Native Evidence of Non-ferrous Metalworking in Early Historic Ireland

Native Evidence of Non-ferrous Metalworking in Early Historic Ireland

Michelle Comber

BAR International Series 1296
2004

Published in 2016 by
BAR Publishing, Oxford

BAR International Series 1296

Native Evidence of Non-ferrous Metalworking in Early Historic Ireland

ISBN 978 1 84171 380 9

Volume Editor: John W Hedges

BAR Publishing is the trading name of British Archaeological Reports (Oxford) Ltd. British Archaeological Reports was first incorporated in 1974 to publish the BAR Series, International and British. In 1992 Hadrian Books Ltd became part of the BAR group. This volume was originally published by John and Erica Hedges Ltd. in conjunction with British Archaeological Reports (Oxford) Ltd / Hadrian Books Ltd, the Series principal publisher, in 2004. This present volume is published by BAR Publishing, 2016.

Printed in England

BAR
PUBLISHING

BAR titles are available from:

BAR Publishing
122 Banbury Rd, Oxford, OX2 7BP, UK
EMAIL info@barpublishing.com
PHONE +44 (0)1865 310431
FAX +44 (0)1865 316916
www.barpublishing.com

CONTENTS

ACKNOWLEDGEMENTS

Sincere thanks are due to all members of the Archaeology Department at NUI, Galway who assisted in the completion of this work. Particular thanks to Dr. William O'Brien who suggested the BAR as a place to publish and served as supervisor of this work when submitted as an MA thesis in 1996. Thanks also to professor John Waddell for reading the thesis text and resultant helpful suggestions, and to Angela Gallagher for her practical assistance in relation to artefact illustration.

I am also grateful for the contribution of Professor Gearóid MacEoin, Department of Old and Middle Irish at NUI, Galway – his translation of the 'Dinnra Clasaige' passage and his guidance through the myriad of early Irish literary sources. Professor Barry Raftery of the Archaeology Department at UCD kindly provided illustrations and information from Rathtinaun, Co. Sligo, prior to his own publication. Thanks also to the Northern Ireland Environment Service, Historic Monuments and Buildings branch, for location information and descriptions of relevant sites in Northern Ireland. And to the staff of the National Museum of Ireland, Antiquities division, especially Andy Halpin and Raghnall Ó Floinn, for their tireless help in the compilation of the Lagore catalogue.

I offer special thanks to my friends, family and parents for their constant support and encouragement throughout this work – *Go raibh míle maith agaibh!*

LIST OF FIGURES

LIST OF TABLES

CHAPTER 1
INTRODUCTION

1.1 Outline

The Early Historic Period in Ireland (fifth to twelfth century AD) has always been a popular subject of study in Irish archaeology, its art and metalwork perhaps being the two most celebrated aspects. The lure of beautifully ornate bronze, silver and gold objects has for several centuries attracted the attention of both antiquarian and treasure-hunter alike. The importance of such well known artefacts as the Tara Brooch and Ardagh Chalice lies both in their immense aesthetic value and their reflection of technical achievement. While these objects are celebrated for their technical excellence, little consideration has been given to their site production and the resulting archaeological record. A study of such evidence is undertaken here, exploring the artefactual and site evidence of non-ferrous metalworking in Early Historic Ireland.

The study of on-site evidence for non-ferrous metalworking in Early Historic Ireland has been hitherto largely unexplored, despite the obvious potential to shed light on various aspects of contemporary society. At a local level, an examination of physical remains may reveal the nature of a site and its internal organisation. Regionally, a study of site evidence may reflect the existence, or not, of a broader infrastructure or 'industry'. Such an examination can also provide information on the economy and status of those communities and settlements associated with fine metalworking.

Another equally important area that cannot be ignored is literary evidence. The earliest Irish literature dates from the latter half of the first millennium AD, providing an opportunity to glean whatever information possible from contemporary written record. This particular source of information does not exist for scholars of Bronze and Iron Age metallurgy. The study of the earliest Irish documents is, however, fraught with hazard. Language used in early Irish texts may be deliberately archaic or propagandistic, with no early technical treatises comparable to the *Divers Arts* of Theophilus (written in the twelfth century AD). Despite these problems, this is an area of study with much potential to uncover evidence of direct or indirect relevance to the study of Early Historic metalworking.

In this study the metallurgical remains from Lagore crannóg, Co. Meath are, in particular, examined to illustrate the nature of the Irish evidence. Lagore, excavated by the Harvard archaeological expedition in the early 1930s (Hencken 1950), has been identified as *Loch Gabair*, an important royal site of Leinster between the seventh and tenth centuries AD (Price in Hencken 1950, 18–34). The high status of this site is confirmed by the large and varied artefact assemblage recovered and the evidence for fine metalworking, with the latter only briefly examined in the published report. The Lagore material has much potential, its largely unstudied nature encouraging a fresh approach independent of an excavator's opinion.

Lagore is unique in many ways, especially in its position as the seat of the kings of southern Brega and of several of the 'High Kings'. From a metallurgical point of view it is important in the discovery of copper ore on-site, as no other Early Historic settlement has produced such obvious evidence of primary smelting. The occupation of Lagore lasted some four to five hundred years, from the mid seventh to the late tenth/early eleventh century AD. This lengthy duration of occupation offers the opportunity to trace developments and changes in all aspects of life, including metalworking.

Hencken suggested that an apparent increase in iron-working on the site in the ninth century reflected a period of unrest in the area, possibly due to Norse attacks. Lagore itself was destroyed in 934 AD by Amlaibh ua h-Imair, probably Olaf son of Gothfrith, the Norse king of Dublin who died in 940 AD. With such historical connections, Lagore has the potential, not only to illustrate the tools and techniques of non-ferrous metalworking used on a high-status site, but also to trace development and change through periods of political unrest.

1.2. Previous research

Recent publications on the Early Historic Period have dealt only briefly with the subject of non-ferrous metalworking in Ireland (e.g. de Paor 1967; Youngs 1989; Edwards 1990) or not all (Ó Cróinín 1995). Research into mining and metal sources is particularly weak, with little firm archaeological evidence to corroborate the sparse literary record (see Whitfield 1993). However, some information on primary copper smelting at a mine site will be provided by the forthcoming excavation results from Ross Island, Co. Kerry (O'Brien 2004). The general technology of non-ferrous metallurgy has been comprehensively covered by Tylecote, his 1986 publication dealing specifically with the metallurgy of the British Isles.

Michael Ryan has explored metalworking techniques largely from the finished artefact viewpoint, tending to concentrate on the techniques of decoration (e.g. Ryan

1983). Specific metalworking techniques from this period have also been studied, for example Niamh Whitfield's 1987 work on filigree, while there are a wealth of publications dealing specifically with individual artefacts from this 'Golden Age'. Although Brian Scott's book, *Early Irish Ironworking* (1990), does not directly deal with non-ferrous metalworking, sections dealing with the early literary period and settlement organisation are obviously relevant.

As already pointed out, little research has been undertaken in the area of early literary references to metalworking. This is a specialised area of study and many of the earliest law tracts and other documents have been translated by scholars over the past century or so, covering various topics from genealogies to liability laws. Scattered throughout these works are brief references to metalworking and metal-smiths. No specific analysis of these references has been undertaken, probably due to their frequent ambiguity. Very often they refer only to the metalworker or 'smith', failing to differentiate between blacksmith and bronzesmith. Even in cases where such distinctions are made, it is very likely that the term used refers to all metalworkers unless specifically stated otherwise. Brian Scott (1990) does bring together those references which may relate to ironworking, many of which may also be relevant to non-ferrous metallurgy. An article by the author demonstrates the potential of the literary sources when used in conjunction with archaeological evidence. In this case, site use at Garranes ringfort was examined (Comber 1998).

Scientific research in this area is very scarce, limited to analyses of crucibles and their residues for the most part. In 1927, Moss carried out an analysis of several crucibles, then in the Royal Irish Academy collection of antiquities, the results of which are relevant to our understanding of the use of such vessels. Unfortunately, the application of scientific analyses has not improved much since then. Recent excavation reports frequently include the results of scientific examination, although where non-ferrous evidence is concerned they are still largely confined to the analysis of crucible residues. Detailed analyses of other items, such as possible furnace remains, have yet to be undertaken despite the obvious value of such work. There has also been very little experimentation on the production side of Early Historic non-ferrous metallurgy comparable to, for example, O'Kelly's work on early iron smelting (O'Kelly 1961).

The main source of published research in this area consists of the excavation reports of the relevant sites. The most recent attempt to collate this information is Proudfoot's 1961 article on the economy of the Irish rath where he briefly discusses the then available excavation evidence. Until recently, the only synthesis attempted for Irish crannógs was Wood-Martin's *The Lake-Dwellings of Ireland* which, published in 1886, can no longer be regarded as comprehensive. Aidan O'Sullivan's more recent *The Archaeology of Lake Settlement in Ireland* (1998) takes a multi-period approach to the subject, including discussions of sites relevant to this study.

The majority of the excavation evidence dates from the 1940s to the present with the more recent of these reports usually highly detailed and often including various scientific analyses (e.g. Gaskell-Brown and Harper's excavation of Cathedral Hill, published 1984). Some excavation reports may not have sections dealing specifically with metalworking or non-ferrous metalworking, though much valuable information can be incorporated in the text, e.g. the Lough Faughan report (Collins 1955). As with most aspects of this study, the various references and pieces of information must be drawn together to form a coherent account. Fortunately, the vast majority of relevant sites have been well documented and published. This study brings together information provided by excavation reports and the results of artefact and technical studies.

1.3. The present study

The present study was undertaken and first written up as a postgraduate thesis (Comber 1996). Primary research was undertaken on unpublished sources in the National Museum of Ireland. The first of these sources was the Museum archive, involving an examination of the Museum registers, excavation records and photographs, and various items of correspondence relating to the discovery of artefacts. The second source of data in the Museum was, obviously, the artefacts themselves. A detailed examination and primary record was made of the relevant material from Lagore crannóg, Co. Meath. This cataloguing involved the description, measurement and illustration of the metalworking artefacts and residues, of which there are over three hundred. Unfortunately, several relevant items could not be located in the Museum and so, where possible, descriptions based on previous work are used to fill gaps in the catalogue.

Viking metallurgy

Due to the intrusive nature of the metallurgy from the Viking towns, Dublin in particular, it is not included in this work dealing with native activity. It would appear, from the scattered reports published thus far, that the relevant assemblage from Dublin is quite large consisting of, amongst other things, many motif-pieces, tools, moulds and crucibles (Fig. 1). The most probable site of a non-ferrous workshop in Dublin seems to have been Christ Church Place during the eleventh and twelfth centuries. These dates place this evidence slightly later than much of the native Irish remains. The evidence from Viking Dublin constitutes an interesting area of future work which might usefully compare the native and Viking non-ferrous evidence, an exploration of possible routes of influence.

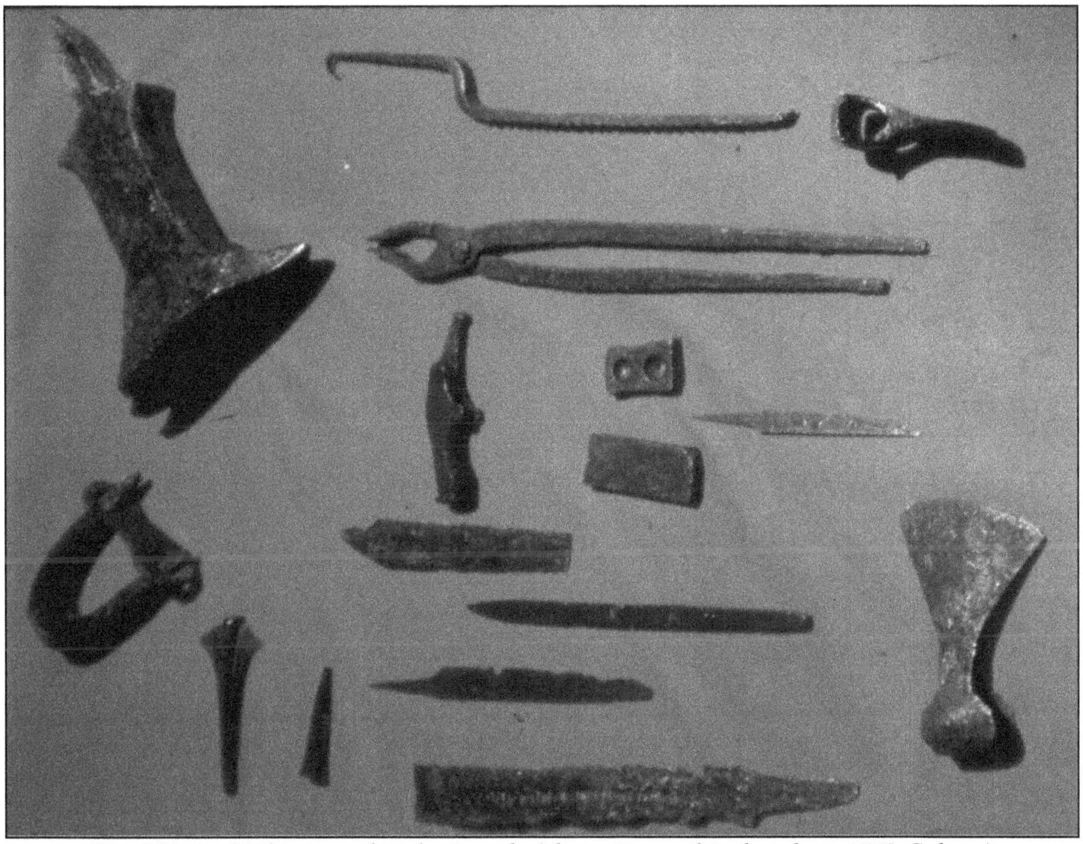

Fig. 1 Typical Viking metalworker's tools (photo; Dept. of Archaeology, NUI, Galway).

1.4. Technological background

An understanding of prehistoric technology is obviously relevant to any study of non-ferrous metallurgy in Early Historic times, as it was in prehistory that the basics of working in copper, bronze, lead, silver and gold were established. The origins of Irish metallurgy lie in the Final Neolithic period, *c.*2500–2100 BC, when objects of unalloyed copper and gold began to be produced on this island. The introduction of metallurgy to Ireland is generally attributed to the early users of Beaker pottery and to Continental connections at that time (Sheridan 1983). The introduction of copper metallurgy to Ireland, followed by the development of tin-bronze technology *c.*2100 BC, was highly successful, a fact which many commentators attribute to a ready availability of raw material (O'Brien 1995). With regard to copper sources, considerable research has been carried out in recent years (O'Brien 1994, 2003, 2004) on Copper Age/Early Bronze Age mine sites in the Cork-Kerry region (Fig. 2).

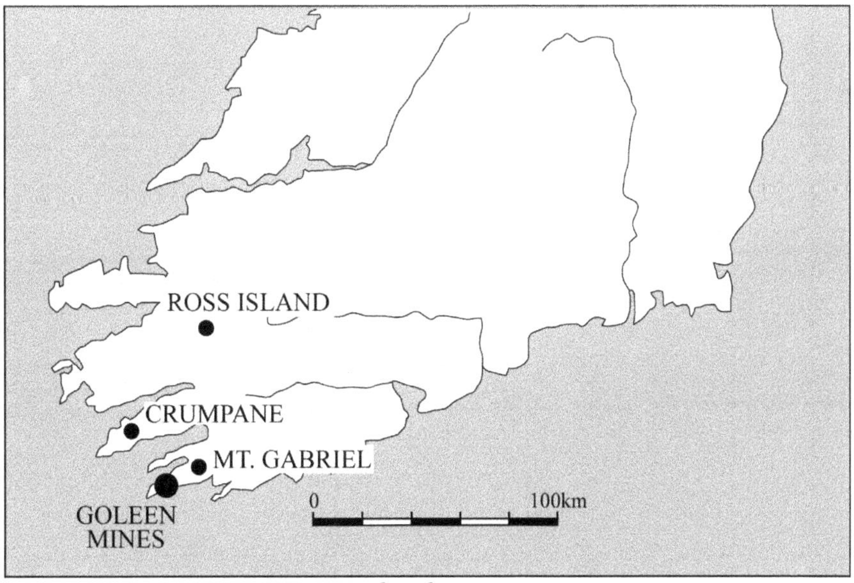

Fig. 2 Excavated prehistoric copper mines.

Fig. 3 Ross Island mine opening (photo; W.O'Brien).

The Beaker associated mines of Ross Island, Co. Kerry and the Bronze Age mines of the Beara and Mizen peninsulas, Co. Cork are excavated examples of early Irish copper mines (Fig. 2). The earliest operations at Ross Island are dated to *c.*2500 BC, the end of the Irish Neolithic, and are the earliest such workings known in western Europe. The mines consist of large, cave-like openings with smooth concave walls – a result of the use of fire-setting and stone hammers (mauls) to remove the ore from the rock-face (Fig. 3). A work-camp was located on an adjacent shelf, with activities here including ore concentration, smelting and temporary habitation. Small pit furnaces used for the roasting and/or smelting of copper ore were identified in this settlement (O'Brien 1996, 2004).

The main period of mining at Mt. Gabriel on the Mizen peninsula occurred during the Early Bronze Age, *c.*1700–1500 BC. Some 32 workings were identified, each with the characteristic smooth, concave profiles produced by fire-setting. Mining equipment discovered during excavation included oak and hazel fuel for fire-setting, resinous pine splints for lighting, alder shovels, hazel picks, and stone mauls. No evidence of smelting was found on site (O'Brien 1994, 1996). Limited sample excavations have also been undertaken at a number of other mine sites on the Beara and Mizen peninsulas. All are of the Mt. Gabriel type and have produced dates in the early/mid second millennium BC (O'Brien 2003).

Prehistoric mines are also known from elsewhere in Bronze Age Europe e.g. south-west Spain (Rothenberg and Blanco-Freijeiro 1981). Here, smelting took place in a bowl-shaped furnace partly dug into the ground. Smelting conditions were achieved by placing the crushed ore into the furnace along with charcoal, and perhaps a fluxing agent, and then inducing a draught with a bellows or blow-pipe (Craddock 1995). The simplicity of this furnace type would make it a logical design in any early metal-using society.

Once smelted, the ore was refined in a crucible in preparation for casting. The extant crucibles known in Ireland date to the Late Bronze Age and are few in number. Excavation on Dalkey Island, Co. Dublin (Liversage 1968) and Raftery's excavation of Rathgall hillfort, Co. Wicklow (O'Kelly 1989, 156) both produced Late Bronze Age evidence. The type of crucible found on both of these sites was a large, shallow oval vessel. Similar crucibles from Britain are also dated to the Late Bronze Age, e.g. Breiddin, Powys (Tylecote 1986, 233).

While smelting evidence remains elusive, we can follow the development of casting technology in the Bronze Age through the evidence of both moulds and finished artefacts. In the Copper and Early Bronze Age the open stone mould was commonly used, largely for casting flat axes. In addition to the axe-head, other artefact types cast in such moulds include daggers, awls, razors, and razor-knives (Coghlan and Raftery 1961). Halberds of Copper Age/Early Bronze Age date were most probably cast in bi-valve or two-piece moulds, though no mould remains have been found. Casting technology developed further in the Middle Bronze Age when the range of artefact types produced was largely the same as in the Early Bronze Age. The moulds used continued to be made of stone although they were now mostly two-part or bivalve moulds. It has also been argued that clay moulds must have been in use (Tylecote 1987), however they do not survive in the archaeological record. One of the most important developments of the Middle Bronze Age was the introduction of the clay core in the manufacture of socketed objects such as spearheads (O'Kelly 1989).

The Late Bronze Age period in Ireland lasted from approximately 1300 BC to shortly after the earliest contact with iron use after 600 BC. This was a period of considerable metal production and metalworking in Irish prehistory, facilitated to some extent by the use of clay moulds. The majority of moulds in the Late Bronze Age were made of clay, while the use of stone went into decline. The use of a clay core was well established and this, in association with the flexibility of casting in clay moulds, allowed the production of a wide range of artefacts including swords and socketed tools (Ó Faoláin and Northover 1998). Late Bronze Age clay moulds are known from sites such as Dalkey Island (Liversage 1968), Rathgall (O'Kelly 1989), and Lough Gur (Ó Ríordáin 1954). It is also highly probable that the *cire perdue* technique of casting was practised in the Late Bronze Age although this must be inferred from the finished artefacts rather than the actual mould evidence. All of the techniques mentioned here are described in detail in chapter 3.

The most spectacular Bronze Age metal artefacts are made of gold, usually displaying a high level of decoration which developed over time (Eogan 1994). As items of personal ornament, more time was spent in their decoration than on weapons or tools. Decorative objects of the Early Bronze Age include sundiscs, earrings and lunulae while Late Bronze Age objects include armlets, torcs, gorgets, tress-rings and dress-fasteners. Sheet gold, and indeed sheet bronze working, was used throughout the Bronze Age, the hammering out of which required the development of a heat treatment technique known as annealing. Wire production was also developed, involving either the twisting of a narrow strip of metal or rolling it with a flat stone on an anvil. It would appear that techniques of gold decoration changed little throughout the period. Decorative techniques included *repoussé* and the use of compasses, as seen on gorgets for example. That other non-ferrous metals were known and worked can be seen in the lead cores of such objects as the ring from Killyleagh, Co. Down (O'Kelly 1989, 176), the bulla from the Bog of Allen (*ibid.*), and the three bracelets of tin from Rathtinaun (Raftery unpublished).

The status of non-ferrous metallurgy in prehistoric Ireland is understandably affected by the introduction of iron-working in the period *c.*600–300 BC. There is very little evidence of non-ferrous metalworking from the pagan Iron Age which follows, up to the advent of Christianity in the fifth century AD (Scott 1990). This largely reflects the limited settlement record from the Iron Age. Recent work at Ráith na Ríg on Tara has, however, produced evidence of the first non-ferrous workshop dated to the Iron Age (Roche 2002). Beneath the enclosing bank of Ráith na Ríg were found the remains of industrial activity including 'precious' metalworking. Evidence consisted of a large metalworking hearth, structural/workshop remains, and an assemblage of relevant finds. Both iron-smithing and bronze-working occurred in this area. The crucibles employed were triangular in shape, similar to examples found in Iron Age Britain. Unfortunately, the clay mould fragments were too small to positively identify the objects cast therein. Bronze waste and finished objects were also recovered from this industrial area.

The hearth at Ráith na Ríg consisted of a shallow scoop cut into the subsoil, its sides and base heavily burnt. Numerous stake-holes around its edge suggest the presence of a reinforced clay superstructure, while the excavator also mentions a possible flue for this hearth. Analysis of the metalworking debris revealed no evidence of on-site copper smelting, causing Crew and Rehren to rule out the use of this feature as a smelting furnace (*ibid.* 83–102). Radiocarbon analysis of the hearth fill produced a date of 200 BC–16 AD (*ibid.* 34). There is abundant evidence in the finished artefact record that bronze continued to be produced in this period (Raftery 1983). Obviously, with the initial introduction of iron technology one would expect the native bronze-workers to have experimented with this 'new' metal. Such experimentation is evident in several early iron artefacts e.g. the Lough Mourne axehead and the Drumlane cauldron, Co. Cavan (Scott 1990, 48) (Fig. 4). Both of these iron objects were made in a manner identical to that of their Late Bronze Age equivalents.

Fig. 4 Drumlane cauldron (photo; Dept. of Archaeology, NUI, Galway).

In addition to such skeuomorphic objects, there is further evidence of the early association between 'new' and established metals. Both the Lisdrumturk cauldron and Derrymacash bucket are composite objects of iron and copper-alloy (Scott 1990, 58). The two activities may, of course, have become distinct from one another with time.

Other metalworking/decorative techniques were introduced in the Iron Age, for example, the use of enamel and the *cire perdue* technique of casting e.g. the three-dimensional bird's head handles from Somerset, Co. Galway (Fig. 5) and the cup from Keshcarrigan in Leitrim (Raftery 1994). Unfortunately, there is no known site or equipment evidence of non-ferrous working in Iron Age Ireland - no workshops, moulds or crucibles. In this period we are forced to infer the techniques and tools of non-ferrous metalworking through the finished artefacts, which are relatively few in number compared to the Late Bronze Age and Early Historic Period.

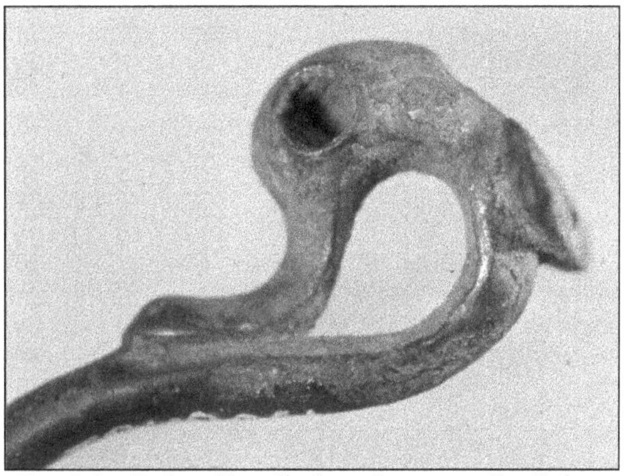

Fig. 5 Bird's head handle from Somerset, Co. Galway (photo; Dept. of Archaeology, NUI, Galway).

In Iron Age Britain, the crucibles used were of a large triangular type whilst the remains of *cire perdue* moulds were found at Gussage All Saints in Dorset (Foster 1980). The largely unknown record from the Iron Age in Ireland leaves open to speculation most of the processes and techniques employed in non-ferrous production.

1.5. Chronological framework

In this study the Early Historic Period, also termed the Early Christian or Early Medieval Period, lasting from the introduction of Christianity in the fifth century to the Anglo-Norman invasion of the twelfth century AD, may be sub-divided as follows;

Period I; fifth – sixth centuries AD (Fig. 6)

The beginning of Early Historic Ireland is generally marked by the introduction of Christianity in the fifth century. The first church in Ireland was episcopal in nature and limited both in terms of its distribution and influence, existing alongside native pagan beliefs and practices (Ó Cróinín 1995).

The fifth and sixth centuries saw a period of disturbance and displacement throughout the country (MacNiocaill 1972). Older tribal groups began to be replaced by new powerful dynastic families causing the displacement of large groups of people, many of whom settled in Wales and western Britain. One of the most important events during this period was the movement of the Uí Néill from Connaught, northwards to Ulster and eastwards to Meath.

Fig. 6 Main political groups of the fifth/sixth century (miscellaneous sources).

In Ulster the northern Uí Néill defeated the dominant Uí Cruithin kings at the battle of Druim Dergi in 516. In a series of battles between 498 and 503, the southern Uí Néill defeated the Leinstermen, particularly the Dál Cormaic of south Leinster. The dominant powers in north Leinster were the Uí Garrchon, followed later by the Uí Failgi, while the Eóganachta were dominant in Munster. There was no strong pattern of provincial or national kingship during this period (Ó Cróinín 1995).

Period II; seventh – eighth centuries AD (Fig. 7)

The seventh and eighth centuries AD saw the acceptance of Christianity into Irish society. In the mid-seventh century the church in Ireland made the transition from an episcopal system to a monastic one. The new monasteries encouraged the development of the arts and facilitated the introduction of literacy. This period was also marked by extensive Irish missions on the Continent and in Britain.

In Connaught the most prominent king was Guaire of the Uí Fiachrach Aidne until the eighth century which saw the emergence of the Uí Briúin. From 764 on, rival branches of the latter contested the provincial kingship. In Ulster the Dál Fiatach provided 52 of the 62 kings of the province.

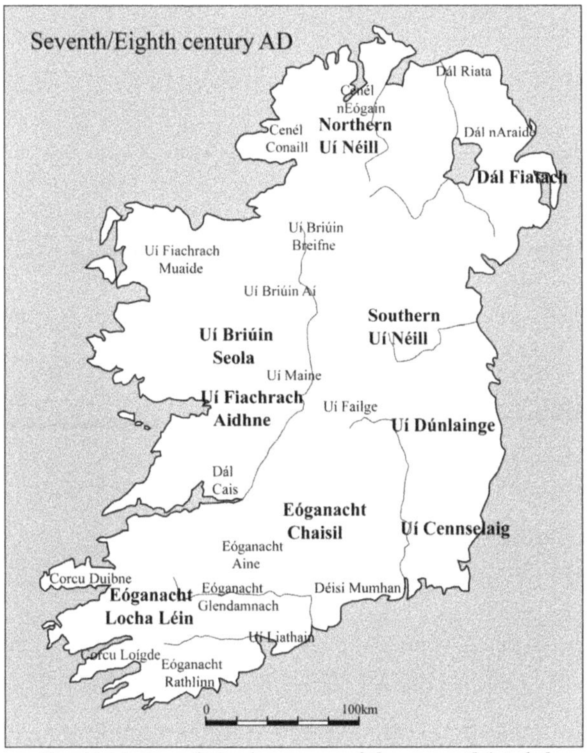

Fig. 7 Main political groups of the seventh/eighth century (miscellaneous sources).

The battle of Áth Senaig in 738 saw the decimation of the Uí Cennselaig of south Leinster by the southern Uí Néill. The latter continued to rule Meath and provide the first high-kings of Ireland. From 738 to 1042 the Leinster kingship was dominated by the Uí Dúnlainge of north Leinster. The eighth century in Munster saw the emergence of rival branches of the Eóganachta. This fragmentation resulted in the inability of Munster to claim the high-kingship (Ó Cróinín 1995).

It was during this period also that, arguably, the high-point of native non-ferrous metalworking was reached, with achievements such as the Tara Brooch and Ardagh Chalice. The end of the eighth century saw the first Viking raids on Ireland, specifically Rathlin Island in 795 AD.

Period III; ninth – tenth centuries AD (Fig. 8)

By this time, the church in Ireland had become totally integrated into society. Large monasteries played host to regional fairs or *oénach*, acted as patrons of craft and learning in addition to their religious function. The growth of monastic communities saw some sites emerge as small towns or monastic 'cities'. A decline in Irish scholarship on the Continent began in the tenth century marking a return to concentration on insular affairs.

In the provinces, the Dál Fiatach were dominant in Ulster, the Uí Dúnlainge in Leinster, the Eóganachta in Munster, the Uí Briúin in Connaught and the Uí Néill in Meath. The high-kingship was still claimed by the southern Uí Néill.

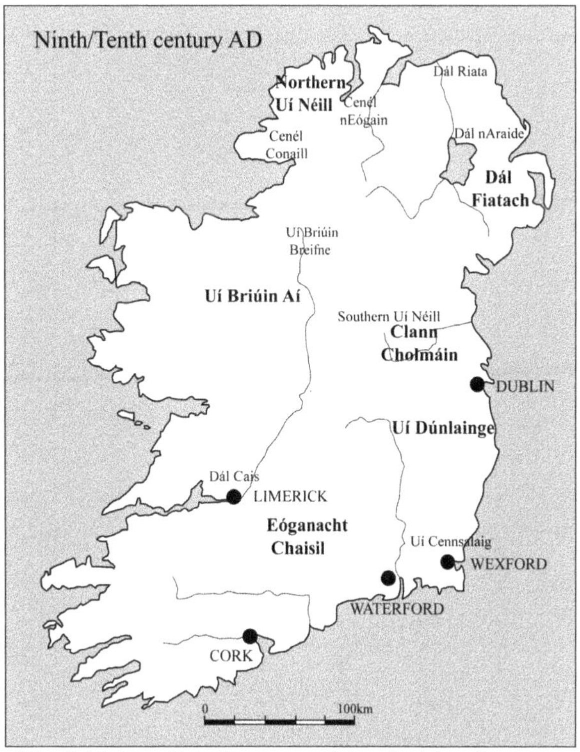

Fig. 8 Main political groups of the ninth/tenth century (miscellaneous sources).

In the 830s sporadic Viking raids became sustained campaigns with the establishment of the first wintering stations or longphorts, including Dublin, in 841 AD. The Vikings were driven from Dublin in 902, returning in 917 AD. Dublin became one of the most important of all the Viking towns, however by the end of the period had lost most of their independent political power. The Scandinavian settlers of Dublin, Waterford, Wexford, Cork and Limerick interacted with their native Irish neighbours and eventually became integrated into Irish society (Ó Cróinín 1995).

Period IV; eleventh – twelfth centuries AD (Fig. 9)

During this period the church underwent a reform which saw the establishment of the diocesan structure, still in use today. The move towards insularity was hastened by the advances of the Normans on the Continent. The twelfth century marks the introduction of Romanesque architecture and new monastic orders, such as the Cistercians, into Ireland.

This period sees the increasing militarization of Irish society with warfare becoming more brutal and prolonged. All kings claiming the high-kingship did so with opposition. In Connaught these included Turlough O'Connor (died 1156) and his son Rory (ousted from the kingship in 1183), in Ulster Muirchertach MacLoughlinn claimed the title of high-king from 1156 to his defeat in 1166. The first claimants to the high-kingship during this period were Turlough O'Brien (died 1086) and his son Muirchertach (died 1119) from Munster.

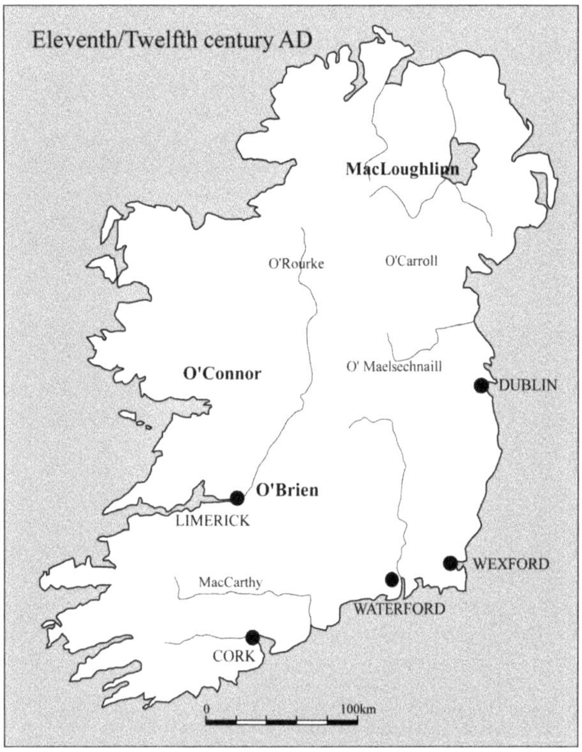

Fig. 9 Main political groups of the eleventh/twelfth century (miscellaneous sources).

By the eleventh century, the Viking towns no longer had any meaningful autonomy, they were thoroughly absorbed into Irish society and were ruled by Irish kings. The banishing of Dermot MacMurrough from Ireland in 1166 by his enemy, Rory O'Connor, resulted in his return the following year aided by the Anglo-Normans. In 1169, Richard Fitzgilbert de Clare or 'Strongbow' arrived in Ireland and took Waterford, Dublin and Meath. Henry II landed in Waterford in 1171 and by 1177 the Anglo-Normans had advanced as far as Cork despite meeting with fierce native resistance. The Treaty of Windsor in 1175 established Henry's title to the conquered territories and Rory O'Connor as ruler of the rest (Ó Cróinín 1995).

1.6. Finished Product Range

The non-ferrous metals in Early Historic Ireland, mostly copper-alloy, lead, tin, silver and gold, were used almost exclusively in the manufacture of luxury items. These objects were made in a society that was fully iron-using – every day work tools and weapons were made of iron – bronze (and, of course, precious metals) was used exclusively for the production of personal ornaments and decorative pieces. One of the best known artefacts from the period is the penannular brooch. Evolving sometime during the Iron Age and Roman period, these pieces consisted of a separate pin and penannular ring, the pin attached to the ring though free to move along it (Fig. 10). A dress-fastener, the brooch was worn with the pin pointing upwards (Edwards 1990, 133). During the earlier part of the period (sixth-seventh century AD), zoomorphic forms of these brooches were popular (Fowler 1960, type F). The terminals of these rings were shaped as elongated animal heads and were often decorated with enamel, millefiori and spiral designs (Kilbride-Jones 1980). Type F forms were most common in Ireland, with a much smaller quantity of types G and H. Type G brooches consisted of small faceted terminals with lozenge ornament, and date from the seventh to the ninth century. Type H forms have flattened and expanded terminals (Fowler 1960). Penannular brooches remained common throughout the Early Historic period becoming larger, heavier and more ornate with time. Eventually, the gap between the terminals was closed, resulting in the pseudo-penannular brooch, as exemplified in the Tara brooch.

Simpler pins were also used as dress-fasteners, probably by a greater number of people. The earliest type of pin, from the beginning of the period, was the hand-pin. These consisted of a straight shank of copper-alloy and a flat semi-circular pin-head with three to five 'digits' on top, often decorated with red enamel and spiral designs (Edwards 1990, 136). Later, during the seventh and eighth centuries, the ringed-pin became common (Fig. 11). Mostly of copper alloy, a small number of iron examples are also known. These consisted of a straight pin with a looped or perforated head, through which a ring was inserted. The earliest sub-type was the spiral ringed-pin where a ring of spiralled wire was passed through a baluster pin-head.

Fig. 10 Bronze penannular brooch from Lagore, Co. Meath (after Hencken 1950).

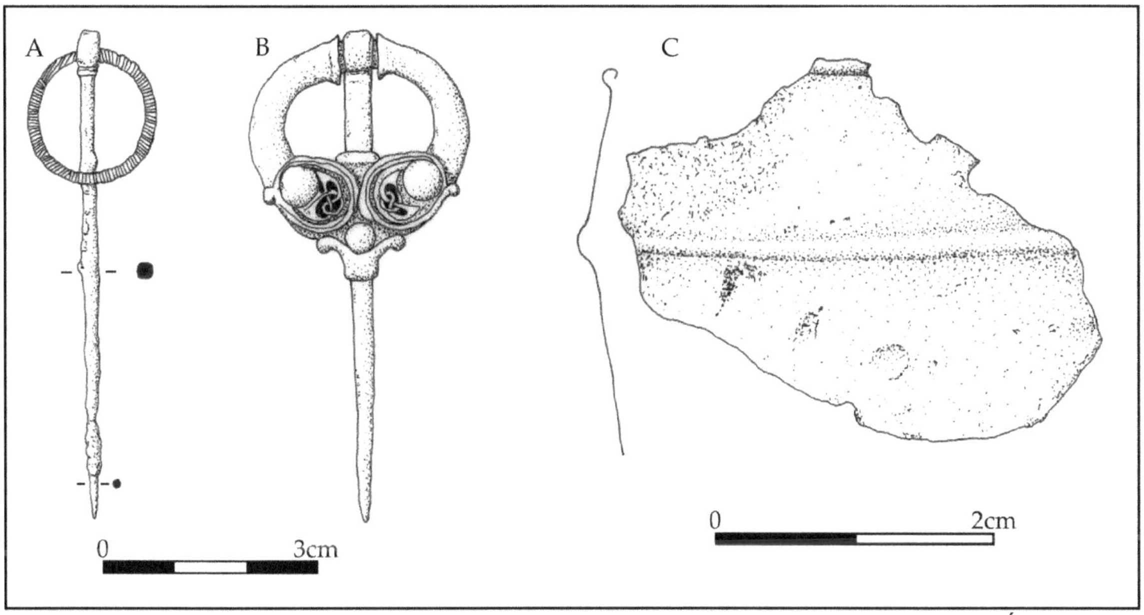

Fig. 11 Bronze pin (A), brooch (B), and vessel fragment (C) from Carraig Aille, Co. Limerick (after Ó Ríordáin 1949).

They bore simple ornament on the pin-head and occasionally on the ring. These were gradually replaced by plain ringed-pins which continued in use into the ninth century. The ring was attached through a looped pin-head, occasionally polyhedral or baluster pin-heads. A related type of dress-fastener, developed in the eighth and early ninth century, was the ring-brooch. This was, essentially, a cross between the pseudo-penannular brooch and the ringed-pin (Edwards 1990, 142).

Other artefacts from the beginning of the Early Historic period include latchets and hanging-bowls. Latchets, possible dress-fasteners, consist of a circular head with an 'S'-shaped tail, and are occasionally decorated with enamel (*ibid.* 136). Hanging bowls were formed from sheet metal (copper-alloy most commonly) and bore decorative escutcheons and mounts. These featured openwork patterns, spirals, interlace, animal ornament, enamel and occasionally millefiori (*ibid.* 136). Whatever their place of manufacture (only two escutcheons have been found in Ireland), the metalworking techniques they display were all practised in Ireland.

While many of these artefact types may have been both secular and ecclesiastical, a smaller number of items can be definitely associated with the latter. A series of shrines decorated with precious metals are known from Early Historic Ireland. These include three small, house-shaped, shrines for carrying portable relics (Emly, Shannon, Lough Erne), the Lough Kinale book-shrine and the Moylough belt-shrine (*ibid.* 138). A small quantity of altar plate has also been found, most notably the Ardagh Chalice (Organ 1973) and the Derrynaflan Hoard (Ryan 1983). Miscellaneous other ecclesiastical items include a fragmentary copper-alloy processional cross from Co. Antrim, decorated with enamel and millefiori, a number of bronze hand-bells, and the crucifixion plaque from near Athlone, possibly once attached to a book (*ibid.* 140).

The ninth century saw the beginning of a decline in Irish fine metalworking, perhaps due to the Viking impact. Artefacts became much coarser and plainer. Amber replaced the use of millefiori and enamel, and both *kerbschnitt* and filigree disappeared. Animal motifs and simple plait-work remained, while spirals, intricate interlace and step-patterns died out (Henry 1967).

1.7. Research Design

This study aims to explore all aspects of native non-ferrous metalworking in Early Historic Ireland, from the initial procurement of raw material through mining to the production of metal by smelting and alloying, to the final fabrication of artefacts by casting and decoration. While much has been written with regard to the finished products, this work sets out to explore the processes and structures which made their production possible.

The present study is undoubtedly limited by an inability to apply modern scientific approaches to the technical study of these finished objects and metalworking residues. Nevertheless, a purely archaeological approach to the study of this evidence has the potential to reveal new information on the practice and organisation of this metalworking craft. A study of the techniques used to fabricate different artefact types will allow us to reconstruct the tool-kit of the Early Historic metalworker. The examination of on-site activity will shed light on the organisation and status of non-ferrous metalworking in this period.

Emphasis is placed on understanding the technology of fine metal production. A knowledge of the technical background is necessary for the identification of artefacts that may have been used in non-ferrous metalworking and the highlighting of any gaps in the relevant assemblages or tool-kits. Questions arising from a technical study include whether or not development can be traced in the techniques employed, what were the outside influences (if any) on the technology of the period, what sources of raw material were exploited, and how those raw materials were processed and manufactured into finished objects. This synthesis of the archaeological evidence of native non-ferrous metalworking will also provide a foundation for future scientific studies.

Fine metal production is also examined within a broader social context. The evidence here is drawn from excavated Early Historic settlements, the study of which offers an insight into the organisation of contemporary Irish society. On a micro level, internal site layout may help identify craft or industrial areas. These might, in turn, reflect the organisation of labour. On a macro level, settlement evidence may provide clues as to the status and wealth of various regions, thus enabling the development of a pattern of settlement hierarchy. An examination of settlement contexts might, in addition, reveal information about trade and the broader economy of Early Historic Ireland.

An attempt is also made to interpret early literary evidence and to correlate it with the archaeological record where possible. Questions regarding the status and role of the smith in society are addressed using this information. It is hoped that, through an examination of literary, technical and artefactual evidence, much can be revealed about the structures and organisation that facilitated native production of fine metal objects in Early Historic Ireland.

CHAPTER 2
LITERARY EVIDENCE

Fig. 12 The Forge of Vulcan by Velazquez, 1630.

2.1. Source criticism

References to metalwork and metalworking are scattered throughout early Irish literature dating from the seventh to the twelfth centuries AD. While most of these references are extremely brief and not very informative, when viewed together as a composite body of evidence they can give an insight into the position of the smith in Early Historic Ireland. In general, the picture presented must be viewed with caution given the problematic dating and origin of this literary record.

The two main relevant areas of early Irish literature are myths and legends, and the first law tracts. Some additional information can occasionally be garnered from the genealogies and the various lives of the saints. While early Irish myths and legends are overtly fictitious they may contain a core or element of truth. Conversely, it must also be remembered that the supposedly 'factual' sources may very well have been creatively embellished by an author wielding the power of exaggeration. In this way, for example, certain grand attributes may have been given to a king in an attempt to increase or improve his status in the eyes of others. These attributes might include a list of tributes/rents due every year from his clients or 'tenants', the number and quality of which being a direct reflection of the king's status. As metal and metal products were much valued, references to them are susceptible to this stretching of the truth.

The use of the term 'smith' is here understood as describing the worker in bronze, copper, gold and silver and not just iron. The terms used in the literature are ambiguous, possibly due to the borrowing of technical terms of non-ferrous metalworking by the new technology. In the majority of cases it is practically impossible to determine the exact meaning of a term. For example, *cerd* originally meant a worker in bronze and precious metals, but was later used to mean craftsmen in general.

Another major problem in interpreting these references, as Brian Scott points out, is that 'the works in which they occur are literary, religious or historical compositions which have no technological pretensions' (1981f, 251). This calls into question the technical accuracy of these references and their interpretation today. The writer in question may have had an understanding of what he/she was describing, on the other hand a considerable amount of guesswork may have been involved. An example of this is found in Lucan's *Pharsalia* in the *Cath Catharda* (Stokes 1909) regarding the definition of electrum. At one point it is described as a mixture of silver and copper-alloy while eleven lines later it is correctly given as a mixture of gold and silver. Such discrepancies immediately lead to suspicions of accuracy. Just as the original authors of a text may have had little knowledge of technology and technical terms, so too the translators of such texts. These translations are undertaken by scholars of early Irish linguistics who may not be very familiar with metalworking terms and techniques.

The fact that some works may have passed through the hands of several scribes who may have made errors in their transcription or merely edited or changed what they were transcribing, could also have resulted in the loss of original meaning. Added to this is the fact that some words may have undergone a change in meaning through time. A prime example of this is the term *findbruine*, originally used to describe a silver-coloured surface but which by the seventeenth century was used in reference to brass. Bearing in mind these limitations, the literary sources must be used with some caution in the study of Early Historic metalworking.

2.2. The smith

From these sources, the metalworker in Early Historic times does not appear to have worked in both iron *and* non-ferrous metals. Such a distinction is made both in fictional and 'factual' sources, so it is reasonable to assume that such a division of labour actually existed. In legends detailing the earliest events on these islands, specifically those of the battle of Moytura (*Mag Tuired*), the god Lugh is aided in his battle by the three gods of craftsmanship: Goibniu the blacksmith, Luchta the wright, and Credne the worker in bronze.

This three-fold distinction very likely reflects the position of craftsmen in society around the ninth century AD when this tale was committed to written record (Gray 1982). This distinction is further reinforced in the early law tract *Lebor Aicle* (O'Donovan and O'Curry 1901), where a section titled 'Blai Ord Indeoin' (see Appendix 1) deals with the blacksmith's forge. Here, it is stated that *gobae* work in iron, *umaide* work in copper and copper-alloy, while *cerda* work in gold and silver. In a further passage, the 'Dinnra Clasaige' (see Appendix 1), a distinction is made between the blacksmith and bronzesmith;

> And the smith never fashions a 'white vessel' (i.e. of precious metal), as the cerd (craftsman) does, but makes only a 'black vessel'. (*Corpus Iuris Hibernici*, V.1571.18–30)

In general, it would seem that the title of *cerd* applied to one who worked in both copper-alloy and the precious metals. This is clearly seen in the 'Blai Ord Indeoin' also, where the relevant passage can be translated as;

> Gold and silver and copper-alloy are unindemnified in the forge of a smith, but not in the workshop of a cerd since they are inappropriate to a blacksmith's forge but not to the workshop of a cerd.
> (*C.I.H.* III:936.1ff)

Indeed, many items of fine metalwork from the Early Historic Period are often composite objects of copper-bronze, gold, silver and related alloys. If made by an individual then that craftsman must have had a knowledge of working in different metals. Although smiths were generally male, this was not always the case. In the tract 'Bretha Crolige', it is stated that if a craftsman did not have a son, he could have his daughter trained in his profession instead (*C.I.H.* VII.2295).

There is also evidence of a three-tier hierarchy within the metalworking trade. At the highest level was the master craftsman (*ollav*), next the ordinary craftsman (*cerd*), and finally the apprentices (*felmac*). The work of a craftsman had to be passed by the local *ollav*, before he could practice in the area. These masters also took on apprentices, for which, if they received a fee, they were held liable. In 'Uraicecht Becc' we find out that the title of master craftsman was conferred by the king of the *tuath*, but that this was only significant if the king ruled several *tuaths*, i.e. was an 'overking' (*C.I.H.* V.1614).

2.3. The workshop

Literary references to iron-workers and artificers in non-ferrous metals would appear to indicate a considerable degree of craft specialisation in Early Historic times, consistent with the high quality of finished objects. Such specialisation may also have extended to the place of work within a contemporary settlement site. This is somewhat supported by the archaeological evidence where evidence of ferrous and non-ferrous metalworking was found to have been spatially separated e.g. Dunmisk, Co. Tyrone (Ivens 1989) and Knowth, Co. Meath (Eogan 1977).

This may only have applied to the wealthiest settlements with smaller sites such as Reask, Co. Kerry (Fanning 1981) providing evidence for the working of both iron and non-ferrous metals in the one area. At other sites, such as Garranes, Co. Cork (Ó Ríordáin 1942), where evidence of ferrous and non-ferrous metalworking is found together, the excavation record does not allow us to assess their spatial relationship (see chapter 5).

While there are no specific references to non-ferrous metal workshops, the forge of the blacksmith is dealt with in one particular law tract, 'Blai Ord Indeoin' (*C.I.H.* III.936.1ff). This lists the injuries which could occur in a forge and outlines where the responsibility should lie in such an eventuality (Appendix 1). It is possible to glean some interesting information about the blacksmith's forge from the tract. It appears to have been home to the smith, one or two people wielding hammers, a bellows-blower, and apprentices. It also confirms the use of a bellows-blown furnace. Some of the tract's provisions include that the person wielding the hammer is exempt from penalty for injuries arising from this work, providing he has made no obvious errors and has no knowledge of any defect in the tool he is using. Similarly, the bellows-blower is not to be held at fault for injuries arising from the showers of sparks emanating from the furnace, unless of course he has blown too hard and caused an excess of sparks. If the smith has urged him to do this then the liability rests with the smith.

Among the list of incidents for which the blacksmith can be held accountable is the injuring of livestock that may wander too close. This obviously suggests that there were no walls or barriers to prevent animals approaching, and may reflect the use of flimsy shelters only. This may explain the lack of archaeological evidence of solid workshop structures. Scattered post- and stake-holes are known from many excavated settlements which may well be all that remain of such a flimsy shelter.

2.4. Metal

The importance of metal in Early Historic Ireland can be clearly seen in early literary references beginning with those referring to its initial extraction or mining. Mining, without reference to a specific mineral, is dealt with in the 'Bla Miand Midchlais' (see Appendix 1) section of the *Lebor Aicle* law tract (*C.I.H.* I.270.15f). Here, it is clearly stated that a mine could be privately owned, and that if ore is stolen from that mine, the thief is fined and the property returned in whatever form it is found (ore, smelted ingots, semi-products, or finished artefacts).

Mining is also referred to in other texts, for example, the eleventh century *Lebor na Cert* which lists one of the tributes due to the Connacht dynasty as 350 blooms of iron (Dillon 1962, 48.681f). This text has been identified as a blatant piece of propaganda, however as Scott points out, 'the fact that it was felt that consolidated blooms were a worthy tribute may be indicative of continuing small scale production' (1994, 176). The value of having a mine on your land is expressed in the eighth century *Tír Cumaile* (Mac Niocaill 1971), a tract on the divisions of land. Here it is clearly stated that a mine of copper or iron adds to the value of the land - the relevant passage is translated as;

> How many things increase the values of these lands?... a mine...
> ... If it be a mine of copper or iron it adds 5 séts.
>
> (*C.I.H.* II.676.6)

These tracts also differentiate between the surface of the land and what lies beneath. The landowner did not necessarily own any minerals beneath it, which could be the property of others providing they could be extracted without damaging the agricultural value of the land. It is also clear that some mines were communal, i.e. those mines situated in or on a sea cliff. Stealing from a mine resulted in distress of five days, while the stealing of smelted copper (refined or otherwise) resulted in distress of three days. The fact that such provisions existed suggests that thieving from mines was common, which in turn, reflects the value of metal in Early Historic Ireland.

The importance of silver can be seen in *Bretha Déin Checht* where a physician's fee is given as a yearling heifer or its equivalent in silver (Binchy 1955). Generally, an ounce of silver appears to have been equal in value to one cow or two séts. An *ocaire*, or 'small farmer', had an honour-price of three séts but could only be paid in cattle as his status was not high enough to allow him possess a precious metal. A powerful king, on the other hand, if wounded was entitled to the worth of a cow in silver. Therefore, the *ocaire* was to be paid in cattle only, the physician, craftsman and local king could be paid in either cattle or silver, while the king of several *tuaths* had to be paid in silver. Silver was obviously a form of currency reserved for the upper echelons in society.

2.5. Craft-working Communities

A number of specific groups associated with mining and metalworking have been identified in early genealogical sources. The metal in question is not usually specified, one exception being Na Cuirc who, as already mentioned, paid their tributes in iron blooms. The most commonly referred to group, however, are the Cerdr(a)ige. 'Cerd' originally meant a bronze worker but was later used to mean craftsmen in general, therefore the exact nature of the associated metalworking is unknown. The second part of the name, -raige or -rige, merely refers to a 'local autonomous community' (Ó Córráin 1974, 66). Ó Córráin identifies at least three distinct groups of Cerdrige, all of whom appear in Munster. One is located in West Cork, another is associated with the Deisi of East Limerick, and the third is closely linked with the Eóganacht, possibly located near Cashel. A further group associated with mining are the Síl mBuinne of Bantry. In the *Dinnshenchas of Ailen Cobthaig* they are referred to in a verse, translated as;

> Men shall come and go between the rocks and the land, quarrying ore with great toil. They who do the crushing shall be Síl mBuinne of Benntraige.
>
> (Ó Córráin 1974, 68)

This appears to suggest that this particular group were employed in crushing the ore in preparation for smelting. Ó Córráin concludes that 'One thing that comes out in the genealogies... is that the bronze workers form a coherent community at a local level, whereas there are no certain references to communities of smiths' (71). However, one could also interpret the literary evidence as referring to a professional or family relationship, not necessarily to a self-contained community. Ó Ríordáin suggested that the large earthen ringfort of Garranes, Co. Cork had been the home of some such community of craftsmen. However, it is not feasible to associate this fort solely with a bronze working community as evidence of iron working and agriculture were also found on the site. Garranes is also tenuously identified as *Rath Raithleann*, a royal dwelling place, which argues against it being the exclusive home of a group of craftsmen (Comber 1998).

In conclusion, it is more feasible that one or two metalworkers lived, either permanently or temporarily, in a patron's fort or crannóg. The various legends would seem to support this idea. One can see this, for example, in the tale where Lugh arrives at Tara wishing to enter into the service of Nuada of the Silver Hand (Rolleston 1994). The gate-keeper asks him what service he can provide and when Lugh lists his many talents, including those of blacksmith and brazier, he is told that such a craftsman already resides within. Thus it would seem that, on the high status sites at least, the role of metalworker was a specialised one.

2.6. The metalworker in society

Linked with this argument of Garranes not belonging to a community of metalworkers, is the position or status of the metalworker in early historic society. Ó Córráin states that 'the status of these, whenever it is possible to establish it, is more or less honourable vassal status as indeed the status of the Cerdrige peoples as a whole seems to have been' (1974, 67). It is true that the metalworker held a respected place in society. According to a law tract which deals with status, 'Uraicecht Becc' (*C.I.H.* V.1614), the blacksmith, coppersmith and silversmith all had a fixed honour-price of seven séts. This was the price (or a fraction of) that had to be paid to the smith for various offences carried out against his person or property. This honour-price was equal to that of the lowest grade of noble allowing the smith to, amongst other things, host feasts for kings. In the *Táin*, Culann the blacksmith entertains King Conchobar and his retinue;

> Culand the cerd dwelt in Ulster. He prepared a feast for the king and went to Emain to invite him. He told him to come with a small number unless he could bring a few genuine guests, for neither land nor domain had he but only his hammers and his anvils, his fists and his tongs.
>
> (*Book of Leinster*, 23.822f)

Evidence of the employment or patronage of metalworkers by kings is also present in early literary sources. In the genealogies the Cerdrige are recorded as having provided metalworkers to the king of Cashel (*Book of Lecan* 38v c26). Scott (1981a) believes such royal patronage is indicated by the concentration of metalworking debris on sites such as Lagore crannóg, Co. Meath. Triad 156 of *The Triads of Ireland* (*C.I.H.* IV) states that no delay is allowed on a manufactured item, in other words work had to be completed within the agreed time. Further evidence of the patronage of the smith is found in 'Críth Gablach' where it states that if a craftsman is illegally injured, the sick-maintenance he is entitled to should equal half the rate due to his present employer (*C.I.H.* IV.332).

Where distraint (punishment for a wrong-doing) is concerned we again see the privileged status of the smith. In 'Uraicecht Becc' we are told that craftsmen, unlike other people, are not subject to distraint for a kinsman's liability (*C.I.H.* V.96–42). They are, however, subject to distraint for their own misdemeanours. A withe was tied around their anvil and nothing could be made on it until the claim against the smith had been settled. If the smith ignored the withe then his property could be seized. Metalworkers were also free to move from tuath to tuath and indeed to stop practising their trade if they so desired, either way their honour price stayed the same. A king, on the other hand, could cease to be a king for various reasons such as travelling without his full retinue. The importance of the smith is also seen in the provision that notice of any lost property found had to be sent to the chief smith as well as to such people as the local king and monastic superior.

The smith was also held in high regard by the Church which became a strong and wealthy patron, as evidenced by the finds associated with metalworking on sites such as Cathedral Hill, Co. Armagh (Gaskell Brown and Harper 1984) and Tullylish, Co. Down (Ivens 1987). Several prominent church figures were associated with metalworking e.g. Amairgen, the father of St. Bairre of Cork, was chief smith to the king of *Rath Raithleann* (remembering the possible association with Garranes). St. Patrick himself had in his household three smiths, one of whom was Fortchern, son of the king of Ireland, and three braziers; Assicus,

Tairill, and Tasach, all of whom were native Irishmen (Joyce 1968). St. Conleth, the first bishop of Kildare, presided over a school of metalworkers, while St. Kieran, of sixth century SeirKieran, was a metalworker as well as a scribe (Joyce *ibid.*, 458).

2.7. The smith and the supernatural

The smith was also linked to supernatural powers and magic. In the *Triads of Ireland* a blacksmith of status is associated with a list of supernatural items, such as the anvil of the mythical god Dagda (*Triads*, 120). A seventh century prayer, sometimes ascribed to Patrick, depicts the smith as a very powerful, magical person; 'I summon today all these virtues ... against spells of women, smiths and druids ...' (*Tripartite Life of St. Patrick* I.50.19ff; Stokes 1887). Some of this is attributable to the fact that the smith manufactured weapons of death and destruction. It also relates to the presence of smith-gods in the legends, for example Goibniu the smith and Credne the worker in bronze. In the later legends the smith retains some of the god-like attributes, attributes which also carried over into real life to some extent.

A similar pattern can be seen in Roman, Greek and Saxon mythology, all of which contain a smith-god; Vulcan, Hephaestus, and Weland respectively (Figs. 12 and 13). As for the representation of metal and metal objects in the legends, they clearly cannot be trusted, being used merely as props to a story and so greatly exaggerated in every aspect. The actual metalworkers themselves, as already pointed out, are portrayed as gods in the earliest legends

2.8. Conclusion

Some aspects of metalworking as portrayed in early literature are confirmed by archaeological evidence, e.g. the association of metalworking with royal patrons. However, many questions still remain as to the position of the smith in Early Historic Ireland. For example, did these craftsmen live in communities together as the genealogies suggest? Did they travel from patron to patron or were they attached to just one? Was the art of metalworking handed down from generation to generation within families, or was it open to all? Was it an age or gender segregated activity? Unfortunately, these are questions which remain unanswered and all we can say about the smith in early Irish literature is that he held a position of high status, slightly tempered with the supernatural. To what extent this reflected the position of the smith in reality remains a matter of opinion.

Fig. 13 Hephaestus, detail from a Greek postage stamp, 1951.

CHAPTER 3
THE TECHNOLOGY OF NON-FERROUS METALS

3.1. Introduction

The technology of most metals can be divided into four main areas; the acquisition and processing of raw material, the initial fabrication of an object, finishing and possibly final decoration. The first step involves obtaining the necessary raw materials. Depending on circumstances these may vary from metal ore and scrap metal, to cakes of smelted metal or ingots. Obviously, acquiring supplies of primary copper ore necessitates the development of prospecting, mining and smelting technology. It would appear that a limited number of people actually carried out these processes, while others relied on secondary sources such as the recycling of older objects or trade to satisfy their needs (see chapter 5).

Evidence of the initial fabrication of objects is plentiful, consisting largely of crucible and mould remains, while evidence of final decoration is more difficult to identify. Working backwards from finished objects such as penannular brooches and decorative mounts, it is possible to identify a range of tools which may have been employed in the decoration of artefacts (see chapter 4). Most of this chapter deals with copper and copper-alloy technology, that most commonly used in the manufacture of fine metal objects in Early Historic Ireland. The technology of tin, lead, silver, and gold is dealt with to a lesser extent, reflecting the relative usage of such metals in the Early Historic Period.

3.2. Acquisition of raw material

From the present evidence it is very difficult to determine the source or form of the raw materials used on any particular Early Historic site in Ireland. Only one site, Lagore crannóg Co. Meath, has produced any finds of copper ore and then only two pieces. Such a discovery indicates the use of a primary source but whether directly (on site) or indirectly (off site but under the control of the Lagore residents) is impossible to ascertain. The inhabitants of Lagore may have had direct control of a mining operation or they may have obtained the ore in some other manner such as trade or tribute. There is no evidence of contemporary mining in eastern/central Ireland. Uncertainty surrounds every aspect of the raw materials exploited, and their source(s) is one of the biggest problems in the study of Early Historic period metallurgy.

Ore

There is very little evidence of metal mining in Ireland during the Early Historic Period. Indirect evidence is provided by early Irish literary references to groups of people involved in, and the laws governing, mining (see chapter 2). A further literary reference, by the British monk Nennius in the ninth century, describes the mineral wealth of the area around Killarney, Co. Kerry (*Historia Britonum*, Stevenson 1838). The Ross Island copper mines are located in this area and, interestingly, two small furnace pits have recently been radiocarbon dated to the Early Historic Period. While this in itself is not definite evidence of ore extraction it is suggestive of mining at this site. These mines were first exploited in Beaker times (O'Brien 1995a and 1995b) and it now seems possible that they were also worked in the Early Historic Period.

Unfortunately, we have no evidence of the techniques which may have been used in such ore extraction. The possibility that ore was imported from elsewhere (most probably Britain) cannot be discounted, especially in the case of tin which was probably brought in from Cornwall. Ireland, however, is rich in copper deposits and it seems likely that at least some of these were exploited in Early Historic times. Both gold and copper are found in native form, i.e. they 'occur in their natural metallic form on the earth's crust and can be used by man without the problem of smelting' (Tylecote 1987, 66). Native gold in the form of stream nuggets is the most probable source of that metal in Early Historic Ireland. Whether or not this native gold was Irish in origin is uncertain. Significant concentrations are known from Co. Wicklow, Cornwall in England, Highland and Leadhills in Scotland, and Dolaucothi and Dolgellau in Wales. Nuggets of native gold are relatively pure and can be used directly without the need for refining (Tylecote 1987).

Although native gold was probably the only type of gold ore exploited, native copper probably played a very minor role in Early Historic Ireland. It is found at a few locations in Cornwall, Scotland and Ireland and can be extremely pure. Small artefacts can be produced from native copper by cold working alone, though a knowledge of annealing would have been needed for the manufacture of larger items. The most likely source of both silver and lead in Early Historic times was the ore mineral, galena, an argentiferous lead sulphide (PbS – Table 1). Galena is a very dense, darkly coloured ore and is commonly argentiferous. In antiquity, the silver content of lead ore was separated by a process called cupellation.

Some Common Metalliferous Minerals		
Mineral	Chemical Formula	Colour
Copper Ores		
Melaconite	CuO	black
Cuprite	Cu_2O	red
Azurite	$(CuCO_3)_2Cu(OH)_2$	blue
Malachite	$CuCO_3Cu(OH)_2$	green
Chalcocite	Cu_2S	black
Covellite	CuS	dark blue
Chalcopyrite	CuFeS2	metallic yellow
Bornite	Cu_3FeS_3	iridescent
Chrysocolla	$CuSiO_32H_2O$	turquoise
Chalcanthite	$CuSO_45H_20$	blue
Atacamite	$Cu_2(OH)_3Cl$	green
Pseudomalachite	$6CuOP_2O_53H_20$	green
Arsenic Ores		
Realgar	AsS	red
Orpiment	As_2S_3	yellow
Arsenopyrite	$FeAsS$	silver grey
Antimony Ores		
Senarmontite	Sb_2O_3	grey
Stibnite	Sb_2S_3	grey
Lead Ores		
Galena	PbS	metallic grey
Cerusite	$PbCO_3$	white
Anglesite	$PbSO_4$	white
Pyromorphite	$Pb_4(PO_4)_3PbCl$	green
Tin Ores		
Cassiterite	SnO_2	grey
Zinc Ores		
Sphalerite	$ZnFeS$	yellow/black
Calamine/Smithsonite	$ZnCO_3$	white
Complex Ores		
Enargite	$3Cu_2S2As_2S_3$	black
Famatine	$3Cu_2S2Sb_2S_3$	grey
Tetrahedrite/Tennantite	$(Cu, Fe, Zn, Ag)_{12}(Sb, As)_4S_{13}$	dark grey
Stannite	Cu_2FeSnS_4	metallic grey
Bournonite	$CuPbSbS_3$	dark grey

Table 1 Some common metalliferous minerals (after Slater 1988).

Lead ores are found in Ireland though again there is no direct archaeological evidence of early mining. Mercury occurs as the red sulphide, cinnabar (HgS), which is found in Tuscany and Spain. Accounts of exploitation at Almaden, Spain date back to AD 912–961.

The main sources of tin were most likely the deposits of Cornwall and/or Brittany, though small quantities are found in Ireland also. Metallic tin does not occur naturally, its most common form being the mineral cassiterite (SnO_2). The process of tin recovery is known as 'tin-streaming', a slightly misleading term. The ore is dug from around rivers and then washed in the river or stream. It is not, as is often believed, 'panned' from the river in gold-prospecting fashion (Penhallurick 1986).

Mining

Mining in this context refers to the deliberate extraction of a metalliferous ore from a bedrock source. Where copper is concerned, ore deposits can be classified into two groups: oxidized ores, and sulphide ores. Oxidized ores include minerals such as malachite, azurite, cuprite, melanconite, and chrysocolla. Sulphide ores include chalcopyrite (copper pyrites), chalcocite, tetrahedrite-tennanite, and bornite (Table 1). These ores may in theory occur in three different zones, though typically they can all be found at the surface. The three levels include a surface or oxidation zone where percolating water results in the oxidation of the metalliferous minerals leading to the formation of oxidised minerals, the most common being the hydroxy-carbonate malachite. The cementation zone, or zone of secondary enrichment, just below this contains sulphide minerals such as chalcocite, bornite, and tetrahedrite. Below this again, primary sulphide minerals such as chalcopyrite and bornite occur in a deep, unaltered zone. The characteristic bright colours of the oxidized ores was the principal surface indicator of these minerals. The most common copper ores, malachite and azurite, are green and blue respectively and so were easily noticed (Tylecote 1987). This is an idealised depth zonation of copper mineralization, which can vary from one mine to the next.

Having located the ore, the next problem was its extraction. As there is no known archaeological evidence of Early Historic mining in Ireland it is impossible to say whether both sulphide and/or oxide ores were exploited, or to identify the method employed in that exploitation. We do, however, have clear evidence from Ross Island for smelting of primary copper ore (chalcopyrite). One would expect the form of the mine to depend on the geology of the surrounding area, on the shape and size of the mineral veins or pockets, and on the encasing rock type. In Britain, no mines have been definitely dated to the relevant centuries. Some, such as Keswick in Cumbria, which was worked before 1250 AD, may have been exploited between the fifth and tenth centuries AD (Tylecote 1986, 13).

In the Bronze Age, fire-setting was the method of ore recovery employed at sites such as Mount Gabriel in Co. Cork (O'Brien 1994), the Great Orme in Wales (Dutton 1990), and the Mitterberg mines in Austria (Pittioni 1951). It involved the building of a large fire up against the rock-face to heat it. This was then allowed to cool naturally or was quenched with water. The fire-weakened rock-face was then susceptible to the use of stone hammers (mauls) and organic picks in the removal of ore. Fire-setting is well documented from historic contexts in Europe, used as late as the nineteenth century in Norway. No archaeological or literary evidence has, however, been found for the Early Historic period in Ireland. It is likely that the ability to mine rock using tempered iron tools did exist in Ireland from the Early Historic Period onwards. This is suggested by the use of iron chisels to dress stonework in early Irish church buildings and by the discovery of rock-cut souterrains with wall-stroke patterns indicating the use of iron tools (O'Brien pers. comm.).

Recycling

The mining of copper ore is only one source of raw material used in the production of fine metal objects in Early Historic Ireland. Metal may also have been acquired by melting down older objects or scrap. This is particularly relevant to gold as it is not susceptible to corrosion and is relatively easy to melt, making it ideal for recycling purposes. A hoard of Viking silver from Carraig Aille II, Co. Limerick (Ó Ríordáin 1949), provides possible evidence for the recycling of silver as does an older find of Roman hacksilver from Balline in the same county (Bateson 1973). The other non-ferrous metals could also have been recycled in this manner. While undoubtedly providing some metal, perhaps on the less wealthy sites, it is unlikely that it was the only source.

Importation

The final possibility is that of importation – of ore, scrap, or smelted cakes or ingots. Importation of raw material from Britain and the Continent would also have allowed for the associated introduction of new ideas and techniques from abroad. Internal trade would then have facilitated their dissemination. However, other mechanisms such as the arrival of Christian missionaries with metalworking experience or the importation of finished artefacts may be relevant. Internal trade or importation from native sources would also have facilitated the spread of new ideas and technology, though on a lesser scale. Of the non-ferrous metals, the importation of tin is likely, given its scarcity in Ireland. The metal was probably imported from the significant Cornish or Breton deposits. Cornwall is also rich in other non-ferrous metals, particularly copper and native gold, so the possibility exists that they too were imported. Mercury had to be imported from the Continent as it does not occur naturally either in Britain or Ireland.

In the twelfth century Geraldus Cambrensis recorded in his work, *Conquest of Ireland*, that some gold was indeed imported. He also records, however, that the main source of gold in Ireland was native ore, whether in the form of sediment or nuggets in rivers and streams. In *Conquest of Ireland*, Cambrensis advises King John to tax the Irish in gold. It is clear from this work that the Normans saw Ireland as a great source of gold. 'From the time accurate historical records were kept, native gold is known to have been exploited. So why should this tradition not stretch back further?' (Whitfield 1993, 23). It would seem more probable, if only for economic reasons and the tradition of native exploitation, that native sources provided the raw material, importation playing a less important role.

3.3. Processing of raw material

Once ore was removed from the mine it underwent 'dressing' or concentration (Fig. 14). This involved crushing the ore, probably with hand-held stone hammers, on large slabs or anvils. The mineralised rock was then hand-sorted to remove gangue or waste materials. Repeated crushings would further improve the mineral quality. There is also a possibility that water flotation may have been used to up-grade the mineral even further. The ore would then be transported to the smelting site for reduction to metal, if indeed this process was not carried out at the actual smelting site.

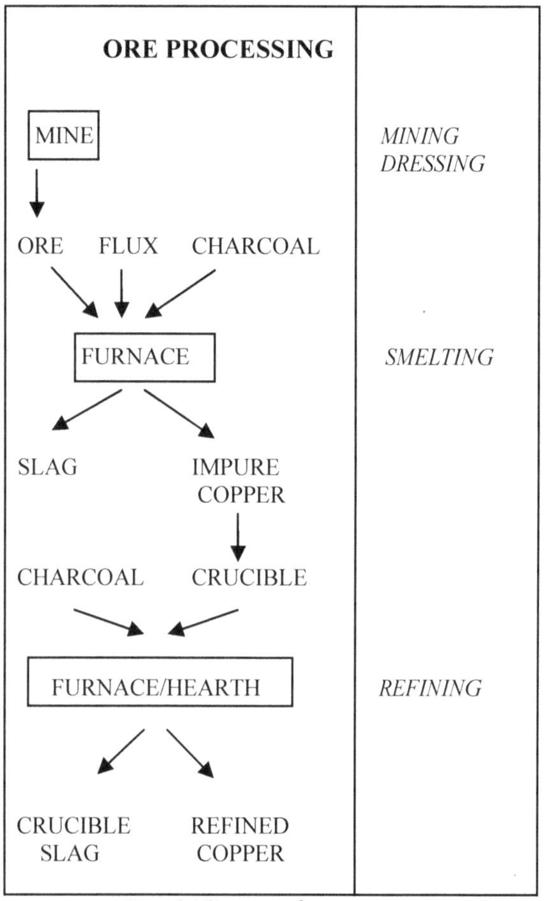

Fig. 14 Stages of ore processing.

Smelting process

The two main non-ferrous metals which require smelting are copper and lead. Gold was available in a native form and did not require smelting, while silver was produced through the cupellation of lead ore (see below). Oxidised copper ores such as malachite and cuprite are relatively easy to smelt and can be reduced without any significant slag production. When ore is placed into a furnace with charcoal at temperatures reaching 1100°C, reducing conditions are achieved. This reducing atmosphere causes the following chemical reaction:

Cu_2O	+	CO	→	$2Cu$	+	CO_2
Cuprite		Carbon Monoxide		Copper Metal		Carbon Dioxide

The carbon monoxide formed by the charcoal and air from the bellows reacts with the oxygen in the cuprite to form carbon dioxide gas, leaving behind the copper metal. The same basic principle of this carbon reduction process can be applied to the smelting of all copper oxides.

Matte smelting

While copper oxides can be smelted without any further preparation, sulphidic ore must undergo pre-treatment. There are two methods of treating copper sulphides, the most common sulphide being copper pyrites. The first method requires the prior roasting of the ore. Chalcopyrite (copper pyrites) when roasted to 700°C is converted first to copper sulphate and then to copper oxide, and the pyrite is converted to pyrrohtite (FeS). These are then smelted under reducing conditions to form copper metal. Alternatively the sulphide ores may be smelted to a matte, which is a 'liquid or solid mixture of sulphides, usually FeS and Cu_2S' (Tylecote 1987, 22) and a coarse slag. The matte is then roasted to an oxide and smelted a second time to produce copper metal.

The most commonly found evidence or residue of smelting on any site is slag. Because copper minerals are rarely, if ever, found in a completely pure state, waste materials are produced in the smelting process. The most common impurities are silica, iron, calcium and magnesium. These generally form silicate slags such as iron silicate or calcium silicate. Copper is virtually insoluble in slag (the most common slag being fayalite $2FeO.SiO_2$) and forms discrete globules during the smelt. These globules sink to the bottom of the furnace and there form an ingot or cake of copper. However, some can remain trapped in the slag. These 'prills' can subsequently be recovered from the slag by crushing it with a hammer stone, and then re-melted together in a crucible. Slag is a relatively common find on Early Historic sites but it is often unclear as to whether it is iron slag or slag from non-ferrous metals or glass (Bachmann 1982).

Tapping the molten slag in early furnaces would have been quite difficult. In a furnace where the temperature is not quite high enough the molten copper and slag will solidify too quickly to be run off. Tapping either slag or metal would cause substantial heat loss and possibly 'freeze' the smelting process. If a furnace was re-used on several occasions the slag would slowly begin to dissolve its lining, tending to build up around the furnace walls. This would prevent the smooth flow of both molten slag and copper to the bottom of the furnace where tapping could have occurred (Craddock 1990). Pure minerals such as malachite, cuprite and chalcocite can be easily smelted in the above manner whereas other minerals require a 'fluxing agent' to facilitate the formation of molten slag. Sulphides generally have an excess of iron and need fluxing with silica (sand) to form fayalite, oxides have an excess of silica and need fluxing with iron oxide. If the excess material or impurity is lime (CaO) or magnesia (MgO) then both iron oxide and silica must be added. The copper metal being of a higher density than, and insoluble in, fayalite will fall through it to the bottom of the furnace. Occasionally some minerals may be self-fluxing in that they contain enough iron and silica compounds to form a molten slag (Tylecote 1987).

Early copper smelting furnaces are known from Huelva province of south-west Spain (Rothenberg and Blanco-Freijeiro 1981) and Mitterberg and Trieben in Austria (Tylecote 1987). The former consist of a very simple bowl-shaped hearth, dug into the ground or placed in a natural hollow. The inside of this furnace pit was often lined only with a layer of clay. The earliest furnaces were not tapped, the slag being removed only when it had solidified after the furnace had cooled down. The lining of such a furnace had to be replaced and repaired regularly ('fettled') due to the damage caused by successive high temperature smelts and the physical removal of the furnace contents once the smelt was completed (Craddock 1995, 198–202).

The examples found in Mitterberg and Trieben are twin furnaces. The evidence from St. Lorenzen near Trieben is dated to the Urnfield culture *c.*1000 BC. Analysis of residues indicates the smelting of copper sulphides in two stages, with one furnace employed in matte smelting and the other in reduction smelting. The furnaces were constructed of clay, placed in a pit and probably had a low superstructure. They measure 0.55m x 0.45m internally and are very similar to the Mitterberg examples (Tylecote 1987, 127–131).

In order to reach the high temperatures required in smelting an artificial draught is needed. Either bellows or hollow blowing-sticks (possibly reeds) were used for this purpose. Bellows were made of organic materials such as wood and animal skin which decay rapidly, explaining the lack of archaeological evidence. An efficient furnace would most likely employ a pair of bellows, sometimes even two pairs, in order to maintain a continuous blast into the furnace. Both bellows and blow-pipes were connected to the furnace with tuyères. A tuyère is a baked clay nozzle which can withstand the high temperatures of the furnace. Pre-fired tuyères would be much more effective and enduring; the discovery of a tuyère on a site indicating the use of bellows or a blow-pipe.

Tin is smelted in a similar fashion, at temperatures reaching 1000°C (Table 2). Iron ore is used to flux the siliceous gangue to produce fayalite. Mercury was produced by the smelting of cinnabar in a furnace which yielded mercury vapour and sulphur dioxide. The mercury was condensed by cooling the vapour, perhaps in a vessel of water. The most common lead ore, galena, is extremely easy to smelt. As a sulphide (PbS), galena must be roasted before smelting. Both roasting and smelting can be undertaken in the one procedure. The low melting point of lead (327°C) and the fact that its oxide can be reduced to metal below 800°C means that a domestic fire of dry wood or charcoal is sufficient for the operation. Once placed in the fire, part of the galena will be roasted to its oxide (PbO). This in turn reacts with the un-roasted galena (PbS) to produce lead (Pb) and sulphur dioxide (SO_2) which is expelled as a gas:

$$2PbO + PbS \rightarrow 3Pb + SO_2$$

This is known as a 'Double Decomposition Reaction', as both the oxide and sulphide decompose to produce lead.

°C	Melting and Boiling Points
183	Lead solder melts (lead 38%, tin 62%)
232	Pure tin melts
327	Pure lead melts
357	Mercury boils
420	Pure zinc melts
779	Silver solder melts (copper 28.5%, silver 71.5%)
878	Gold solder melts (copper 18%, gold 82%)
907	Zinc boils
1000	12% tin bronze melts
1063	Pure gold melts
1083	Pure copper melts
1150	Lead boils
1525	Pure iron melts

Table 2 Melting and boiling points of some metals and alloys (after Hodges 1964).

Cupellation

With very limited sources of native silver, the primary source of silver in Early Historic Ireland was probably the cupellation of argentiferous lead ore. This involves the oxidation of all base metals in the ore (lead, copper etc.) to a litharge which rises to the top of the furnace. Silver and gold are noble metals and as such remain un-oxidized. The litharge can be removed by skimming it off, through absorption in a bone-ash hearth, or given off in gaseous form. This process takes place in a shallow hearth of charcoal or wood at a temperature of 1000°–1100°C. A bellows is needed to achieve such a temperature and to facilitate oxidation. To separate the gold from the silver, the impure silver must be granulated in water and then mixed with salt and clay in a charcoal furnace. Here, the silver is converted to silver chloride which is absorbed by the clay while the gold is unchanged and can be washed out. To recover the silver from the clay it must be mixed with lead and cupelled once again (*ibid.*).

Furnace evidence

Definite Early Historic furnace remains have recently been discovered at the Ross Island copper mines near Killarney, Co. Kerry. Excavations in 1995 within the mine complex revealed four concentrations of copper smelting slag, two of which can be associated with the use of two small metalworking furnaces. Three of these slag deposits are dated to the late sixth/early seventh century AD. The remains were found on the periphery of an early Beaker work-camp and mining site (O'Brien 2004).

Fig. 15 Early Historic Period copper-smelting furnace from Ross Island prior to excavation, showing dark slag-rich sediment on top of white fuel ash pit-fill (photo: W.O'Brien).

Charcoal-rich sediments containing local spreads of comminuted plate slag marked the location of the two furnaces, both of which were revealed by magnetic susceptibility survey prior to excavation. Small amounts of vitrified ceramic and slagged sandy clay were also found, possibly the remains of clay tuyères and furnace wall material (O'Brien 2004).

The first furnace consisted of a small circular, bowl-shaped pit [Context 784] with a maximum diameter of 0.50m and depth of 0.15m. This pit had a lower side opening, perhaps to facilitate the use of bellows or blow-pipe at this point. The sides of the furnace were intensely fire-reddened and the pit contained three undisturbed layers of fine silty sediment, probably fuel ash deposits. No metallurgical residues were recovered from the furnace, though it was covered by the slag and charcoal spread (Fig. 15). Immediately adjacent to this furnace was a second pit most likely employed in some aspect of the metallurgical operation, however its exact function is unknown.

The second furnace was found approximately 26m to the north of the first. Again, a black sediment containing large amounts of plate slag covered the actual furnace pit. This pit [Context 788] is oval-shaped and measures 0.44 x 0.39m with a central depth of 0.12m. Associated with this feature also were fuel ash deposits and small quantities of vitrified ceramic and slagged clay (O'Brien 2004). Scientific analysis of the metallurgical remains undertaken at the Institute of Archaeology, London revealed that these furnaces were employed in the smelting of sulphide ores (chalcopyrite) (Meyerdirks *et al.* 2004). Both of these bowl furnaces originally had a low shaft superstructure, which was destroyed on completion of the metalworking operation (*ibid.*). There is no evidence of metal casting on the site (no mould or crucible remains), therefore this evidence appears to indicate an Early Historic primary processing site. Radiocarbon dates place the smelting activity in the seventh/early eighth century AD. This somewhat substantiates the slightly later, ninth century, reference by the British monk Nennius to the mineral wealth of the Killarney area. Although this evidence does not confirm Early Historic mining at Ross Island, it does strongly suggest the possibility.

Identical plate slag was also found during nearby rescue excavations at Scrahane on the outskirts of Killarney town (Fig. 16). This slag was associated with a number of pit features, though no actual furnaces have been found (O'Donnell 2000). There are no available radiocarbon dates from Scrahane and the archaeological record there remains largely unpublished. The large number of slag deposits suggests an early metal production centre of considerable size. Possible smelting furnaces are known from a few Early Historic settlements. At Movilla Abbey, Co. Down (Ivens 1984) a shallow scoop had been cut into the subsoil, the fill of which contained much burnt material (Fig. 88).

Fig. 16 Scrahane, Co. Kerry (photo: W.O'Brien).

The excavator identified the feature as the base of a furnace, associated with a considerable quantity of slag of an unspecified nature, together with crucible fragments and scrap bronze. The latter may indicate the recycling of metal although the large amounts of slag would suggest primary smelting.

A more convincing example of a smelting furnace was uncovered at Moynagh Lough, Co. Meath (Bradley 1991). This furnace was bowl-shaped with a circular mouth. Constructed of clay, the base was carefully stepped creating different levels in the interior. It measured 0.14m deep with a maximum diameter of 0.35m and was constructed in a prepared hollow. The furnace yielded what the excavator considered to be a deliberately clean fill of sand and tiny pebbles. The use of this furnace feature at Moynagh Lough for non-ferrous metallurgy is suggested by its location in a metalworking area (Fig. 114) which produced finds including a tuyère, 67 crucible sherds (both pyramidal and oval-mouthed), over 600 clay mould fragments, and one kilogram of slag. Due to the relatively small amount of slag found, Bradley suggested that the furnace was employed in melting metal rather than smelting ore. It must be remembered, however, that a relatively pure ore does not produce a large amount of slag, while the melting of smelted and refined metal should produce very little.

Clay tuyère fragments have been found on Cathedral Hill, Co. Armagh (Gaskell Brown and Harper 1984), Dunmisk, Co. Tyrone (Ivens 1989) and Reask, Co. Kerry (Fanning 1981). While these finds indicate the presence of a furnace, no such structures have been definitely identified in the archaeological investigation of these sites.

Refining

Refining is a vital step in the preparation of a metal for casting, excepting native copper and gold which are relatively pure. The impurities remaining in smelted copper would hamper the casting process, if not totally prevent it. Such impurities can include large quantities of iron, some residual slag, and smaller amounts of arsenic, antimony, lead, oxygen and hydrogen. The refining of copper may be simply carried out by re-melting the metal in a crucible under oxidizing conditions. Impurities which are less noble than copper will oxidize first, e.g. iron, arsenic, antimony, zinc and tin. The main impurity is usually iron which, due to its lower density, will rise above the molten copper where it can be oxidized off. Tylecote suggests that any hydrogen remaining in the smelt was reduced with cuprous oxide or malachite (1987, 193). Another suggested method of refining is 'poling', the stirring with a stick to evolve hydrogen gas and carbon monoxide. As Tylecote points out, however, there is 'little evidence that it was done in early times' (1987, 193). Similarly, there is no direct evidence for the use of malachite.

The process of refining metal using malachite helps avoid the 'gassing' of an object during casting. 'Gassing' describes the release of dissolved gases on solidification, leading to porosity of the metal. The method of poling, however, can leave behind hydrogen which leads to the problem of gassing. To remove gold and silver from the copper, lead or lead ore is added during smelting or refining. The precious metals dissolve in it and the mixture later removed by melting it out from the cakes of impure copper. Lead has a much lower melting point than copper and will melt first, carrying the precious metals with it. These can then be recovered from the lead by cupellation (Tylecote 1986).

Alloying

An optional, and final, step before casting is alloying, the 'mixing' together of two or more metals. Tin-bronze was obviously the most common alloy, but others were commonly used in Early Historic times. In the production of bronze, tin was most likely added in the form of its main ore cassiterite (SnO_2) though tin metal could also be added directly. Lead may occasionally have been added to lower the melting point of the alloy. This makes it easier to cast as it reduces the speed required to pour the metal into the mould (it remains molten longer). Although used in the Late Bronze Age, lead was not widely used as an alloying agent in the Early Historic Period as it is not soluble in bronze. It remains as globules which may join up if too many are present, forming weaker areas of lead 'lakes'. Both tin and lead can be added to the molten copper once it has been refined.

The alloying of gold with silver was practiced in the Early Historic Period, producing electrum. This consists of gold with a high percentage of silver, though sometimes a small percentage of copper was also added. The addition of the latter two lowered the melting point of the metal, increasing the ease with which it was worked and its wear resistance.

3.4. Initial fabrication

If a secondary source of raw material, such as scrap or ingots, was being used, then the only preparation necessary for casting was the melting of the stock and, perhaps, alloying. The stock was most likely broken up into very small fragments, placed in a crucible and heated in either a hearth or the top of a smelting furnace.

The initial fabrication of most non-ferrous artefact types involves some form of casting, though objects could also be made by forging. Simple objects can be hammered out of ingots in a process that is archaeologically invisible. Some composite objects were made by joining cast and forged pieces e.g. the Ardagh Chalice is composed of beaten sheet-metal and cast handles. Integral to the casting process was the mould which determined the shape and form of the object to be manufactured.

Moulds

Three different types of mould were employed in the Early Historic Period: the open mould, the bivalve mould, and the *cire perdue* ('lost wax') investment mould. The open mould is the simplest type and was made of stone, most commonly sandstone. The shape of the required object was cut into the stone which was probably covered by a flat cap-stone to reduce the formation of air bubbles on the surface during cooling. Objects cast in such a mould have flat upper surfaces and therefore need further cold-working and annealing. Moulds without a cover would have been inefficient as much metal would be lost through surface oxidation. In Early Historic Ireland, open moulds were used mainly for casting small metal ingots and unfinished artefacts. Examples of open stone ingot moulds are known from Garranes, Co. Cork (Ó Ríordáin 1942), Lagore, Co. Meath (Hencken 1950), and Cathedral Hill, Co. Armagh (Gaskell Brown and Harper 1984).

Most commonly employed was the bivalve mould, made of clay, stone, and possibly metal. Used in conjunction with an internal clay core, these moulds could be used to manufacture socketed objects. Stone bivalve moulds had to be carved by hand, while clay moulds were made by impressing a (usually) wooden pattern into each half. When the pattern was removed, the interior of the mould half may have been dusted with soot to prevent the metal sticking to the clay.

In the production of these clay moulds, dowels or moulded depressions were added as a method of registering both halves accurately. The first valve to receive the impression was dried and baked. When completed, the pattern was replaced in it and the second valve created by impressing the first (with the pattern *in situ*) into the wet clay. This ensured a completely accurate surface match of both halves. The second half is then baked in the same manner as the first. Before baking, vents (to allow the escape of gases) and a feeding cup or 'gate' were added to the valves. Examples of bivalve clay moulds are known from Lagore, Co. Meath (Hencken 1950), Moynagh Lough, Co. Meath (Bradley 1991), Garranes, Co. Cork (Ó Ríordáin 1942).

The *cire perdue* or 'lost wax' technique involves the use of one-piece clay moulds. A pattern or model of the required item was first carved in wax and then covered with a layer of fine clay to form the mould. As with all moulds, a hole or gate was left in the clay. This was then heated, baking the mould and melting the wax. The mould must be broken to retrieve the cast object (Hodges 1964). This may be the reason for the lack of identifiable *cire perdue* mould remains. The neck of the Ardagh Chalice and the shank of the Tara

Brooch pin were cast using the *cire perdue* method.

Lead, due to its low melting point (327°C), can be used in a similar fashion. Lead models are known from Lagore, Co. Meath (Hencken 1950), and Moylarg, Co. Antrim (Buick 1893). At the Iron Age site of Gussage All Saints, Dorset, 7380 broken clay mould fragments were uncovered. These moulds were used in the lost wax process 'and the uniformity of the objects which would have come from the moulds, suggests that master moulds were used to produce the wax models' (Foster 1980, 185).

Both open and bivalve stone moulds, if treated correctly, could be used for numerous castings. To prevent the mould cracking and shattering due to thermal shock when the molten metal is poured in, it must be slowly pre-heated to at least 200°C. When required, a clay core was used. The parting line of a bivalve stone mould allowed for the escape of gases, some valves have crude incisions or grooves on their surfaces to aid this. Bivalve stone moulds were tied to secure them together.

The invention of the clay core, which could be used with both clay and stone moulds, was a very important step in the development of casting technology. A good core had to be permeable to allow gases released on solidification to pass through. Cores were best made of clay or a mixture of clay and sand tempered with straw and were held in place by core-prints and/or chaplets.

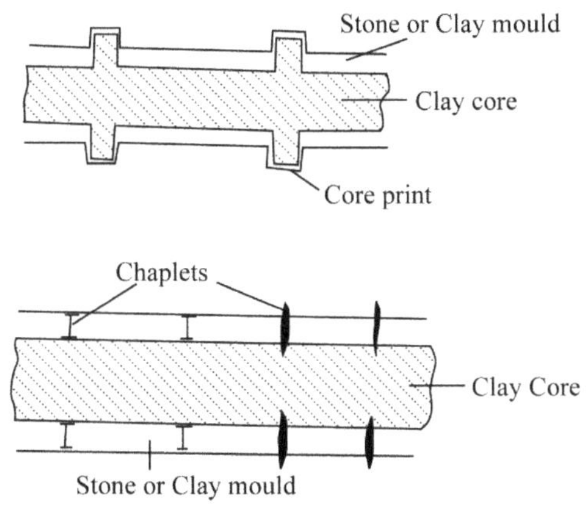

Fig. 17 Core prints and chaplets (after Tylecote 1987).

Core-prints are holes made in the moist clay of the mould before firing, holes which would be filled by clay projections from the core. They help secure the core in place and also create rivet holes in the finished object (Fig. 17). Also used for this purpose were chaplets. Coghlan describes a chaplet as a 'small piece of metal which is pressed into the surface of the core, and projects the required distance so as to maintain a true register between the core and surrounding mould' (1951, 57). These chaplets should be of a higher melting point alloy than the metal being cast so as to fulfil their function before melting into the cast. In addition to their primary function, they also maintain a uniform thickness in the metal of the object being cast (Fig. 19).

The biggest problem experienced in casting is gassing. Water vapour, formed by the combustion of hydrogen in the fuel, is reduced by the metal being melted:

$2Cu$	+	H_2O	$\rightarrow$	Cu_2O	+	$2H$
Copper		Water		Copper Oxide		Hydrogen Gas

The hydrogen so formed enters the metal only to be released as gas bubbles when that metal solidifies, spoiling the cast. To prevent this, large feeders (gates) and risers were added to moulds. A riser is an excess part of a casting and is discarded. Therefore, the formation of gas bubbles on its surface will not affect the finished artefact. The feeder or gate is the hole through which the molten metal is poured into the mould. To compensate for the shrinkage of metal on solidification, additional molten metal can be poured in via the gate.

Both bivalve and *cire perdue* moulds were made of clay. Unlike their stone counterparts, most clay moulds could only be used for one casting. Large-scale output of replicated shapes was achieved through the use of an enduring pattern of wood, or metal. The clay used in making moulds was usually of two types: a carefully worked fine clay mixed with sand to minimise shrinkage (on drying) on the inside, and an external or outer layer of coarser clay. Clay moulds were made by 'investing' the wooden, wax or metal pattern in wet clay. Both the mould and core had to be thoroughly dried before casting to prevent gassing. A clay mould develops small cracks during drying, cracks which act as vents during casting.

Moulds were also made of metal, although there are far fewer finds of such than their stone and clay counterparts. In fact, there are only 23 known examples from Britain, and none from the Early Historic Period in Ireland. Metal moulds are direct reproductions of clay moulds and are instruments of mass production, capable of numerous castings if handled correctly. The mould is usually made of the same metal as the objects being cast. For this to be effective, the mass of the mould must be great enough to avoid melting when the molten metal is poured in. Again, a coating of soot is used to insulate and to prevent sticking. Tylecote raised the 'possibility that they were often used for wax or lead patterns and not always for direct casting' (1962, 124).

Crucibles

Used in conjunction with the mould, another vital piece of equipment in casting is the crucible. A crucible is a small vessel used for melting and pouring metal (Fig. 25). The vast majority used in Early Historic Ireland were made of clay, however some stone crucibles are known from Garranes, Co. Cork (Ó Ríordáin 1942). The type of crucible in use at any given time depended to a large extent on the level of ceramic technology and the methods of lifting available. Some vessels have handles or projections to facilitate gripping with either iron tongs, a pair of freshly cut green withies (sap preventing burning), or two pieces of charred wood.

The smelted metal is broken up (unless in the form of prills) into the crucible which is then placed in the melting furnace. Once melted, the metal had to be poured very quickly, within approximately five seconds of removing it from the furnace, before it solidified. The use of vessels with lids or the use of leaded bronze allowed more time. The re-melting of metal may, of course, have taken place in the smelting furnace and would result in traces of slag on the exterior of the crucible.

It has been stated by numerous authors that Early Historic crucibles from Ireland are all very small when compared to the items produced. This, in fact, is not the case. Firstly, the vast majority of cast fine metalwork (e.g. ring-pins, penannular brooches) are actually quite small and light, not requiring vast quantities of metal. Even larger objects are usually of a composite nature, their individual components not requiring large amounts of molten metal. Secondly, larger crucibles are known e.g. Appendix 3, no. 3. Such a crucible could easily hold the volume of molten metal necessary to cast larger objects, such as the silver brooches from the Ardagh Hoard.

In the Early Iron Age in Britain (there is no Irish evidence for the same period), the crucibles used were mainly large triangular shaped vessels. During the Roman Period in Britain, native sites continued to use these crucibles while circular, conical, and 'D'-shaped vessels are found on Roman sites. Pyramidal, bag-shaped and flat-bottomed crucibles were used in Early Historic Ireland (see chapter 4 for detailed description). In Saxon England, bag-shaped vessels were used, while in Medieval times a wide variety of crucibles were in use. Graphite and refractory clay crucibles were used in the eighteenth-nineteenth century period, right up until the introduction of the modern silicon carbide crucible (Tylecote 1987, 184–189).

Cast Decoration

Although a form of decoration, *kerbschnitt* must be included as part of the initial fabrication of a metal object as it is executed prior to casting. *Kerbschnitt*, or 'chip-carving', is a technique borrowed from the Anglo-Saxons and was originally used in wood-carving. A chisel and hammer are used to chip out areas of the material being worked, generally the background to a design, leaving the latter in relief (Ryan 1993).

The Early Historic Irish adapted chip-carving by applying it to the wax pattern prior to the manufacture of the mould. Irish *kerbschnitt* is a form of cast decoration, undertaken before, not after, casting and is indicative of *cire perdue* casting. A good example is found on the neck of the Ardagh Chalice (Organ 1973). Other decorative elements, such as animal heads, were also cast, being added to the pattern before mould manufacture, e.g. the protruding figures on the Tara Brooch (Ryan 1983b).

Sheet Metal

Sheet metal was integral to the formation of numerous artefacts in Early Historic Ireland. Sheet metal is made by casting the metal into as thin a form as possible. This is then carefully hammered with rounded bronze hammers to prevent causing damage, while continuously annealing the metal. The sheet was then cut and twisted into the required shape, joined and decorated if necessary. Small pieces of sheet metal were integral to the fabrication of many decorative panels, acting as a base for filigree, *pressblech* or *repoussé* work. Such panels may then have been attached to larger objects of sheet metal. The Ardagh and Derrynaflan chalices were made in this way. Sheet metal was also used to manufacture items including tweezers, rings, bracelets and some brooch terminals.

Lathe-turning

A lathe is a piece of equipment used to hold and turn an object being shaped with a separate tool (Ryan 1992). The most likely type of lathe employed is a pole-lathe. This consists of a revolving wooden pole with a former of the required shape at one end, a treadle used to drive the pole. In the case of a vessel, the thin sheet of metal is placed over a bowl-shaped former. However, true spinning or turning (resulting in a fully worked metallurgical structure) was probably beyond the capabilities of the pole-lathe. It seems likely that a vessel was beaten into shape and then polished on the lathe to finish it. Polishing affects the surface of the metal and not its full thickness. The foot and bowl of the Ardagh Chalice are lathe-polished (Organ 1973) as are those of the Derrynaflan Chalice, and the dish of the Derrynaflan Paten (Ryan 1983a).

Joining techniques

Several methods of joining metal sections together were employed in Early Historic metalworking. 'Pressure welding' is the hammering together of two pieces of heated metal, whereas 'burning' is the joining of heated solid metal with molten metal. 'Casting-on' is a term used to describe the making and attaching of a piece in one operation. This was done by constructing the mould of the additional piece around the pre-existing object (Fig. 18).

'Soldering' involves the use of an alloy with a lower melting point than that of the metal(s) being joined. Copper-alloy, for example, is used as the solder for gold. To solder two pieces of gold together, a strip of copper-alloy is placed between them. The whole is then heated. When the copper-alloy melts and flows over the edges of the adjoining metal, the heating is stopped. The copper-alloy re-solidifies securing the two gold pieces together.

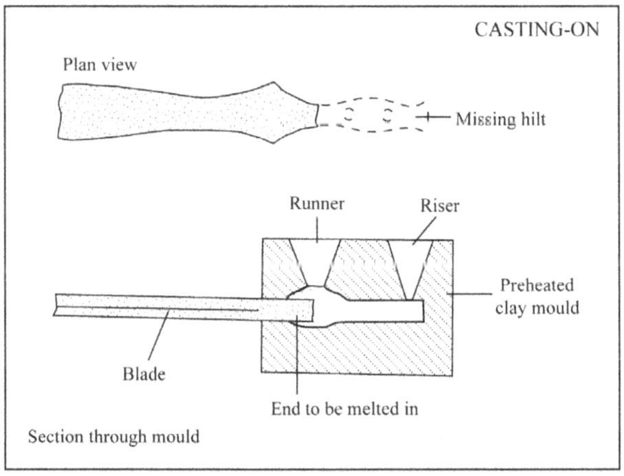

Fig. 18 Casting-on (after Tylecote 1962).

Rivets were used for joining, usually on less ornate items or where they could be disguised and integrated into a design, e.g. the handles of the Ardagh Chalice. Fine objects from Early Historic Ireland are very often composite items consisting of individual panels of decoration. A peculiarly Irish method of attaching such panels to the larger object is known as 'stitching'. The setting which will receive the panel is cast into the object. A tiny vertical cut downwards is made in the wall of the setting. The cut section automatically curls downwards to lap over the edge of the panel. A series of such tabs in the surrounding setting walls successfully secures the panel in place. Due to their tiny size they are virtually invisible to the casual observer (Fig. 19).

The attaching of the filigree panels on the front of the Tara Brooch is a classic example of this method.

3.5. Artefact decoration

The technical expertise evident in the decoration of fine objects is one of the most celebrated aspects of Early Historic Period metalwork. One of the initial steps undertaken in the decoration of some objects was the alteration of their overall appearance either through tinning e.g. two bronze studs from Garranes (Ó Ríordáin 1942) or gilding e.g. the panels of the Moylough Belt-Shrine, the Tara Brooch and the Cavan Brooch.

Tinning involves the application of tin-rich layers to the surface of an object. Tin was applied either by rubbing heated tin directly onto a surface or by dipping the whole object in molten tin. This layer of tin was later transformed into a tin-rich layer by continued heating after application. This causes the tin to inter-diffuse with the copper of a bronze object (Tylecote 1987). The method of gilding used in Early Historic Ireland and Anglo-Saxon Britain is known as mercury or fire gilding, of which there are two different techniques of application. The particular technique used on any specific artefact is impossible to identify, even by scientific means as the end products of both are identical (Oddy 1980).

One technique of mercury gilding involves the thorough cleaning of the base metal object to be gilded. When this is done, an amalgam or alloy of gold and mercury is applied to the surface or desired part of the surface. The object is then heated which causes the mercury to evaporate off, leaving behind a gilded surface ready for polishing. The amalgam or alloy is created by adding gold leaf or gold filings to boiling mercury, the most probable ratio being one part gold to seven parts mercury. When the amalgam cools down it is applied to the desired surface (*ibid.*).

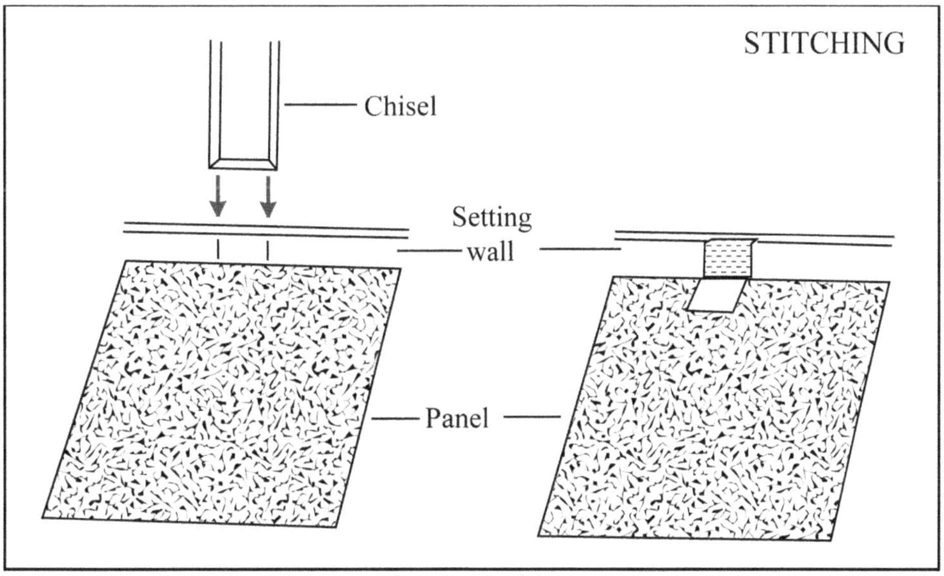

Fig. 19 Stitching.

The second method of application also involves the use of mercury and gold leaf. In this case, the surface of the object to be gilded is first amalgamated by rubbing clean mercury onto it. As Oddy points out (*ibid.* 131), this is not an easy process to carry out; the amalgamation of an object's surface is much easier if the whole object is dipped into a soluble mercury salt solution such as mercuric nitrate. However, the date of introduction of this method is uncertain.

After the surface has been amalgamated, gold leaf is placed on top and immediately dissolves in the mercury. A smooth piece of leather is used to press layer after layer of gold leaf onto the surface until the mercury's saturation point is reached and no more gold will dissolve in it. Again, the object is heated to evaporate off the mercury, leaving the gold. Mercury or fire gilding can be identified by scientific means, using a combination of microscopic examination and either emission spectroscopic analysis or x-ray fluorescence testing. Mercury gilding has been identified as the technique used on items from Anglo-Saxon Britain such as some of the objects from the Sutton-Hoo burial (Bruce-Mitford 1983). The best known object from Early Historic Ireland which has been mercury gilded is probably the early eighth century Tara Brooch.

A vital aspect of gold-working, in particular, was the creation of gold foil or leaf. Foil was manufactured in exactly the same way as sheet metal. The malleability of gold makes it possible, with due care and caution, to produce very thin gold foil or leaf. Gold foil was clearly used in gilding, while metal foil or very thin sheet-gold was used extensively as the basic component of individual decorative panels.

Filigree

In both Early Historic Ireland and Anglo-Saxon Britain many of the best examples of fine metalwork are decorated with filigree. Filigree is a type of ornament formed by soldering metal wires (usually gold) onto a metal background. An essential element of filigree is the metal wire used to form the design. There are many different ways of manufacturing metal wire, the technique employed on these islands in the relevant centuries was that of 'block-twisting'. This involves the twisting of a square or rectangular sectioned metal rod and its subsequent rolling between two plane surfaces. This produces a solid, even, round wire with a double helical seam.

'Celtic' filigree describes the type of filigree used to decorate the surface of objects, to decorate and define the edges of stud cells, and may also be mounted on the surface of the stud itself. Anglo-Saxon filigree also sees the edging of stud cells, while the closest parallel to the mounting of wire on the surface of a stud, as done in Irish work, are the enamel settings with inset gold wire on two Merovingian buckles from Kaiseraugst and Amiens (Whitfield 1987).

A common characteristic of both 'Celtic' and Anglo-Saxon filigree is the use of small foil sheets as back-plates onto which the wires were soldered. Craddock (1989, 173) suggests that the reason for their use is to reduce the amount of heat needed to melt the solder. Clearly, it will take less time and lower temperatures to heat a thin sheet of foil and its solder than a whole object such as a penannular brooch. Also, the application of complex filigree to individual components allows for greater quality control. In the treatment of these back-plates, Whitfield (1987) points out many shared 'Celtic' and Anglo-Saxon characteristics. An example of this is the mounting of filigree on a relief design, executed by stamping or *repoussé*.

The use of a 'hollow platform' is another technique of application common to both 'Celtic' and Anglo-Saxon work. The technique involves the use of two back-plates, one of which is flat, the other one embossed with the design (stamping or *repoussé*). The background to that design is then cut away to leave a raised openwork ornament resting on the flat back-plate. The filigree is soldered on top of this. 'Celtic' objects showing this technique include the Hunterston Brooch (Stevenson 1983), the Ardagh Chalice (Organ 1973) and the Derrynaflan Paten (Ryan 1983a), Anglo-Saxon objects include the Faversham bird-buckles (Whitfield 1987). Again, the technique is employed for the same types of motif, suggesting a connection between 'Celtic' and Anglo-Saxon work.

The central element of filigree, the wires and patterns of application employed, reflect to some extent the similarities observed in the treatment of back-plates. Plain round wires were rarely, if ever, used in either 'Celtic' or Anglo-Saxon work. Instead, the most commonly used type of wire is the 'beaded' wire. Beaded wire (milled as opposed to truly segmented) can be made in several ways, but it is interesting to note that both 'Celtic' and Anglo-Saxon beaded wire was hand-made, using a grooved tool applied at right angles.

The use of twined wires is common to both islands, as is the use of granulation, triple bands, and carpeting with twined and/or beaded wires. Granulation is found on the brooches from Hunterston, Dunbeath and Killamery (all three in Youngs 1989), and on the Sutton Hoo shoulder-clasps (Bruce-Mitford 1983) and large Taplow buckle (Whitfield 1987). A triple band consists of three parallel wires, the centre one usually different to those on either side. Two types of triple band are common to 'Celtic' and Anglo-Saxon filigree; a central beaded wire between two finer beaded wires, and a two-ply twined, beaded wire flanked by single beaded wires e.g. the large triangular panels on the 'terminals' of the Tara Brooch. Carpeting is found on the Hunterston Brooch (Youngs 1989), the Sutton Hoo strip (Bruce-Mitford 1983), and the bird buckles from

Faversham (Whitfield 1987), the first mentioned generally considered of Irish manufacture.

The use of twisted gold ribbon and flattened beaded wire is not found in Anglo-Saxon work. The 'Celtic' smith clearly was not content with merely copying techniques from elsewhere, and added a uniquely Celtic aspect to filigree work. This took the form of a three-dimensional element in the use of superimposed wires. Beaded wire on top of a ribbon set on edge can be seen on the Hunterston, Tara, and Westness brooches (all three in Youngs 1989). Various new combinations of wires and the use of cones of beaded wire were also a 'Celtic' innovation and can be seen on such objects as the Tara Brooch, Ardagh Chalice (Organ 1973), Derrynaflan Chalice and Paten (Ryan 1983a), and the 'Queen's' Brooch (Youngs 1989).

Tools used in filigree work would have included tweezers with which to manipulate the wire, e.g. Lagore (Appendix 3, no. 337) and possibly shears, used to cut the metal sheet or foil. Iron shears are found on several of the relevant sites, e.g. Lough Faughan (Collins 1955). Hammers, like that found at Lagore (Hencken 1950), and chisels as at Garryduff (O'Kelly 1963), were required to cut the wire and sheet.

When the panel of filigree was complete and the wires soldered onto the back-plate, the panel was then attached to the larger object. On the Hunterston and Croy Brooches (Youngs 1989) rivets were used, the heads of which are each collared by a small ring of decorative wire. On the Ardagh (Organ 1973) and Derrynaflan (Ryan 1983a) chalices, and on Anglo-Saxon buckles (Jessup 1950), metal frames are used to hold panels of gold ornament in place.

Stitching, on the other hand, appears to be a uniquely Irish technique of securing panels in place. There is no evidence of gold wire stitching as seen on Late Bronze Age gorgets. It has been suggested that 'the Celtic goldsmith learned many of the basic techniques of filigree manufacture from their neighbours, but they also appear to have developed new techniques and combined established types of ornamental wire in novel ways' (Whitfield 1987, 82).

Miscellaneous decorative techniques

Pressblech is a term used to describe a certain method of decorating sheet-metal. It involves the use of a carved die or pattern upon which the thin metal is impressed or beaten. The use of *pressblech* can be seen on the silver panels on the underside of the Ardagh Chalice foot flange (Organ 1973). A possible *pressblech* die was found at Rathtinaun, Co. Sligo (Raftery unpublished; see chapter 4), while it has been suggested that the designs on some bone motif-pieces may have been used in this way (Ó Meadhra 1987b).

Trichinopoly is an important, if somewhat rare, technique employed on the Ardagh Chalice (Organ 1973), the Derrynaflan Paten and Strainer (Ryan 1983a), and Tara Brooch (Ryan 1983b). The technique is also known from as early as the Iron Age, in the necklet from the Broighter Hoard. It involves the weaving or knitting of metal wire to produce a seamless, chain-like tube. To be used as a flat panel of decoration (as on the Ardagh Chalice), it was cut down one side and then laid flat.

Fig. 20 Athlone crucifixion plaque (photo; Dept. of Archaeology, NUI, Galway).

Chasing is a technique employed in decorating either sheet-metal or whole objects. A hammer and punch of the type found at Lagore (Appendix 3, no. 323) are used to create lines and areas of decoration composed of a series of oblique punched indentations. This technique was often used in conjunction with *repoussé*, finishing from the front designs raised from the back. Both are found on the Athlone crucifixion plaque dated to the eighth century (Fig. 20; Ryan 1983b, 120). With this form of decoration no metal was lost or removed from the object. The opposite is true of engraving where lines are created by removing metal with a hammer and chisel, e.g. the compass-drawn marigolds on the reverse of the Ervey brooch terminals (Fig. 21).

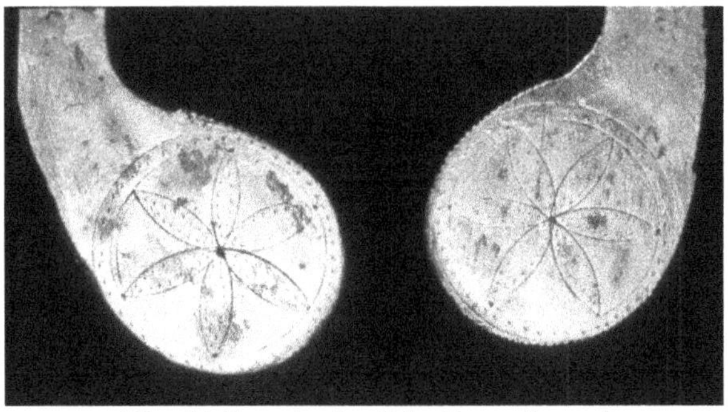

Fig. 21 Terminals of the Ervey brooch (photo; Dept. of Archaeology, NUI, Galway).

Repoussé is an important technique used in the decoration of sheet metal. It involves the hammering of a design into a thin sheet of metal from the back, leaving the design embossed on the front. The design may occasionally have been further enhanced by additional work from the front.

3.6. Non-metal ornament

Although not strictly part of non-ferrous metalworking technology, many objects of fine metalwork were enhanced by the addition of other decorative materials. While finely decorated objects of bronze, silver, and gold were highly valued, the presence of glass, enamel, millefiori, or amber ornament indicates an object of the highest status. The quality of technical execution evident in the application of metal ornament is paralleled in the application of the non-metal ornament.

Enamel

The most common form of non-metallic ornament applied to fine metalwork was probably enamel, a type of glass. 'By glass, we usually understand a hard material, often transparent or translucent, which is made by heating together a mixture of materials such as sand, limestone and soda at a very high temperature' (Frank 1982). Enamel has much the same composition with the added ingredient of an opacifier. The opacity of red enamel is caused by the presence of copper, the opacity of white enamel by the presence of tin oxide, and that of yellow enamel by lead-tin oxide.

The resulting vitreous material may have been applied to an object in powdered form. This was then fused, by heating, to produce the familiar glass-like surface. Alternatively, powdered enamel could have been melted in a crucible and then poured into the prepared setting. Both white and yellow enamel, when heated, also fuse to the metal object. Before applying red enamel, the metal surface must be keyed to enable the enamel to adhere.

There are two different types of enamel application; *champlevé* and *cloisonné*. *Champlevé* (literally translated as 'raised field') describes the use of molten enamel to fill cut or cast recesses, very often areas forming a background against which a metal design was left visible. Good examples of this can be seen on the Dowris latchet and on the terminals of the Arthurstown penannular brooch (both in Youngs 1989).

Cloisonné refers to the use of enamel to fill small cells or compartments formed by wire or strips of metal, as used for decorative studs. This was often applied in the form of enamel chips, similar to gem-setting or garnet-setting in Anglo-Saxon work. Pseudo-*cloisonné* enamel is very often found in association with blue glass in the construction of studs. These studs are manufactured by inserting enamel into a prepared grille from the back not clasping chips of enamel in a setting. Examples are found on the Ardagh Chalice (Organ 1973), Moylough Belt-Shrine (O'Kelly 1965), Derrynaflan Paten (Ryan 1983a), and the Tara Brooch (Ryan 1983b).

There are two main types of glass stud – those with a silver grille inlay and those with an enamel grille inlay. Studs were made by either casting the glass in a clay mould and then applying the metal or enamel inlay or, as O'Kelly (1965) suggests for the Moylough studs, the initial manufacture of a metal grille. This is placed in a clay mould, and then the enamel or glass applied, from behind, to the relevant sections of the grid. Cast glass ornament can be seen on the Tara Brooch in the form of two cast glass human heads on the hinged tab and the purple stud with petal-like ornament in centre-front (Ryan 1983b).

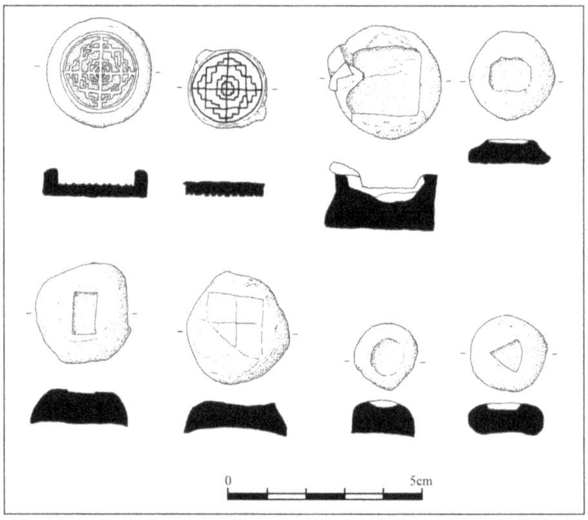

Fig. 22 Moulds and glass stud from Lagore, Co. Meath (after Hencken 1950).

Evidence of enamel and glass working is known from Lagore (Hencken 1950), Moynagh Lough (Bradley 1991), Cathedral Hill (Gaskell Brown and Harper 1984), and Garranes (Ó Ríordáin 1942). At Lagore a clay mould with a glass stud still *in situ* was found. The stud bears a cast sunken geometric pattern which would, presumably, have later been filled with a metal or enamel inlay or grille (Fig. 22). Moynagh Lough produced a block of yellow enamel, Cathedral Hill two sticks of enamel, and Garranes two fragments of crucibles with droplets of red enamel still attached. A stray find of a large block of red enamel is known from the Tara area, Co. Meath. Its hardness suggests it was used as a source of enamel chips (Youngs 1989, 201).

Millefiori

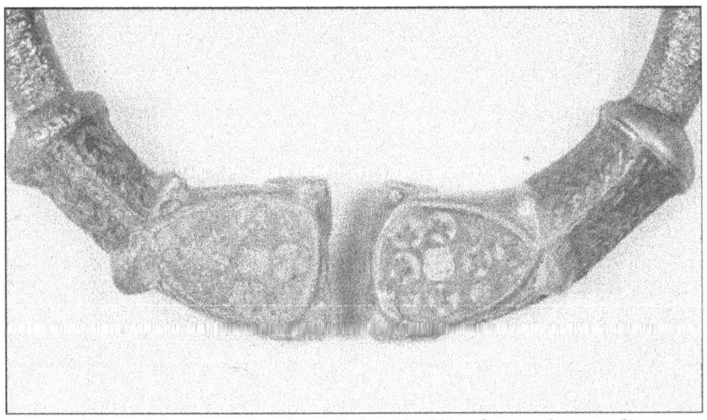

Fig. 23 Terminals of the Ballinderry brooch (photo; Dept. of Archaeology, NUI, Galway).

This is a type of ornament commonly found in association with enamel. Literally meaning 'thousand flowers', as the name suggests this originally consisted of floral motifs (Fig. 24). It involves the arranging together of a group of coloured glass rods so that their cross-section forms the required pattern. The rods are then fused to hold them together and drawn out to reduce their diameter. A thin cross-section or 'slice' of this was used as decoration, usually floated in a bed of red enamel, e.g. the terminals of the Ballinderry Brooch (Fig. 23; Ryan 1983b). In some cases several rods of millefiori are combined to form a larger, more complex pattern.

Millefiori is an ornament used in both the Saxon and Celtic worlds, though its treatment differed. The distinction can best be seen in the objects from the Sutton Hoo ship burial (Bruce-Mitford 1983). The gold regalia (two sword-mounts, purse-lid, and shoulder-clasps), of native Germanic origin, treats the millefiori as it would a precious stone, by cutting and shaping it to fit individual cells. These millefiori inlays were of the larger, more complex type of pattern, and contained antimony as an opacifier and colorant. From the same burial are two hanging-bowls of the Celtic tradition. This millefiori is of single rod type and is embedded in red enamel. Unlike the Saxon work, it employs tin compounds as opacifier and colorant. Fragments of millefiori rods are known from Garranes (Fig. 24; Ó Ríordáin 1942), Lagore (Hencken 1950), and the Scotch Street excavations in Armagh (Lynn 1988).

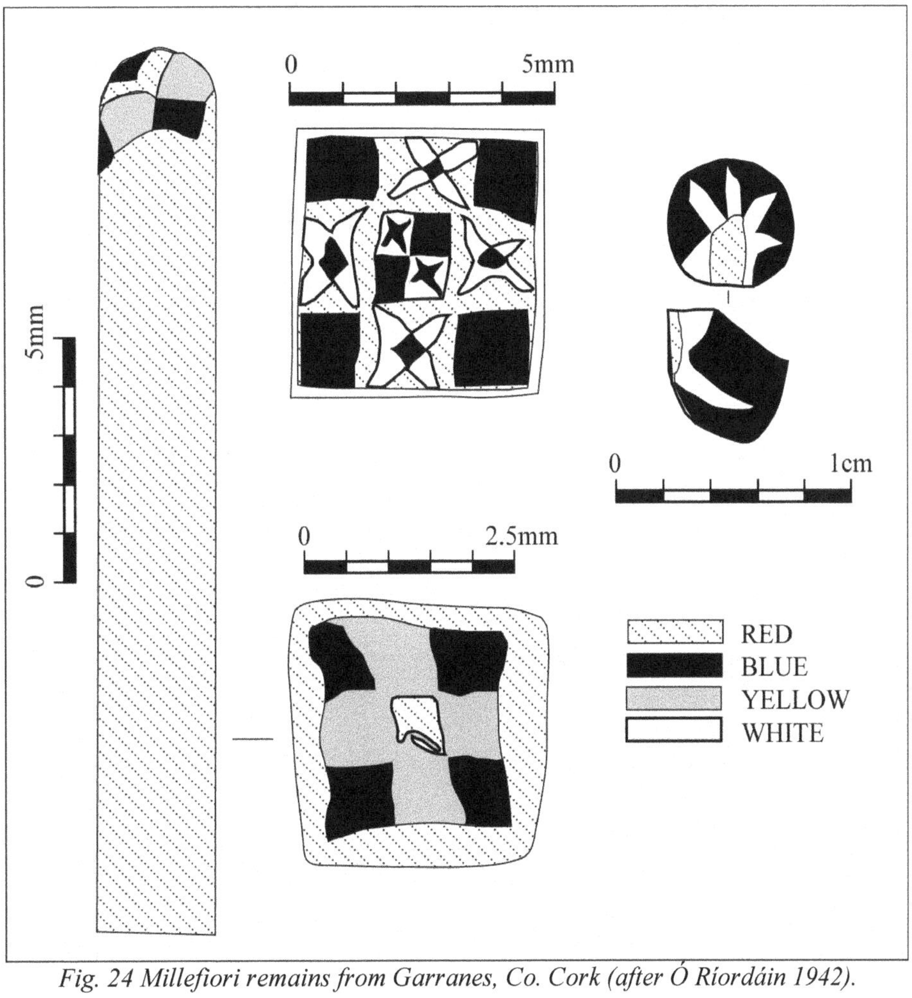

Fig. 24 Millefiori remains from Garranes, Co. Cork (after Ó Ríordáin 1942).

Amber

Amber is the final type of non-metal ornament used to decorate fine metalwork of the Early Historic Period. Amber is a fossilised tree resin not known in Ireland, but which can be found along the eastern shores of Britain and more famously, the sea shores of the West Baltic region. The use of amber during the Early Historic Period was not as extensive as it had been during the Bronze Age, nonetheless it was an important form of ornament. Amber was applied in two ways, either by heating it and then shaping the malleable substance with the fingers, or by grinding and polishing it into shape like a precious stone.

Amber studs can be seen on such items as the Tara Brooch, one of the Ardagh Hoard brooches (Ryan 1983b), the Hunterston Brooch (Stevenson 1983), and the Loughmoe 'Tipperary' Brooch (both in Youngs 1989). However, the largest collection of amber on any individual object from Early Historic Ireland is found on the Derrynaflan Chalice (Ryan 1983a) which bears 57 amber studs. All of these studs appear to have been shaped by grinding and polishing and are clasped in simple settings. Amber-working evidence is known from the Scotch Street excavations in Armagh (Lynn 1988), Viking Dublin (Ó Ríordáin 1971 and 1973) and possibly from Lagore (Hencken 1950) where an unfinished stud and a finished example were found. Amber appears to have taken the place of the more intricate and ornate polychrome glass studs in the ninth century AD.

The tools used in glass, enamel, millefiori and amber working may, of course, have been used in other activities such as metal, leather, or wood-working. None have been definitely associated with the application of non-metal ornament, though a few have useful excavation contexts. For example, an iron plate with a projecting strip at either end was found in a pit containing glass-working debris at Armagh (Scotch Street). This was possibly used for heating and/or rolling glass rods (Lynn 1988). Similarly, iron ladles have been found on sites producing evidence of glass-working and may have been used for melting glass (Craddock 1989).

The use of polychrome glass, enamel, and millefiori marks something of a high point in the fine metalwork of Early Historic Ireland. The replacement of this ornament by amber sees the beginning of a decline in the manufacture of metal objects; from this point on they lack any polychrome ornament.

4.7. Conclusion

Previous studies of non-ferrous metalworking technology from Early Historic Ireland tended to concentrate on the actual fabrication of objects, especially their techniques of decoration. This is understandable, considering the appeal of the finished objects and the lack of definite archaeological evidence regarding the acquisition and preparation of raw material. However, the discoveries at Ross Island provide the first definite evidence of primary copper smelting from Early Historic Ireland and the likelihood of copper mining. This certainly lends credibility to the theory of native ore exploitation and will, perhaps, generate more interest in the earlier processes of non-ferrous technology. Despite the lack of archaeological evidence, one would expect the high quality of technical expertise evident in the fabrication of an object to be reflected in the initial acquisition and processing of the raw material.

Early Historic metalworking initially drew inspiration from a rich tradition of native metalworking in the preceding Iron Age (Raftery 1994). In the decorative techniques employed, a connection between Early Historic Ireland and Anglo-Saxon Britain is evident. Such skills and ideas may have been transmitted along religious routes, by people travelling from one monastic site to another. The large monastic sites like Nendrum were equipped with non-ferrous workshops and were probably important centres for the specialist skills required. It is also plausible that these technical connections are related to the circulation of raw materials or finished artefacts. Viking silver was one of the former, though the Norse appear to have had little affect on native precious metalworking technology. From this admittedly one-sided examination of non-ferrous metallurgy, it seems that the Vikings made more of an impact on styles of decoration than the techniques employed. The Irish did not depend totally on technology or motifs inherited from abroad in the fabrication of objects nor, probably, on imported raw materials.

CHAPTER 4
THE ARTEFACTUAL EVIDENCE

4.1. Introduction

Archaeological evidence of non-ferrous metalworking in Early Historic Ireland includes a wide range of artefacts, many of which are specific to this technology. There are some, however, that are more ambiguous in purpose and could have been used in other craft activities such as carpentry, leather-working and ironworking. When evidence of both ironworking and non-ferrous metalworking are found in close proximity on a site, it is often difficult to identify separate tool-kits.

This also applies to furnaces that are typically of a primitive bowl type that could have been used for smelting or re-melting a range of metals and alloys. Both ferrous and non-ferrous smelting and refining can produce fayalite, an iron silicate and so the presence of slag is not an obvious indicator of the type of metal produced. While many metalworking tools may have been used for both copper and iron production, early documentary sources suggest that these activities were kept separate on many settlements, with different craft-workers involved (see chapter 2).

One artefact type which is commonly found and is specific to non-ferrous metalworking is the crucible. Other metalworking tools include anvils, awls, punches, gouges, chisels, files, hammers, tongs and shears. Also relevant are moulds, heating trays, trial or motif pieces, metal ingots, scrap metal, miscellaneous lumps of waste, slag, ore and tools.

4.2. Crucibles

Crucibles are the most easily recognisable indicator of non-ferrous metalworking as they were not used in Early Historic iron smelting (which was a solid-state bloomery process). Basically, a crucible is a vessel in which metal was melted and poured into a waiting mould. Crucibles may also be used to reduce ore minerals as part of a controlled smelting process, or in the refining and alloying of metal. Recently, a crucible from Moynagh Lough, Co. Meath, was found to have been employed in smelting copper matte directly with a tin ore (Bradley 1991, 23). The vast majority of crucibles from Early Historic Ireland are made of clay although some stone examples are known. Generally speaking, the prepared metal, whether in the form of scrap, cake or ingot, was placed in the crucible which was then heated in a furnace to melt the metal.

Three main forms of crucible have been identified on Irish sites (Fig. 25), with the pyramidal type, or more accurately triangular as they have three sides not four, being most common. These have a triangular mouth with the sides coming to an apex at the base. Good examples were found at Lagore crannóg, Co. Meath (Hencken 1950; Appendix 3) and Garranes ringfort, Co. Cork (Ó Ríordáin 1942). Somewhat resembling these are the bag-shaped crucibles which are deeper and have elliptical, not triangular, mouths. The third type of crucible is flat-bottomed with low sides. The examples found at Lagore are similar to what Ó Ríordáin calls semi-spherical at Garranes.

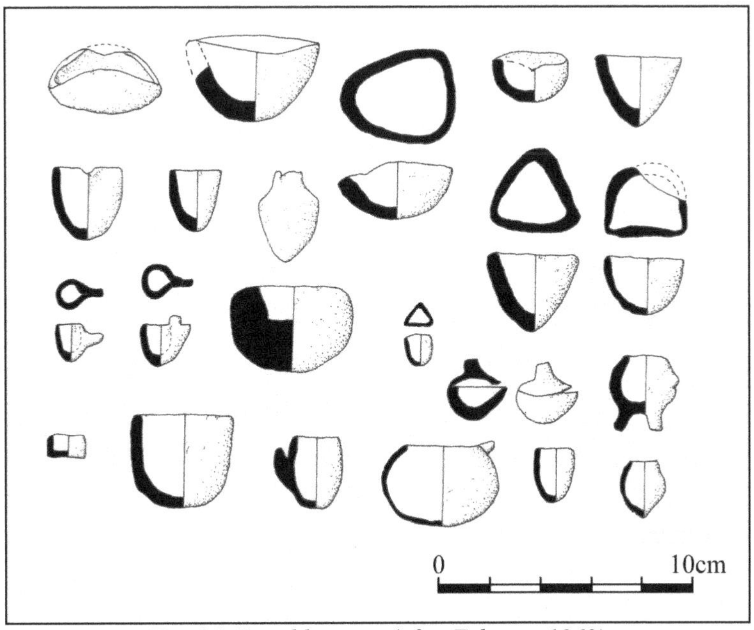

Fig. 25 Crucible types (after Tylecote 1962).

While the simple clay crucible of the three types mentioned above is by far the most commonly found, some variants are known. A small number of seventh-century crucibles have the added feature of a lid e.g. Garryduff, Co. Cork (O'Kelly 1963), Dunadd in Scotland (Craw 1929/30) and Dinas Powys in Wales (Laing and Laing 1995). Such a lid would allow more time for the pouring of the molten metal, slowing down the cooling process and subsequent solidification of the crucible contents. It has been suggested that such 'crucibles were intended for making glass, the covers being necessary so as to exclude impurities, small amounts of which could affect the colour of the finished product' (O'Kelly 1963, 96). However, as he readily admits, there is no residue of such a process evident on these crucibles.

Some flat-bottomed crucibles found on Early Historic settlements are composed of several layers of clay indicating their re-use. After each use the residual material (on the interior surface) was coated over with a layer of clay. No. 109 from Garryduff was used three times as evidenced by two additional clay layers (O'Kelly 1963). Also, a polypod crucible, i.e. a crucible provided with legs to enable it to stand free of a support, is known from Lagore crannóg, Co. Meath (Appendix 3, no. 271).

While clay was the material most commonly used to make crucibles, other raw materials included stone and crushed fragments of previously used crucibles. Crucibles of sandstone, semi-spherical in form, were found at Garranes, Co. Cork (Ó Ríordáin 1942). The use of tongs to handle crucibles can be seen from marks left on clay vessels (e.g. excavation numbers 122, 127, 128, 129 from Cathedral Hill, Co. Armagh). Other examples have handles or clay projections to facilitate their handling with tongs (Appendix 3, no. 272).

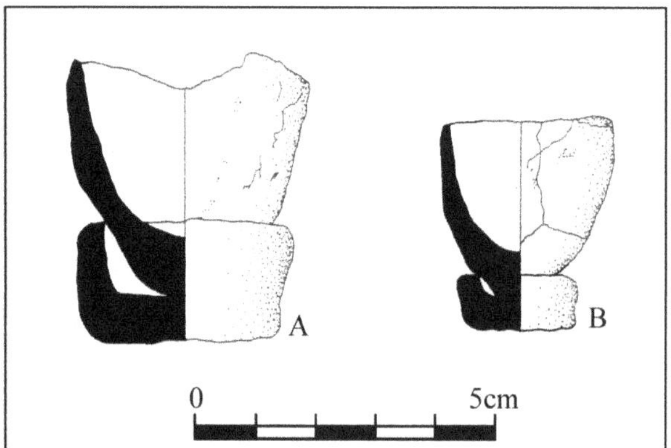

Fig. 26 A and B, pyramidal crucibles 7 and 9 standing in flat-bottomed vessels 71 and 61 from Lagore, respectively.

Crucible residues

When a crucible is heated a glaze forms on the surface in direct contact with the heat. Both the pyramidal and bag-shaped crucibles are glazed on the exterior and exterior base, indicating heating from below by placing them on or over a fire. Flat-bottomed crucibles, however, are glazed internally indicating that they were heated from above. It has been suggested (Tylecote 1986) that heat was directed into them from above with the aid of a blow-pipe or bellows. This would require immersion in a hearth or furnace which would also result in a tell-tale residue around the exterior of the vessel. As no such residues have been found on flat-bottomed crucibles, it is more likely that the latter were used as supports in which a pyramidal or bag-shaped crucible would stand (Fig. 26). The placing of a hot pyramidal vessel into a flat-bottomed crucible would result in the residue from the exterior base of the former adhering to the interior of the latter. It would also result in the heating of the flat-bottomed vessel from above causing the internal glaze.

It has also been suggested that flat-bottomed crucibles from Garranes may have been used for enamel and glass manufacture (Ó Ríordáin 1942). At Lagore, however, a flat-bottomed crucible (Appendix 3, no. 65) was found to contain silica with traces of iron, copper and sulphur, 'presumably from the partial reduction of the native pyrites in a furnace' (Hencken 1950, 237).

Analysis of the glazes and residues apparent on most crucibles has been carried out on several examples. In 1927, R.J. Moss analysed several crucibles and his general findings have been more or less reproduced in more recent work (Table 3). The glaze covering the heated surface of a crucible can range in colour from black to green to red, this last colour often leading to the assumption that the vessel was used for enamel-working. One such example is a pyramidal crucible from Craigywarren crannóg, Co. Antrim. The red glaze of this vessel was identified as consisting mainly of copper, iron, calcium and magnesium, i.e. a copper smelting slag not red enamel as was previously thought (Moss 1927, 178).

In addition to a heating glaze, some residue of the molten metal itself may remain within these crucibles. This can take the form of particles of metal adhering to the interior e.g. bronze-working crucibles from Glastonbury Tor, England (Wilson 1981), a'Cheardach Mhor in the Hebrides (*ibid.*) and Lagore, Co. Meath (Appendix 3, no. 39). Crucibles with a bead of gold still adhering to the interior surface were discovered at Knowth (Eogan 1977) and Clonmacnoise (King 1992). Similar crucible evidence is known from Buston crannóg, Ayrshire, and Cadbury Congresbury, England (both in Wilson 1981).

Locality	Shape	Volume/cc	Residue Type	Copper %	Tin %	Silica %	Iron oxide %	Lime %	Alumina %	Copper oxide %	Lead %	Magnesia %
Craigywarren	Pyramidal	50	Vitreous slag	Trace			11.2	12.8	10.6			6.5
Cavan crannóg	Pyramidal	10	Slag	23.4	9.7	45	7.7					
Ballyjamesduff	Pyramidal	33	Slag			62.8	9.4	15.4	9.4			
Dunshaughlin	Pyramidal	33	Metallic bead	85.05	11.6							
			Slag	28.2		43.9	8.7					
Lagore	Fragment	40	Metallic bead	85.6	4.2							
Loughrea	Not stated	12	Metallic bead	78.8	3.1	Zinc-15.5						
			Slag	20	10.1	20.6	23.1	7.7				
Unknown 158	Bag-shaped	93	Metallic bead	92.5	2.3							
			Slag			43.2	15.2		18.2	10.4		
Unknown 15	Not stated	43	Metallic bead	80.7	2.2						Trace	

Table 3 Crucible residue analyses, compiled from information in Moss 1927.

4.3. Heating trays

An artefact type associated with crucible finds is the heating tray or 'hot-plate', fragments of which were found at Tullylish, Co. Down (Ivens 1987), Moynagh Lough, Co. Meath (Bradley 1991) and Knowth, Co. Meath (Eogan 1977). These are all made of fired clay and were used for heating objects that could not be placed directly into a fire or furnace. They would have been most useful during the final fabrication of an object, enabling the melting of solder during filigree application or the evaporation of mercury during gilding for example.

The 'hot-plates' at Tullylish consisted of alternate layers of sandy clay and charcoal. The uppermost pad of one of these was burnt yellow-red and formed a shallow, vertical-sided dish. At Moynagh Lough, one of the heating trays bears the annular impression of an object worked on it. Unfortunately, there is no available description of the examples from Knowth.

4.4. Moulds

Stone moulds

Another artefact type commonly associated with non-ferrous metalworking is the mould. Both stone and clay moulds were used in Early Historic Ireland, with the latter apparently more versatile than its stone counterpart (Fig. 27). There are no metal moulds known in Ireland from this period. Generally, the vast majority of stone moulds found are ingot moulds. The ingots cast in such moulds are oblong in shape and would clearly have been the simplest to carve in a block of stone. Such moulds are also known in Great Britain e.g. Dunadd (Craw 1929/30) and the Mote of Mark (Laing 1973) in Scotland and Kiondroghad on the Isle of Man (Gelling 1969).

Stone ingot moulds may have been either one piece or bi-valve moulds, uncertainty arising from the fact that two-piece examples are known e.g. Ó Ríordáin's mould no.1 from Garranes, Co. Cork (Ó Ríordáin 1942). It is therefore possible that some single moulds may have originally had a corresponding half. Many of these moulds had more than one casting area or 'matrix' carved into them. The aforementioned mould from Garranes, for example, had four casting grooves. A bronze ingot from this site, Ó Ríordáin's no. 99, fits perfectly into one of these matrices.

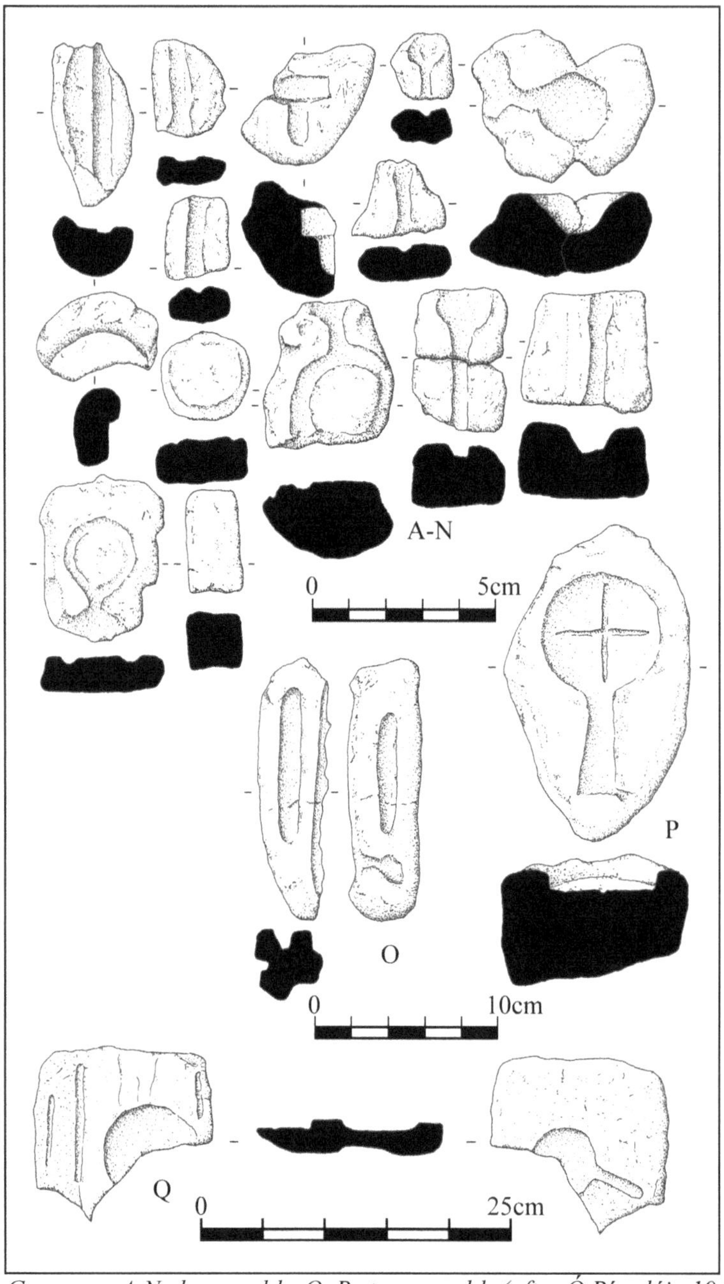

Fig. 27 A-P moulds from Garranes: A-N clay moulds, O–P stone moulds (after Ó Ríordáin 1942); Q stone mould from Cathedral Hill (after Gaskell Brown and Harper 1984).

Seven stone moulds were found on Cathedral Hill, Co. Armagh, producing bar-shaped ingots up to 1cm thick (Gaskell Brown and Harper 1984). Made of Carboniferous sandstone, the moulds also included examples for the manufacture of circular and rectangular objects with bar-shaped extensions (Fig. 27, Q). An example from Garranes with an added cross-like feature, was interpreted by Ó Ríordáin as a matrix used in the hammering out of a sheet-metal (Fig. 27, P). Gaskell Brown and Harper (1984) suggest that the smaller Cathedral Hill examples may have been used in the manufacture of ladles similar to those found on contemporary sites. Stone ingot moulds have been found at sites including Lagore crannóg (Hencken 1950), Dooey sandhills site Co. Donegal (Ó Ríordáin and Rynne 1961), Moylarg crannóg Co. Antrim (Buick 1893), and Ballinderry II crannóg (Hencken 1942).

Clay moulds

Clay moulds were used in the manufacture of finer objects such as penannular brooches and ringed pins. Despite the immense suitability of the *cire perdue* technique to the manufacture of such items, it appears that only bivalve clay moulds were used (Fig. 27, A–N). The *cire perdue* technique was, however, being used in contemporary late Celtic Britain, e.g. at Dunadd (Lane and Campbell 2000) and the Mote of Mark (Laing 1973) in Scotland. However, much of the Irish evidence is extremely fragmentary and therefore very difficult to interpret.

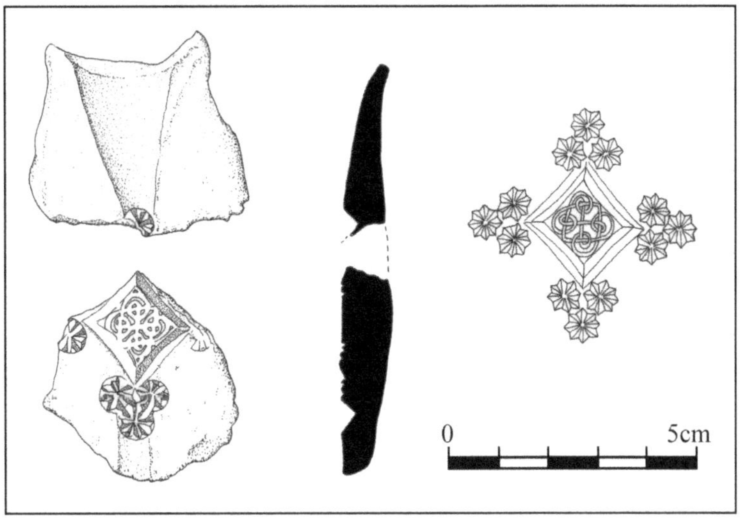

Fig. 28 Clay mould from Cathedral Hill and reconstruction of pattern (after Gaskell Brown and Harper 1984).

It is possible to cut the design of a mould into clay by hand though it is more likely that a pattern of some sort was used. This pattern may have been of carved wood, bone or metal. The latter was suggested for a mould from Cathedral Hill, no. 92 in the excavation report (Gaskell Brown and Harper 1984). It may have been formed using a metal pattern made from a wax original. The use of such an enduring master pattern would have allowed large scale reproduction of items. Also, the intricate detail with which wax is capable of being carved can be transferred to the object being cast. Moulds no. 1214–28, 3191 and 3193–4 from Moynagh Lough were made in this way and were all used in casting interlaced mounts of a similar design and size, suggesting the use of a master pattern or model (Bradley 1991, 20).

The objects manufactured in such moulds include rings, pins, penannular brooches, decorative metal panels, book fittings and studs. Fragments of clay moulds are known from many relevant sites. The examples from Cathedral Hill are particularly impressive, especially moulds 92 and 93. No. 92 was possibly used to cast a decorative metal panel intended for the head of a crozier or some such item. The Lismore crozier has panels of such a shape.

Mould 93 (Fig. 28) bears an ornament consisting of a double heart-shaped interlace within a diamond-shaped setting with three star-shaped bosses at each corner. Similar designs are found on the pin-head of the Killamery Brooch and the brooches from the Ardagh Hoard. The interlace design itself is found on a bone motif piece from Lagore (Appendix 3, no. 302) with only minute differences between it and the Cathedral Hill mould, the border of the Lagore motif is hatched whereas the Armagh example is not.

4.5. Lead models

Moulds may also have been made using a lead model. This model is cast in an enduring master mould and then used to create the new mould matrix. Due to the low melting point of lead (327°C), lead models could be used in roughly the same way as wax *cire perdue* models. Lead brooch-pins, both complete and fragmentary, occur at Dooey (Ó Ríordáin and Rynne 1961), Moylarg (Buick 1893), Clogher (Warner 1973), and Garranes (Ó Ríordáin 1942). Some of these may have acted as patterns for clay moulds. A lead penannular brooch, of an Irish type, from Dinas Powys in Wales has also been interpreted as such (Youngs 1989).

4.6. Motif-pieces

> A motif-piece is a small portable raw piece, or scrap, of any material or waste fragment of a disused artifact, with carved or incised, discrete, positive, patterns comprising art motifs, sporadically positioned over its surface.
> (Ó Meadhra 1987, 11)

Motif-pieces were carved in stone, bone or antler, with examples known from several Early Historic sites including Cathedral Hill, Movilla Abbey (Ivens 1984), Moynagh Lough, Garryduff, Lagore, Clonmacnoise (King 1992), and Ballinderry II. A stone trial piece for a penannular brooch is also known from Dunadd in Scotland (Lane and Campbell 2000, 189). The quality of these motif-pieces can differ greatly. Some are well executed (e.g. Appendix 3, no. 302 from Lagore), while others appear to be the idle scratchings of an amateur (e.g. Cathedral Hill, Fig. 29).

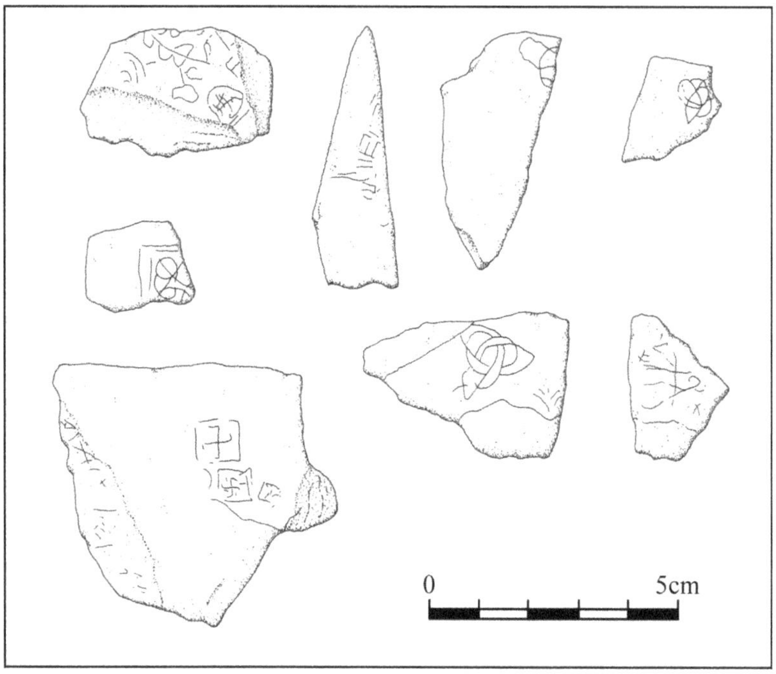

Fig. 29 Stone motif pieces from Cathedral Hill (after Gaskell Brown and Harper 1984).

The purpose of these motif-pieces is much debated. Suggestions relating to metalworking include their use as trial or experimentation pieces upon which designs could be worked out before application to a metal object (or to other media such as vellum, wood or bone). They may also have acted as dies or stamps upon which metal foil may have been pressed, or as patterns from which clay moulds may have been made. There is, however, no mould or artefact evidence of their having been directly involved in metalworking. It seems likely, though, that they may have been trial-pieces, possibly used in both the composition and circulation of designs.

Whatever their purpose, the intricate designs of many motif-pieces resemble, though do not exactly replicate, those found on the contemporary metalwork. 'The motif-pieces are artistic work, and there is enough evidence of skilful work to indicate that some might be the sketchpads of the metalworkers when found on metalworking sites' (Ó Meadhra 1987, 162).

4.7. Raw materials

The almost complete lack of ore from Early Historic settlement sites in Ireland would suggest that metal was smelted elsewhere and transported in the form of ingots or cakes. One site, Lagore crannóg Co. Meath, did produce finds of copper ore. The two pieces of ore were of copper pyrites, such ore having been commercially worked in the past at both Beauparc and Brownstown in Co. Meath, twelve and eight miles (respectively) from Lagore (Hencken 1950).

Several pieces of the lead ore, galena, were found at Ardcloon, Co. Mayo, the nearest sources of which are Ballisodare Bay, Co. Sligo and Ballinrobe, Co. Mayo (Rynne 1956). However, it is impossible to accurately source either copper pyrites or galena to a particular mine as both are very common in Ireland. The lack of any tin ore from Early Historic Ireland may reflect the importation of the metal in smelted form. Two lumps of metallic tin are known from Garranes ringfort, Co. Cork (Ó Ríordáin 1942).

If ore was not the direct source of the metal worked, then scrap metal and ingots must have been, with both ingots and ingot moulds found on several sites of this period. This may indicate the smelting of ore, or alternatively the casting of metal into convenient forms for storage or trade purposes. Ingots of bronze are known from Movilla Abbey, Moynagh Lough, Garranes and Moylarg. A gold ingot was also found at Moynagh Lough, while silver ingots formed part of the silver hoard from Carraig Aille II (Ó Ríordáin 1949). These precious metal ingots are generally bar-shaped with an average length of between 0.05m and 0.1m, and are usually approximately 0.01m thick.

The known ingots do correlate well with the contemporary mould evidence, both in terms of shape and size. In addition to their possible use for metal storage or circulation, ingots may have acted as 'blanks' from which various objects could have been forged. Items fabricated from sheet-metal or metal rods/bars, such as simple rings, bracelets and vessels, may have been manufactured in this way.

The hoard from Carraig Aille II contained fragments cut from silver bracelets. With the ingot association this would suggest that the hoard was possibly a stock-pile of silver intended for use in the manufacture of items on site. Carraig Aille II also produced considerable evidence of bronze working (Ó Ríordáin 1942). A fragment of a Viking silver bracelet was found at Lagore, along with scraps of sheet and wire bronze (Appendix 3, nos. 354, 190–196 and 204–208, 348–353 respectively). Short lengths of gold wire were found at both Movilla Abbey and Moynagh Lough. All of these may represent items of scrap intended for re-use or merely fragments of objects broken after loss.

4.8. Waste

Discarded fragments of bronze have been found on a number of Early Historic settlements. These include a lump of bronze from Lagore (no. 342) that the excavator identified as having come from a flat-bottomed crucible. This is unlikely considering the probable use of such crucibles as stands. This fragment may have been a circular ingot, cast in a stone mould such as no. 316 from Lagore. Solidified drops of molten bronze, spilled while pouring, are known from Lagore (nos. 338 and 339) and also from Southampton in England (Wilson 1981). Irregular fragments of bronze were also found at Carraig Aille II, while a bronze jet discarded from a cast was found at Garranes.

3.9. Tools

Very few tools were required in the actual smelting of ore or the melting of metal. Obviously a bellows or blow-pipe was required to achieve the high temperatures necessary, however these were generally made of organic materials and do not survive. Another tool likely to have been widely used in an Early Historic metal workshop was an iron tongs (Fig. 97). Clay-tipped bronze tongs could also have been used though there are no known examples of such. The clay tips would have protected the metal from the high temperatures encountered when handling crucibles containing molten metal or when holding metal during annealing. Many crucibles (e.g. Appendix 3, no. 47) bear the marks of the tongs used to handle them. Iron tongs are known from Moynagh Lough, Garranes and also from Sibertswold and Shakenoak in Anglo-Saxon England (Wilson 1981).

While the smelting tool-kit was limited, a wide range of tools was employed in the final fabrication of an object, from the initial casting to various finishing techniques. One important piece of equipment in this process was the anvil that served as a stable base for hammering and other metalworking techniques. Surprisingly, very few anvils are known from archaeological sites, the only definite example being that from Garryduff (Fig. 98). This is a very small iron anvil with a working surface of 3.2cm square, probably employed in the working of very fine items. A small spike projects from its base to facilitate insertion into a timber work-bench (O'Kelly 1963). Another possible example is recorded as coming from Randalstown crannóg (Wood-Martin 1886).

Bronze and iron awls (Fig. 69), gouges (Fig. 68) and punches (Fig. 67) have been recovered from Cathedral Hill, Garranes, Garryduff, Lagore, Carraig Aille II, Clea Lakes (Collins and Proudfoot 1959) and Clonmacnoise. All three may have been used in the decoration of fine objects. The awls are simple, one-piece, pointed iron rods. They bear no evidence of having been hafted and may have been used in boring holes. They may also have been used as engraving tools or as punches. Punches, however, are generally identified by a striking platform and such a feature is not evident on the awls.

The known gouges are of iron, some of which were hand-held while larger examples were hafted. These could have been used to gouge out areas of unwanted metal. Most iron gouges found in Early Historic settlements, however, were probably used in carpentry as they would have been most effective when employed in wood-carving. Iron chisels (Fig. 30, D) are known from Garryduff, Carraig Aille and Clonmacnoise. Such tools were required for cutting and decorating purposes, though once again it is hard to be certain about their specific use.

A corroded iron implement from Moylarg has been tentatively identified as a file (Buick 1893), with a similar object found at Thetford, England (Wilson 1981). Files were used in the removal of excess or unwanted metal such as the casting flash created by the seepage of molten metal between the two halves of a bivalve mould. Stone hammer-heads were found at Carraig Aille II and Craigywarren (Knowles and Coffey 1907), while iron versions were found at Lagore (Fig. 30, C) and Randalstown (Wood-Martin 1886). At Moynagh Lough an iron stake was uncovered which the excavator suggested was used in the beating of sheet bronze (Bradley 1991, 18).

Iron shears, which may have been used to cut sheet metal and metal wire were found at Cathedral Hill, Garranes, Garryduff, Lough Faughan (Collins 1955), Lagore and Moylarg. This is but one possibility, as most iron shears were more likely to have been used for domestic or agricultural work. These shears are usually simple, one-piece implements with a single-loop spring. Two examples, one from Lagore (Appendix 3, no. 336) and one from Cathedral Hill (no. 68 in Gaskell Brown and Harper 1984), have a double looped spring (Fig. 30, E).

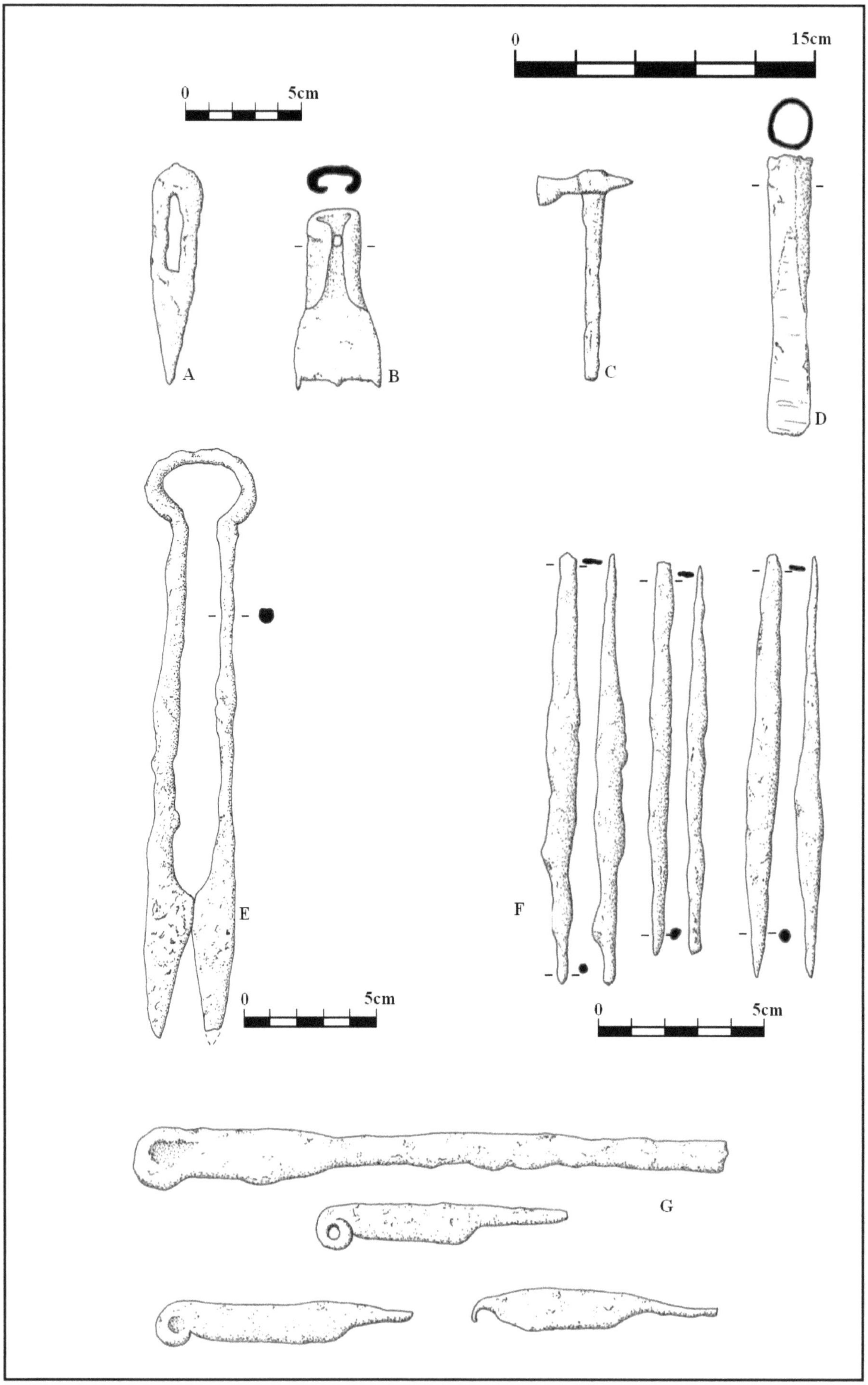

Fig. 30 Iron tools.

A curious type of iron tool, possibly related to metalworking, was found at Lagore, Garryduff, Lough Faughan and Carraig Aille II. This implement takes the form of a socketed pronged tool (Fig. 30, B), averaging approximately 0.05m in length. The small size of these objects must be taken into account when considering their function. They may have been employed in some aspect of metal decoration or perhaps in leatherworking or some other activity.

Three possible dividers (Fig. 30, G) for scribing circles were found at Garryduff (O'Kelly 1963). These consisted of a long flat piece of iron with a hole at one end and a tang at the other. By joining two such pieces together, using a rivet through the holes, the tangs could act as the points of a compass or dividers. At Kilpatrick, Co. Westmeath, the excavator interpreted three tiny needles of bronze as possible compass points (Swan 1994/95).

Bronze tweezers have been found at Lagore, Garryduff, and Kilpatrick. These consist of a strip of bronze, approximately 0.07m wide and 0.1m long, bent in two. Two bronze *styli* are known, from Cathedral Hill and Carraig Aille II, while a wooden example was discovered at Rathtinaun (Raftery forthcoming). The example from Cathedral Hill is 0.06m long (broken), while the Carraig Aille II stylus is 0.07m in length, and the wooden example from Rathtinaun is roughly 0.08m long. All three have a flattened head and may have been used in carving wax *cire perdue* models.

A unique find from Rathtinaun crannóg consists of a rectangular piece of bronze bearing a sunken interlace pattern. This has been interpreted as a possible *pressblech* die for decorating sheet metal (Raftery forthcoming). Pointed and slotted iron implements found at Carraig Aille II and Lough Faughan may have been used in metalworking, with a small piece of wood or metal fitted into the slot serving as a handle of sorts (Fig. 30, A). This would enable the tool to be effectively used as a boring implement (Ó Ríordáin 1949, 79). Whetstones used for the sharpening of metal blades and other implements are commonly found on Early Historic sites in Ireland, however they can rarely be linked specifically to metalworking practices.

The metalworker's tool-kit

The function of several of the above tool-types is uncertain and many could have been used in activities other than metalworking, for example carpentry or leather-working. When found on a site with evidence of non-ferrous metalworking it is probable that they were employed in such work, albeit not exclusively so.

The equipment used by metalworkers in this period varied according to their specific duties. Those involved in smelting and/or casting would have required such items as bellows, crucibles, tongs and perhaps master moulds or patterns from which to make clay moulds as required. The tool-kit utilised in the decorating of a fine object would have included such items as chisels, hammers and punches of varying size and weight.

By examining a finished object of fine metalwork from this period, it is possible to infer the range of tools used in its fabrication. The Ardagh Chalice (Organ 1973) is a suitable example, exhibiting as it does several different, high quality, metalworking techniques. This object (Fig. 31) is, in fact, 'a compendium of the skills of the metalworker in use in Ireland at that time' (Ryan 1983b, 125). The bowl and foot of the vessel are made of beaten sheet silver polished on a lathe. Around the bowl is a girdle consisting of alternating panels and large circular studs. This band comprises an openwork silver frame cut from sheet metal with a hammer and chisel. The framed panels of this girdle were made up of three layers. Firstly, the design was pressed into sheet metal using a *pressblech* die and the background area cut away with a hammer and chisel leaving the design in openwork relief. Secondly, filigree consisting of a treble layer of beaded wire was applied on top of this using a tweezers. Finally, a gilt copper plate was used to back the three-dimensional openwork pattern.

The two silver handles are decorated with gold filigree and polychrome glass studs. When examined in detail it was discovered that the handles were actually cast in moulds, as opposed to having been soldered together. The fact that the two sides of one handle are different from each other but identical to the corresponding side on the other handle indicates that both were cast using moulds made from the same pattern or model. The two roundels or circular medallions were made in the same way. Running around the bowl just beneath the bowl girdle is a Latin inscription of the names of eleven of the original apostles and St. Paul. These letters, very similar in style to the capitals in the book of Lindisfarne, were scribed freehand using a point that may also have been used to stipple the background (Fig. 31). This point may have been a very fine punch or awl. The three components of the neck, the upper and lower ring mountings and the cylindrical collar, are very good examples of cast, imitation *kerbschnitt* ornament (Fig. 31).

The series of decorated panels on the underside of the foot flange are technically very varied. There are eight of these panels, separated one from the other by square blue glass studs. Two of the panels are of copper and are decorated with punched swastikas. Four of the panels are of silver *pressblech*, while the other two are of trichinopoly – one copper, the other silver (Fig. 31). The weaving or knitting of trichinopoly would require the use of a wooden or metal frame around which the metal wire could be wound.

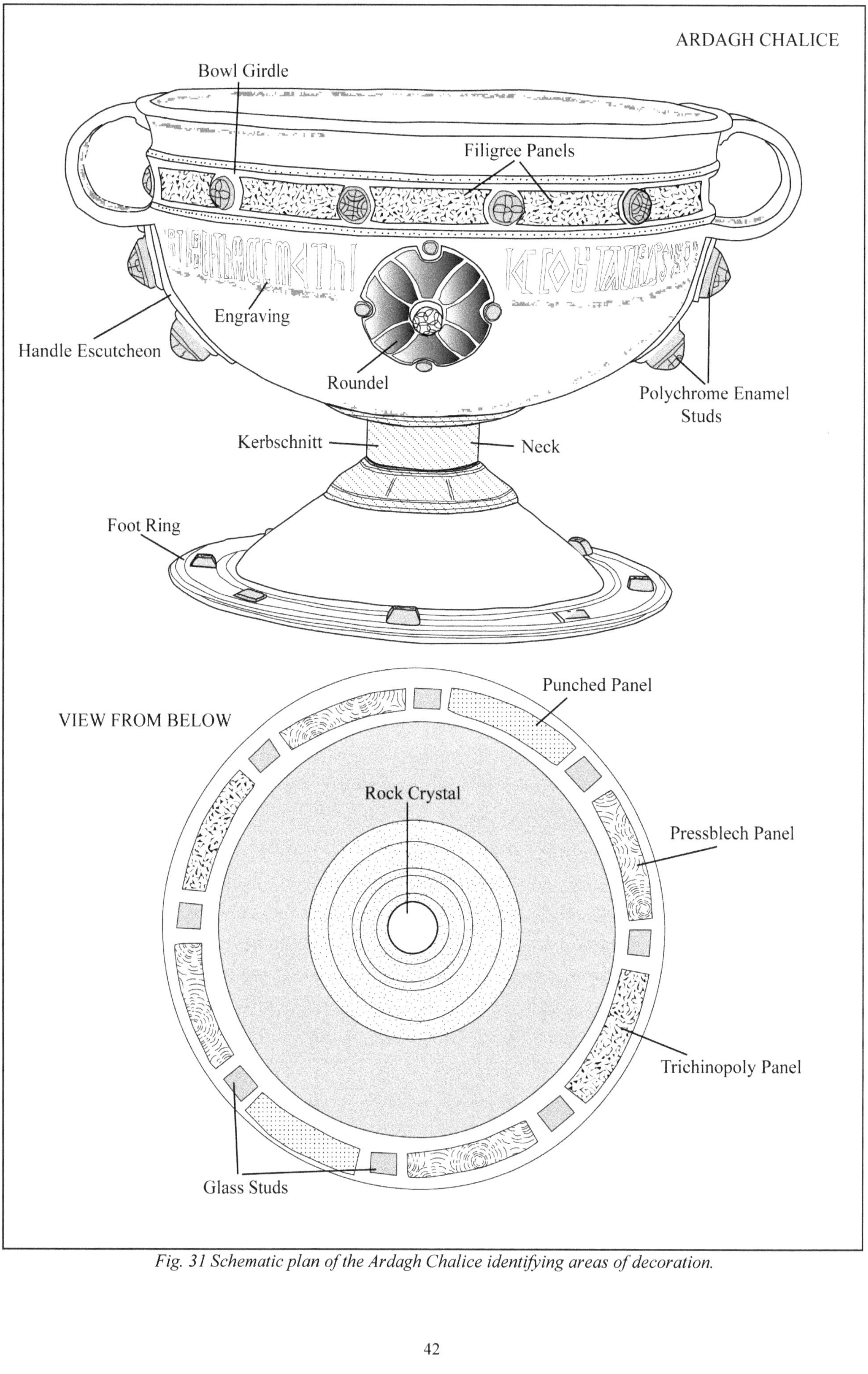

Fig. 31 Schematic plan of the Ardagh Chalice identifying areas of decoration.

The equipment needed, then, in the construction of the Ardagh Chalice includes a lathe, an anvil upon which to carry out all of the work, heating trays to facilitate soldering, a range of hammers, chisels and punches of varying size and shape, a trichinopoly frame, tweezers, *pressblech* dies, crucibles, moulds and mould patterns. Other items such as motif-pieces and *styli* may have been indirectly involved in the manufacturing process. An object like the Ardagh Chalice could only have been produced in a well-equipped workshop by specialist craftsmen with access to a furnace or hearth of some form.

4.10. Conclusion

A difference in status is evident in the metalworking assemblages from certain of the relevant sites. Some such as Cathedral Hill and Lagore do appear to be more impressive than sites like Reask (Fanning 1981) or Lough Faughan. The difference between large ringfort enclosures like Garranes and smaller sites like Ardcloon (Rynne 1956) is particularly striking. A site producing a few crucible fragments is, clearly, not on the same level as a site where a whole range of non-ferrous metalworking artefacts has been discovered (see Appendix 2 for quantitative difference between sites).

The literary sources (chapter 2) provide very little evidence regarding the equipment used. Any references to tools are incidental and many occur in the myths and legends where they act as props to a story. The anvil of the god Dagda is mentioned in several tales though no description of it is given. The anvil is also referred to in the law tract 'Uraicecht Becc' where the distraint of a smith is marked by tying a withie around his anvil though, again, no details of the object are provided. The most useful source of information is the 'Blai Ord Indeoin' tract which deals with the workshop of a smith and mentions such tools as hammers, sledge-hammers and anvils. None of these are described though the anvil is mentioned as having been fixed to a block, whether of stone or wood is unspecified.

The tool-kit which survives in the archaeological record is not completely representative of that employed in the Early Historic Period. The decay of organic materials and the difficulty in identification and interpretation of several objects may account for much of this. The range of recovered items may also be limited to broken and discarded tools which, combined with problems of preservation and interpretation, produces a biased view of the Early Historic tool-kit.

However, in the absence of complete tool-kits, the range of implements available to the smith can be inferred by both manual and scientific examination of finished artefacts. Unfortunately, very few such investigations have been undertaken and those that have usually lack detailed scientific analyses and are confined to a small number of very fine pieces, e.g. the Moylough Belt-Shrine (O'Kelly 1965), the Ardagh Chalice (Organ 1973) and the Derrynaflan Hoard (Ryan 1983a).

CHAPTER 5
SETTLEMENT CONTEXTS FOR NON-FERROUS METALWORKING IN EARLY HISTORIC IRELAND

5.1. Introduction

The settlement contexts for non-ferrous metalworking in Early Historic Ireland take largely the same form as those for ironworking, though there are certain important differences. Evidence of iron production is found on a wide range of sites, from the small earthen ringfort to the supposedly royal crannóg. This reflects a more widespread use of iron implements in domestic and agricultural life, and in warfare.

With a growing dependence on iron implements, the average ringfort farmer required the facility to make and repair equipment. Excavation evidence indicates that many of these small-scale settlements had their own smithy workshop for this purpose. It is also likely that some settlements traded for raw iron and/or finished tools. Ironworking in these smaller settlements may have been largely non-specialised, whereas larger ringforts and crannógs, particularly those which represent tribal or dynastic centres, probably had groups and/or individuals continuously engaged in this work. They were employed in both the manufacture of necessary implements and weapons, and also in manufacturing some items of personal adornment and prestige.

The evidence of non-ferrous metalworking in Early Historic Ireland presents a different picture due largely to the different status of metals like copper-bronze, silver and gold which now played a non-utilitarian role. Artefactual evidence of non-ferrous metalworking in this period commonly takes the form of crucibles, moulds, items of scrap metal to be recycled, slag, ore, structures such as furnaces, and tools such as anvils, awls, chisels, files, and tongs (see chapter 4). This evidence is generally absent from settlements of small to average size. This association with large, presumably high status, settlements would appear to indicate the activity of specifically skilled smiths. Working in copper, copper-alloy, lead, silver and gold would seem to be something that the common landowner had neither the ability nor the wealth to practice.

5.2. Settlement form

One of the most important settlement contexts for non-ferrous metalworking in the Early Historic Period is the ringfort in its various forms. The earthen ringfort or rath is a roughly circular area delimited by an earthen bank and outer ditch, with some sites having up to three bank and ditch combinations. Other ringforts may be built predominantly of stone and are often referred to as cashels. Both rath and cashel probably functioned as farmsteads, though there is obvious ranking of individual sites in both economic and political terms. Ringforts which have produced evidence of non-ferrous metalworking include Garranes (Ó Ríordáin 1942), Garryduff (O'Kelly 1963) and Ballycatteen, Co. Cork (Ó Ríordáin and Hartnett 1943), Carraig Aille II (Ó Ríordáin 1949) and Raheennamadra, Co. Limerick (Stenberger 1967), and Ardcloon, Co. Mayo (Rynne 1956) (Fig. 32).

As with ringforts, the majority of crannógs which have been excavated in Ireland are dated to the Early Historic Period. Crannógs are generally regarded as defended settlements located on an artificially built lake edge, platform or on a modified natural island. Initially regarded as refuge locations, crannógs were already being viewed as settlements by Wood-Martin in his classic 1886 study. They were constructed by throwing down layers of brushwood, stone, timber, and heather, generally consolidated with piles driven in around the edges to prevent slipping. A wooden palisade erected around the perimeter enclosed various hut structures and habitation facilities within.

Many crannógs have produced evidence of non-ferrous metalworking usually in the form of stray finds of crucibles and moulds. This activity is confirmed by the excavation of several sites which appear to be of high status. Crannógs producing evidence of non-ferrous metalworking include Ballinderry II, Co. Offaly (Hencken 1942), Lagore (Hencken 1950) and Moynagh Lough, Co. Meath (Bradley 1991) (Fig. 32).

A third type of site giving evidence of non-ferrous metalworking is the monastic enclosure, many of which were comparable in status to the larger ringforts and crannógs. Not all monastic enclosures operated at the same level with the larger sites, the so-called monastic 'towns', were extremely wealthy and powerful receiving gifts and tributes from local rulers. The standing of these important church settlements led to the establishment and promotion of specialised craft activities, among them non-ferrous metalworking. As a patron of the arts the church was very active. Relevant sites include Reask, Co. Kerry (Fanning 1981), Clonmacnoise, Co. Offaly (King 1992), Cathedral Hill, Co. Armagh (Gaskell Brown and Harper 1984) and Nendrum (Lawlor 1925), Tullylish (Ivens 1987) and Movilla Abbey, Co. Down (Ivens 1984) (Fig. 32).

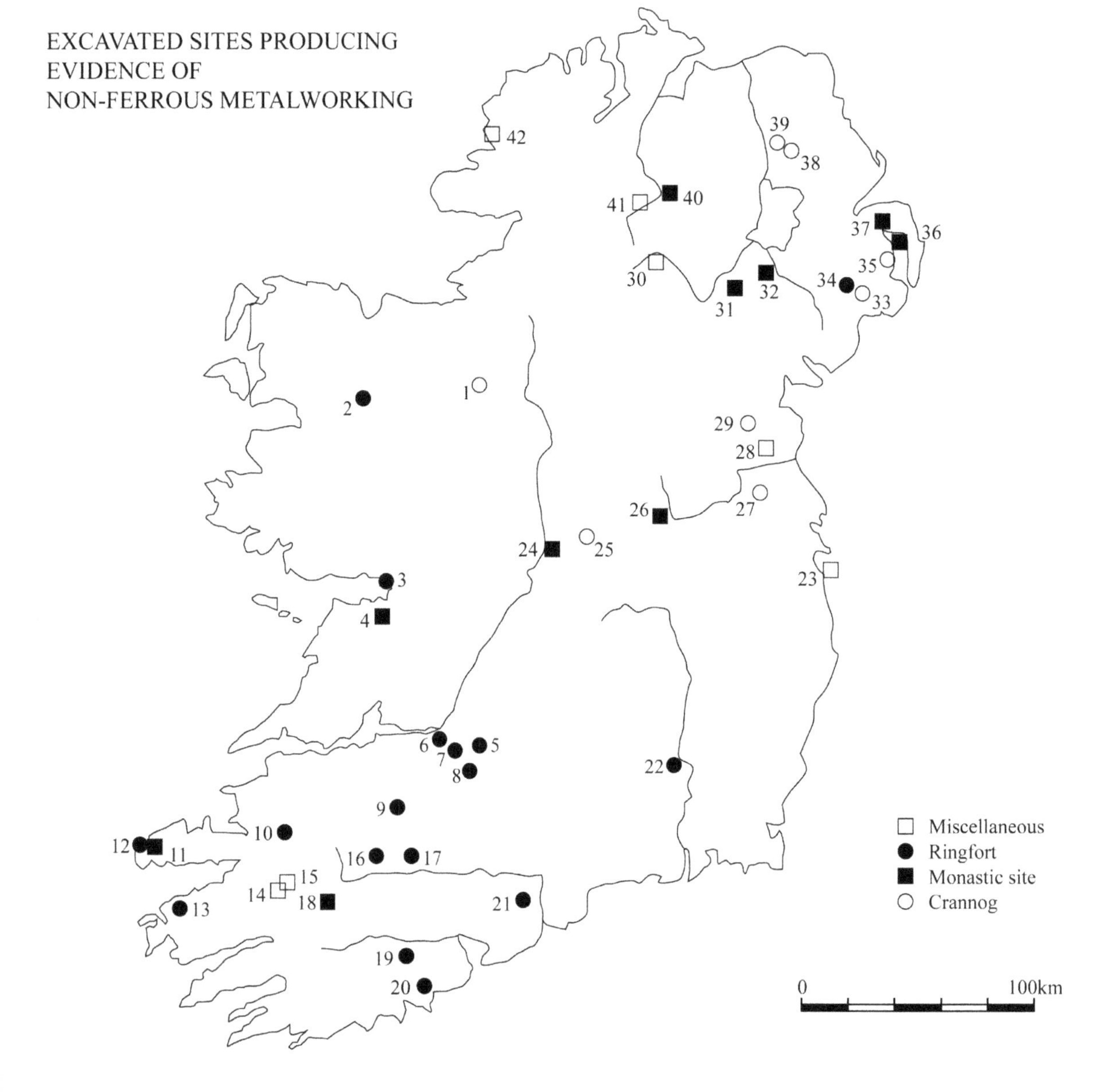

1. Rathtinaun, Sligo
2. Ardcloon, Mayo
3. Rathgurreen, Galway
4. Kiltiernan, Galway
5. Carraig Aille, Limerick
6. Sluggary, Limerick
7. Knockea, Limerick
8. Cush, Limerick
9. Raheennamadra, Limerick
10. Ballyegan, Kerry
11. Reask, Kerry
12. Cathair Fionnurach, Kerry
13. Leacanabuaile, Kerry
14. Ross Island, Kerry
15. Scrahane, Kerry
16. Lisduggan, Cork
17. Raheens, Cork
18. Ballyvourney, Cork
19. Garranes, Cork
20. Ballycatteen, Cork
21. Garryduff, Cork
22. Dunbell, Tipperary
23. Dalkey, Dublin
24. Clonmacnoise, Offaly
25. Ballinderry II, Offaly
26. Kilpatrick, Westmeath
27. Lagore, Meath
28. Knowth, Meath
29. Moynagh Lough, Meath
30. Clogher, Tyrone
31. Cathedral Hill, Armagh
32. Tullylish, Down
33. Lough Faughan, Down
34. Castleskreen, Down
35. Clea Lakes, Down
36. Nendrum, Down
37. Movilla Abbey, Down
38. Moylarg, Antrim
39. Craigywarren, Antrim
40. Dunmisk, Tyrone
41. Island MacHugh, Tyrone
42. Dooey, Donegal

Fig. 32 Non-ferrous metalworking sites of Early Historic date in Ireland.

Other settlement contexts besides ringforts, crannógs and monastic centres are known for non-ferrous metalworking in the Early Historic Period. These include sandhill settlements (Dooey) and the re-use of established sites such as passage-tomb mounds (Knowth) and hillforts (Clogher). It is possible that something of the significance of the original sites may have lingered and prompted their re-use. These sites include Knowth, Co. Meath (Eogan 1977), Dooey, Co. Donegal (Ó Ríordáin and Rynne 1961), Clogher, Co. Tyrone (Warner 1973) and Dalkey Island, Co. Dublin (Liversage 1968) (Fig. 32). There are few, if any, non-site, stray finds of metalworking artefacts in the Early Historic Period unlike in the Bronze Age.

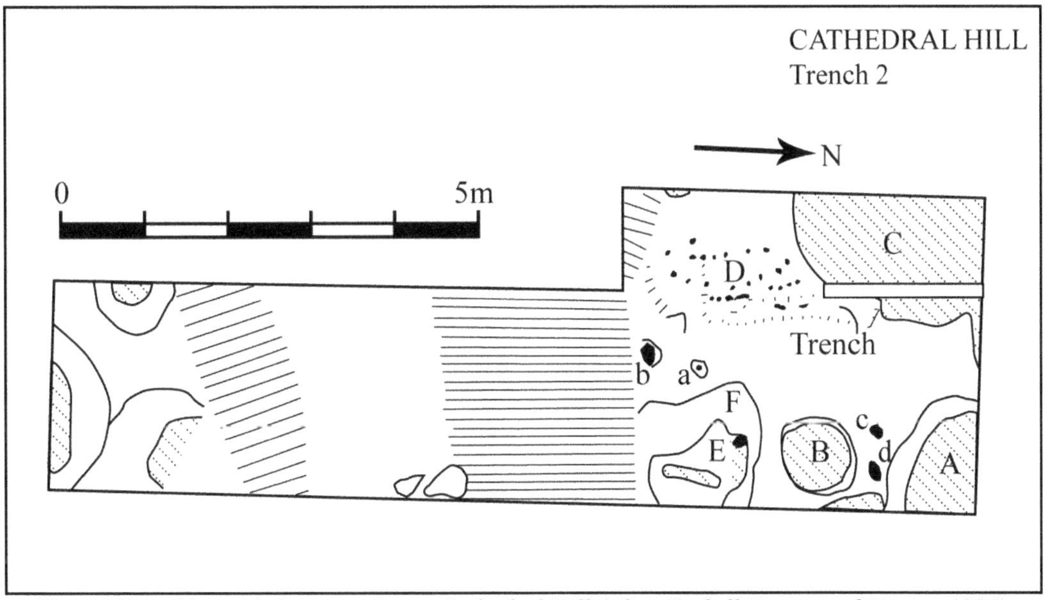

Fig. 33 Possible workshop remains at Cathedral Hill (after Gaskell Brown and Harper 1984).

5.3. Relevant structural remains

Furnaces obviously bear witness to metalworking on a site, though whether or not that work was in iron or non-ferrous metals requires a closer examination of the structure and its associated residues. Where there are no associated finds and all that remains is a spread of burnt clay and charcoal, it is very difficult to interpret the function of these features. In such circumstances, it is easy to confuse this evidence with activities connected to the use of domestic hearths and pottery kilns.

No furnaces from the relevant sites have been definitely associated with non-ferrous *smelting*. The fragmentary nature of the remains (see Appendix 2) and the lack of scientific analysis of associated residues may explain this. An example of a possible copper smelting furnace was found at Movilla Abbey (Appendix 2, no. 8). A feature interpreted as a hearth or furnace base was built upon an infilled shallow scoop in the ground. It contained much burnt material which, unfortunately, was not analysed (Fig. 88). This feature could have had many purposes, however nearby finds of crucible fragments, slag and scrap copper alloy point to a possible copper smelting and/or re-melting furnace (Ivens 1984, 77).

The Moynagh Lough example (Appendix 2, no. 44) was better preserved, consisting of a bowl-shaped furnace dug into the ground, with a maximum diameter of 0.35m and a maximum depth of 0.14m. It was clay-lined with a stepped base. The furnace contained a deliberately clean fill associated with a number of charcoal spreads, presumably cleaned from the furnace at regular intervals. This suggested to the excavator that the furnace had been used recurrently (Bradley 1991, 18).

Found in association with the Moynagh Lough furnace were fragments of a baked clay nozzle lining, crucible sherds, a motif piece, moulds and fragments of heating trays. Due to the lack of ore on the site and relatively small amounts of slag, Bradley concluded 'that the furnace was used for melting metal rather than smelting ores' (1991, 22). He suggests that cooling moulds may have been stacked on a nearby pebbled area where a dump of mould fragments was found (Fig. 116). Also, a pink clay spread probably marks the area where both moulds and crucibles were made as their fabric is very similar to that of the spread.

Indirect evidence for metalworking furnaces is provided by tuyère fragments, which have been found on Cathedral Hill, Lagore, Garranes, Dunmisk and Clonmacnoise (Appendix 2, nos. 2, 38, 18, 5 and 4 respectively). Though indicative of the presence of a furnace on a site, tuyères obviously do not reveal either its form or function and may in some cases have been connected with glass-working or some such activity. In Anglo-Saxon England, possible non-ferrous furnaces have been identified at Cheddar, York, Thetford, and Glastonbury (Wilson 1981).

Workshop locations on Early Historic sites are identified through a concentration of relevant finds in association with structural evidence. Possible workshops have been identified on several of the above sites. At Cathedral Hill, Co. Armagh, four large post-holes (a–d) forming a rectangular plan were identified as the remains of a workshop structure (Fig. 33). This workshop had been burnt twice, with the layers of burning overlying clay floors and relevant finds. Unfortunately, these layers have not been dated so it is not possible to relate them to any historically recorded events at the site. Adjacent to this structure was a large hearth which, due to its location here, may have been employed in non-ferrous metalworking. The probable source of the clay used in manufacturing the moulds and crucibles was located in this area also (Appendix 2, no. 2).

At Kilpatrick, Co. Westmeath (Swan 1994/95) the evidence consists of a semi-circular foundation trench, the open side of which may have been partially enclosed by a line of very fine post-holes. These remains were associated with a forging area and two iron-smelting furnaces as well as the relevant non-ferrous metalworking equipment (Appendix 2, no. 6). The features were stratigraphically identical, cut into the underlying boulder clay.

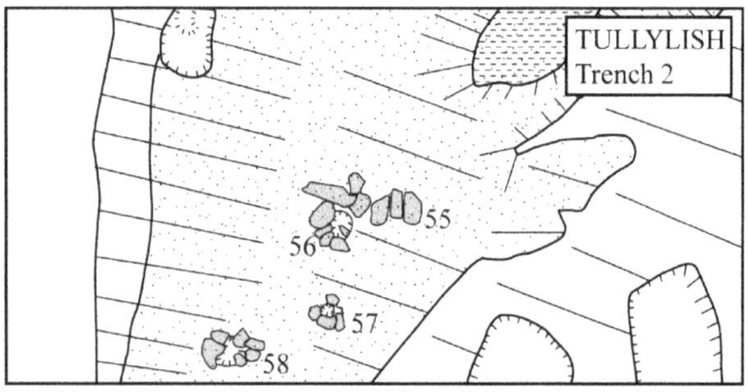

Fig. 34 Workshop at Tullylish (after Ivens 1987).

A timber workshop at Tullylish, Co. Down, was marked by stone packed post-holes and settings. These features were cut into the surface of the ditch surrounding the site (Fig. 34). A layer of industrial debris consisting of charcoal-rich earth and relevant finds accumulated while the structure was in use. Near these remains were found a large hearth, hot-plates and many mould fragments. This area was later (date uncertain) infilled while industrial activities continued in the vicinity (Appendix 2, no. 11).

A possible workshop at Garranes, Co. Cork, consisted of the stone foundation arc of a hut. Though the remains are very fragmentary, their location in the industrial area (site D, Fig. 39) would suggest the possible function of the structure in metalworking (Appendix 2, no. 18).

Finally, the relevant evidence from Reask, Co. Kerry came from hut G indicating its use as a workshop, though one involved in both iron and non-ferrous working (Fanning 1981). The hut is sub-circular in plan with a maximum internal diameter of 2.75m. In the interior were found two pits, one dug after the other went into disuse (Fig. 91). It appears that they were both initially used in the smelting of iron ore before being filled in and re-used as a large hearth. The small number of relevant remains were probably associated with these hearths (Appendix 2, no.10).

5.4. Internal site organisation

The metalworking area within these Early Historic settlements is generally marked by a high concentration of relevant finds, features and residues. Excavation evidence indicates that this activity was generally located within the settlement enclosure, but away from the domestic habitation area. This reflects the desirability of locating the flammable and noisy metalworking zone as far away as possible from the living area, while still taking advantage of the security and shelter afforded by the enclosing element.

Extra-mural activity areas are rarely subjected to archaeological investigation and, as yet, there is no evidence of metalworking having ever occurred in such places. At Ballinderry II, the excavator did find a hearth situated between the water's edge and the palisade enclosing the site (Fig. 35).

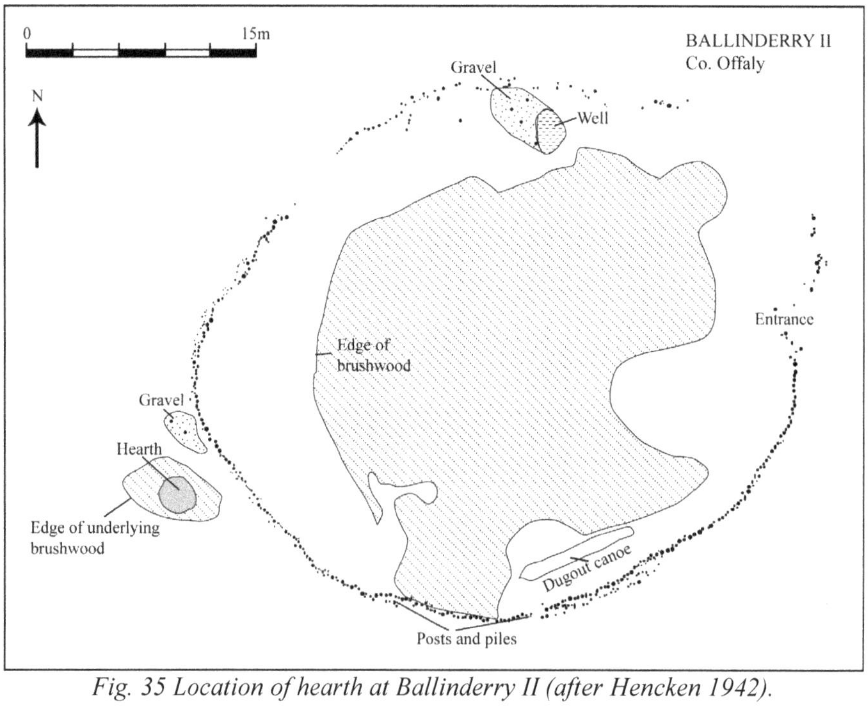

Fig. 35 Location of hearth at Ballinderry II (after Hencken 1942).

It is clear, therefore, that such areas were being utilised. We might also note in this regard that metalworking areas do not have to be associated with enclosed or defended sites. The settlement in the sandhills at Dooey, Co. Donegal, produced substantial evidence of both iron and non-ferrous metalworking (Appendix 2, no. 50).

Where sites have been completely excavated, some pattern is evident regarding the relationship of non-ferrous metalworking to other site activities. Unfortunately, several of the relevant sites were investigated under rescue conditions and only provide a partial record of the metalworking activity. The location of craft activities within these settlements generally depends on the size of the site. Where space was limited, it appears that all metalworking is located in the one place. At Reask, Co. Kerry, both iron and non-ferrous metalworking were carried out in the same area (Hut G) although evidence of iron-working alone is also found on the opposite side of the enclosure outside huts C and D (Fig. 36).

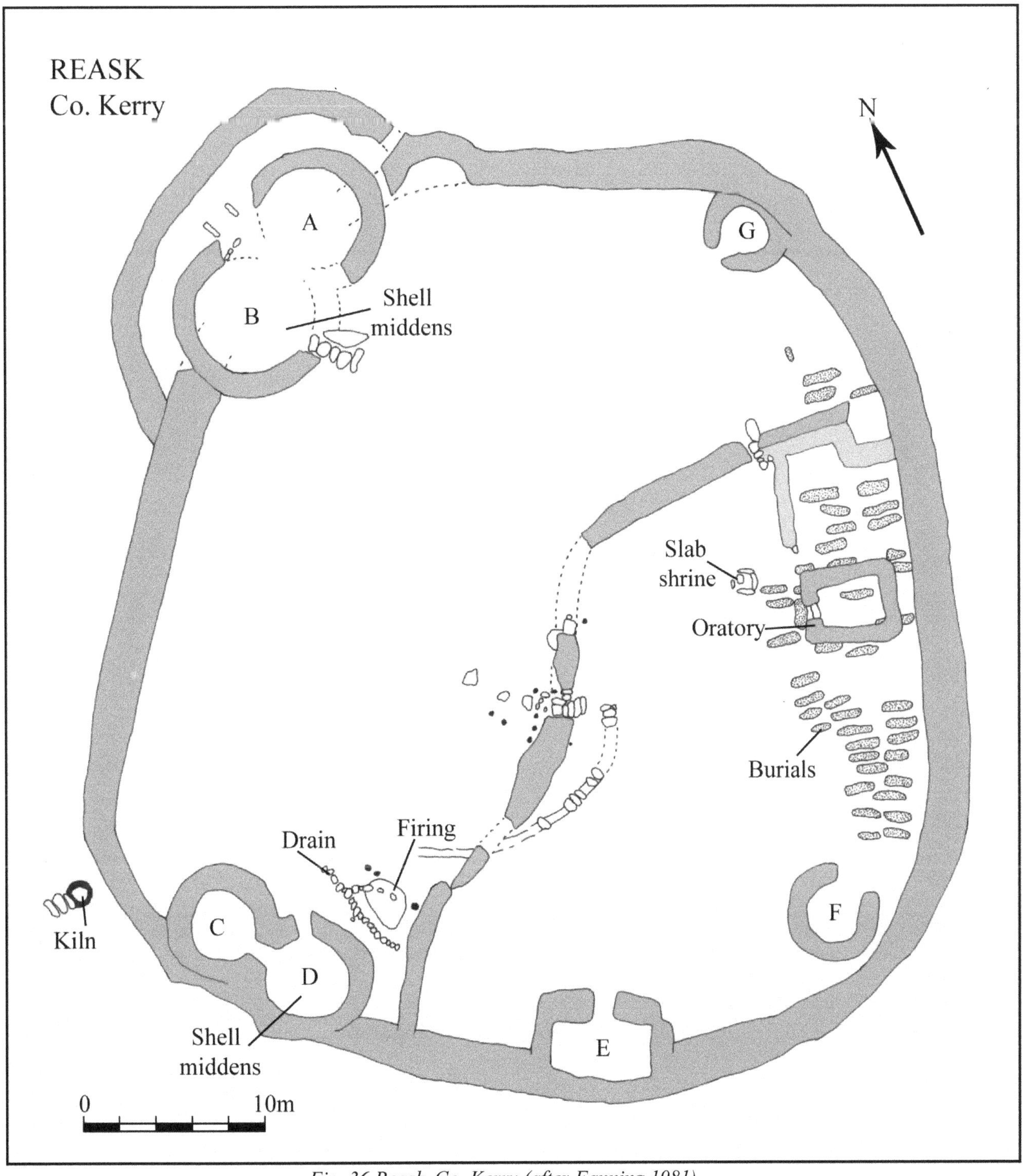

Fig. 36 Reask, Co. Kerry (after Fanning 1981).

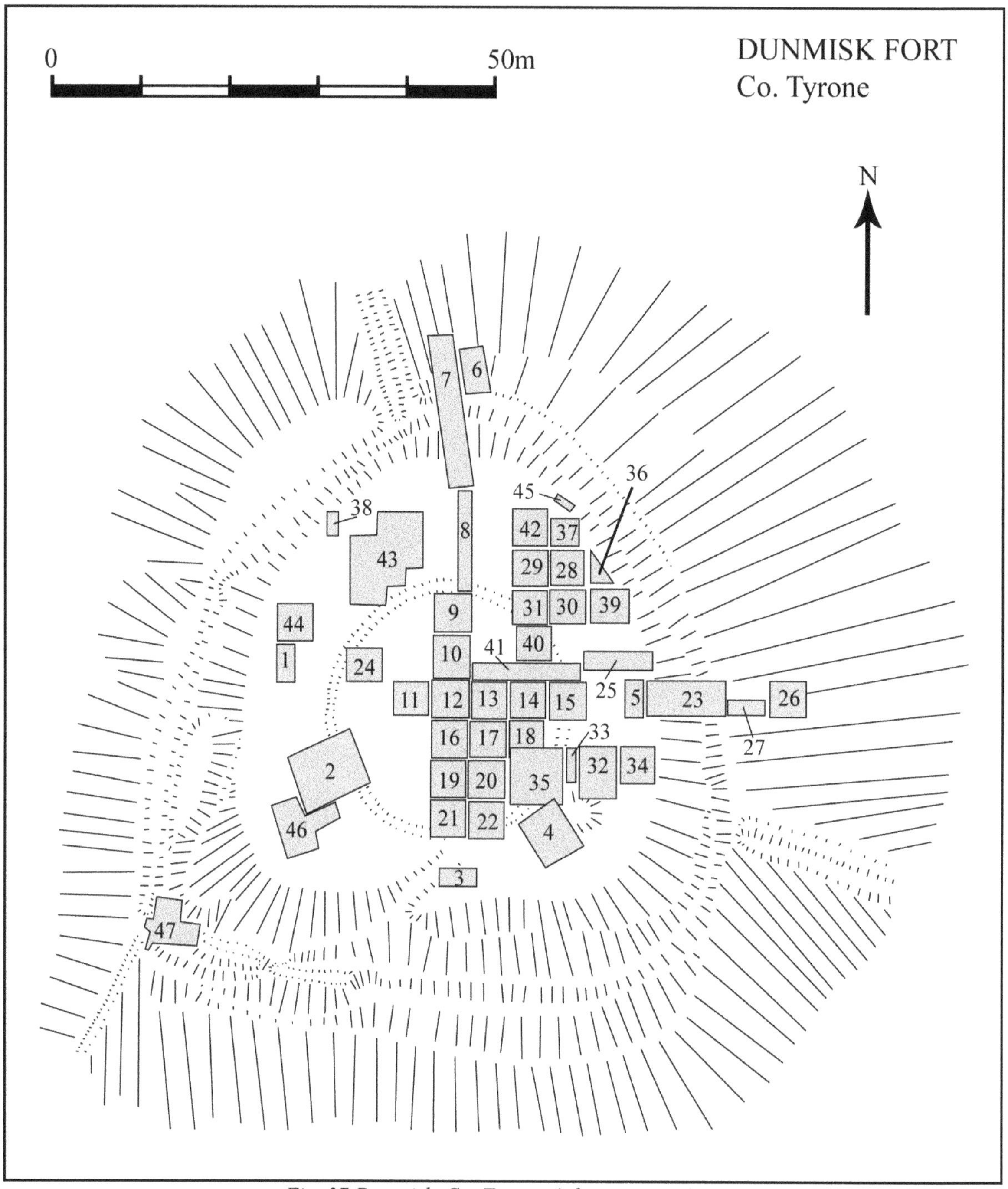

Fig. 37 Dunmisk, Co. Tyrone (after Ivens 1989).

On the larger sites non-ferrous and iron working are usually well separated. At Dunmisk Fort, Co. Tyrone the north-east quadrant of the enclosure was utilised in non-ferrous working, while ironworking was carried out in the south-west area (Fig. 37). Such a separation can also be clearly seen at Knowth where the two took place on opposite ends of the site (Fig. 38). This distinction is also made in the contemporary literature, between blacksmith and worker in precious metals and their places of work (see chapter 2).

Physical separation may, of course, reflect chronology rather than contemporary site organisation. However, at Dunmisk the excavator treats the evidence as contemporary on the basis of C-14 dates, though he does not specify what these are (Ivens 1989, 20). The settlement evidence at Knowth belongs to one phase, dated between the ninth and eleventh centuries AD. Because the site spans roughly three centuries, it is possible that the metalworking areas were not contemporary.

5.5. Economy

The economies of most Early Historic settlements were based on agriculture, the scale of which partly determined the social standing of the occupants and thus their consumption of ornate metalwork made from non-ferrous metals.

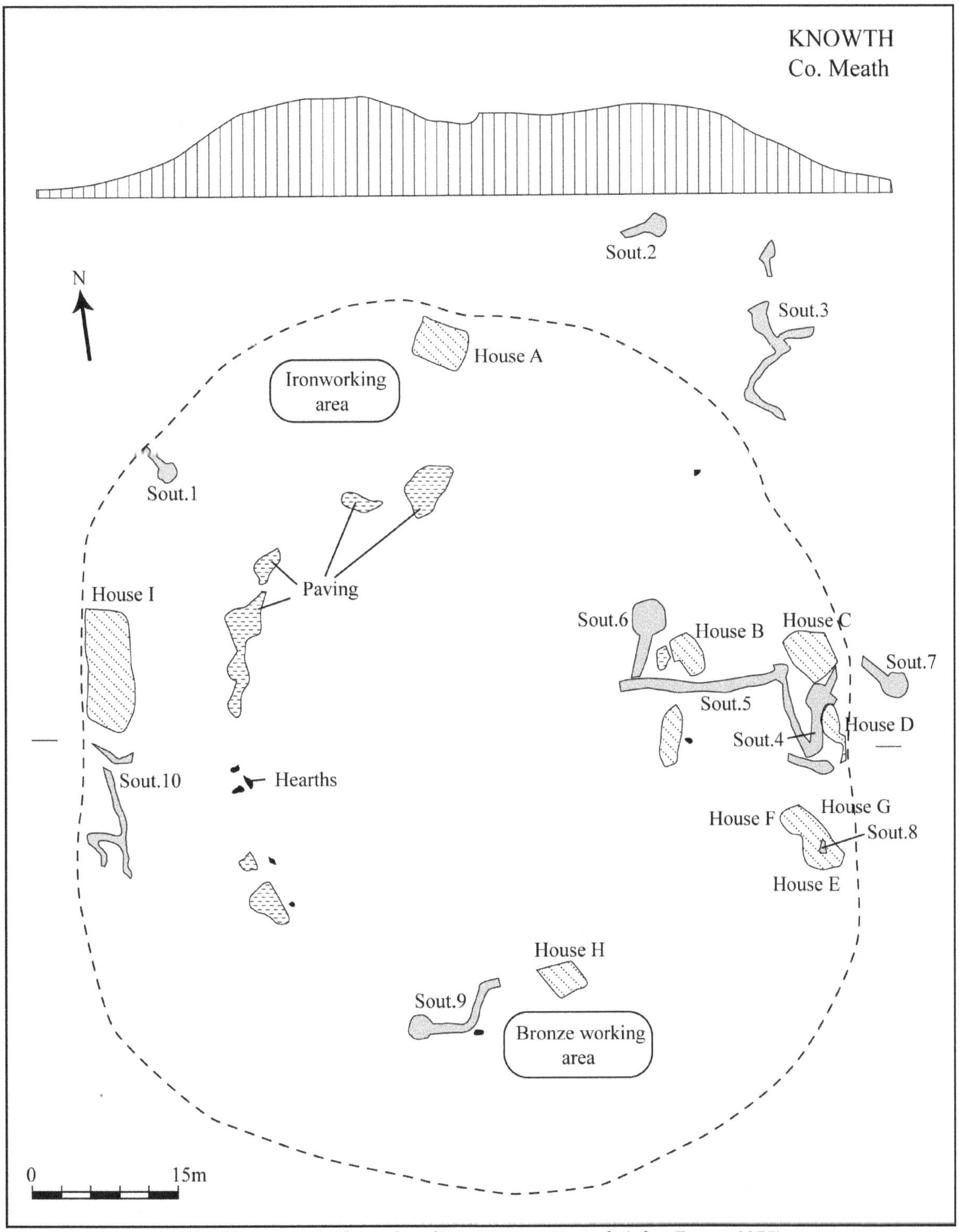

Fig. 38 Location of metalworking areas at Knowth (after Eogan 1977).

Some ringforts and crannógs clearly functioned as important tribal or dynastic centres where craft specialization, including non-ferrous metalworking, was practiced. Garranes, Co. Cork is a good example of the latter. This trivallate bank-and-ditch enclosure (Fig. 39) has produced definite evidence of non-ferrous metalworking dating to the later fifth/early sixth century AD. The excavator saw the metal finds as pointing 'to a specialised occupation – of craftsmen engaged in metalworking and allied pursuits' (Ó Ríordáin 1942, 141). This is clearly at odds with the general interpretation of ringforts as farmsteads. It must be remembered that only between a quarter and one third of the interior of Garranes was excavated, thus producing a biased impression of its economy and internal site organisation. The status of the site, discussed below, also argues against the opinion of the excavator.

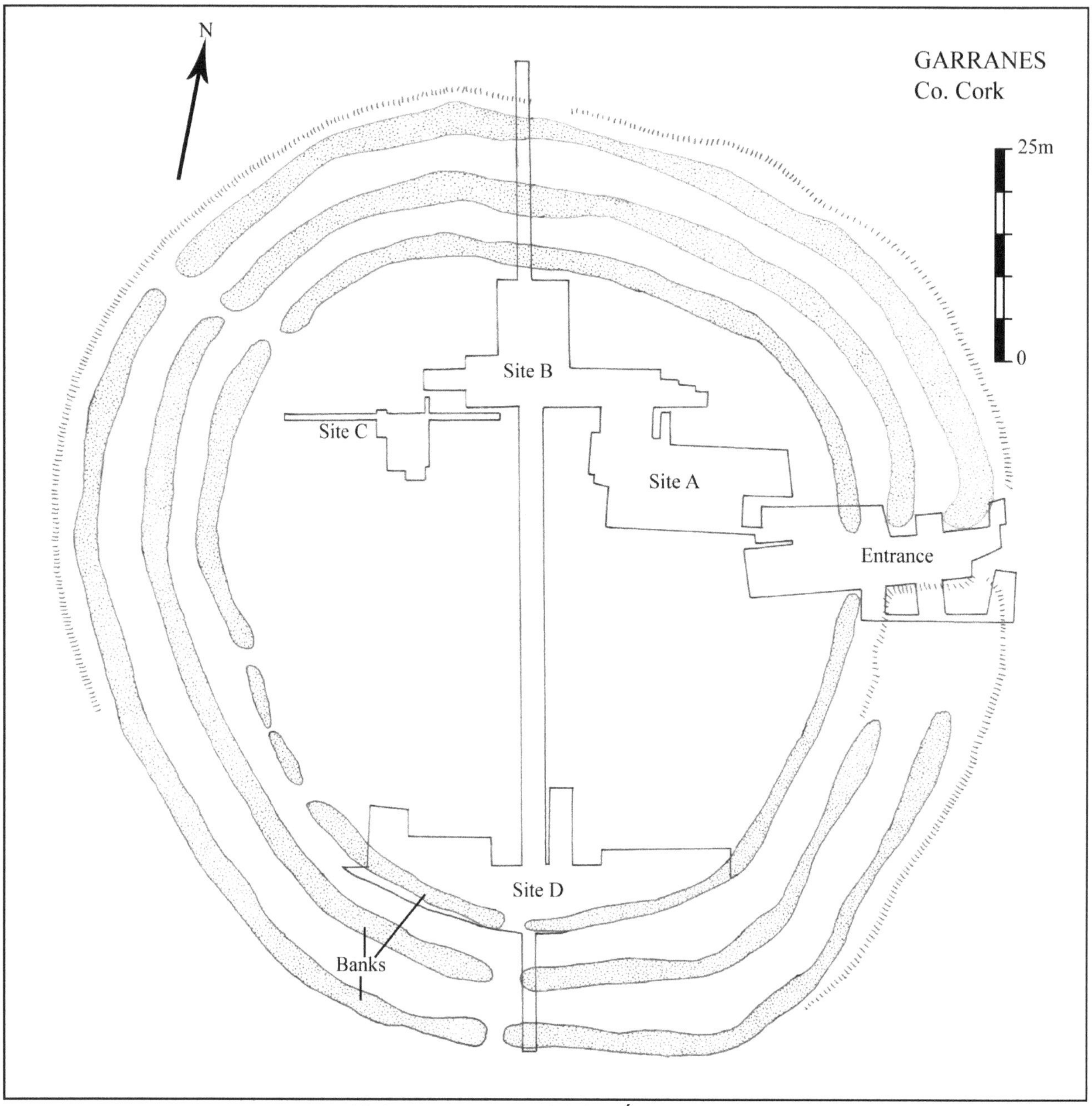

Fig. 39 Garranes, Co. Cork (after Ó Ríordáin 1942).

As with Garranes, the excavator of Garryduff ringfort (Appendix 2, no. 19) felt that the site was not occupied by a warlike people. Unlike Ó Ríordáin, he did not go so far as to say that the occupants were exclusively metalworkers. He does describe them as a 'peace-loving people, interested in trade and in the arts and crafts of the time' and that the 'community included amongst their number artist-craftsmen of considerable capability, as well as farmers and huntsmen' (O'Kelly 1963, 112–114). The defensive nature of a ringfort may be interpreted as a reflection of social disquiet, arguing against O'Kelly's description of the inhabitants as 'peace-loving' (Fig. 40).

Other forts producing evidence of non-ferrous metalworking include Carraig Aille II, Co. Limerick (Appendix 2 no. 14, Fig. 41), directly analogous to Garryduff where a peaceful community practiced metalworking among other activities within the security of an enclosure. The raths at Castleskreen Co. Down (Appendix 2, no. 15) and Ballycatteen Co. Cork (Appendix 2, no. 13, Fig. 42), both of which produced evidence of non-ferrous metalworking, are described by their excavators (Dickinson and Waterman 1959; Ó Ríordáin and Hartnett 1943 respectively) as having had economies based on farming.

For the ringforts, at least, it appears that non-ferrous metalworking was part of a larger economy, whether or not Garranes and Garryduff can be positively described in the same way is uncertain. There is no definite evidence at the present to support a belief that separate communities of metalworkers inhabited some ringforts. The possibility also exists, though, that the non-ferrous metal produced on some sites may not have been exclusively intended for internal consumption.

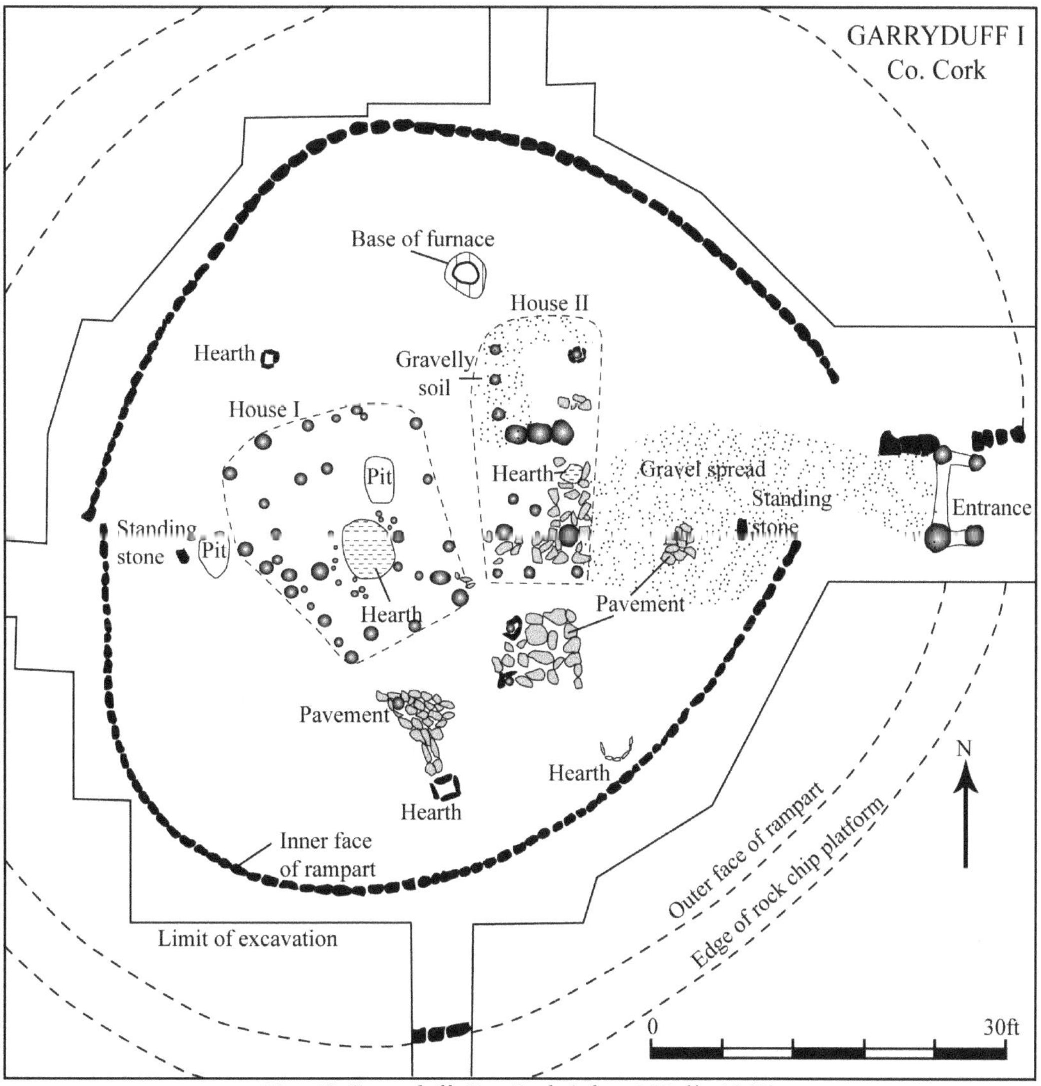

Fig. 40 Garryduff, Co. Cork (after O'Kelly 1963).

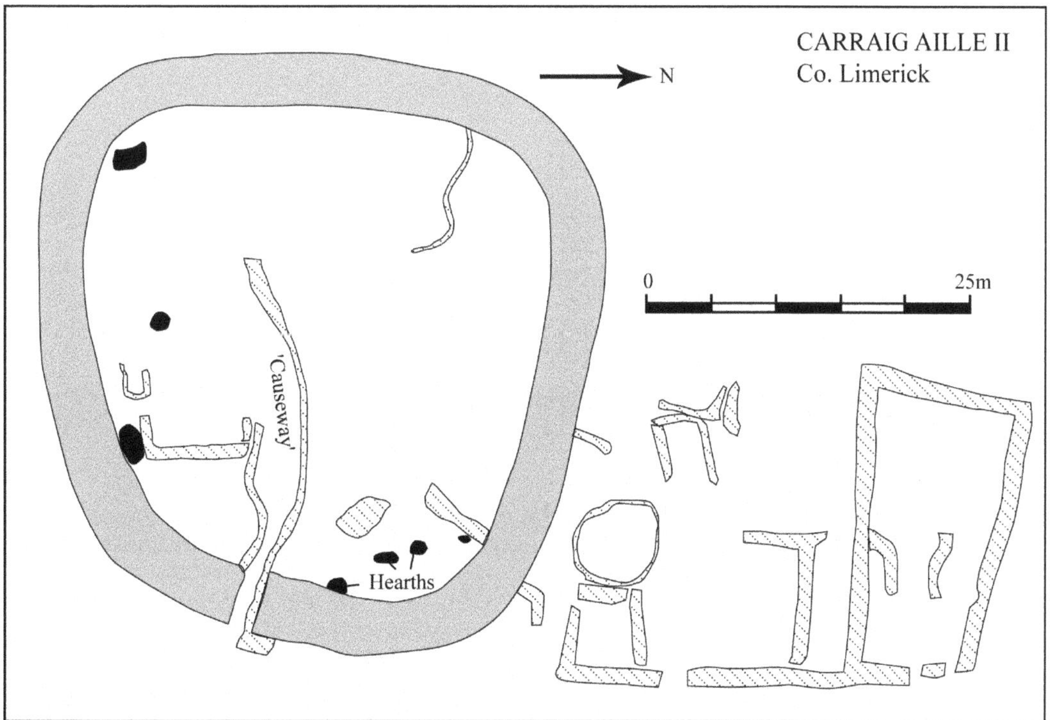

Fig. 41 Carraig Aille II, Co. Limerick (after Ó Ríordáin 1949).

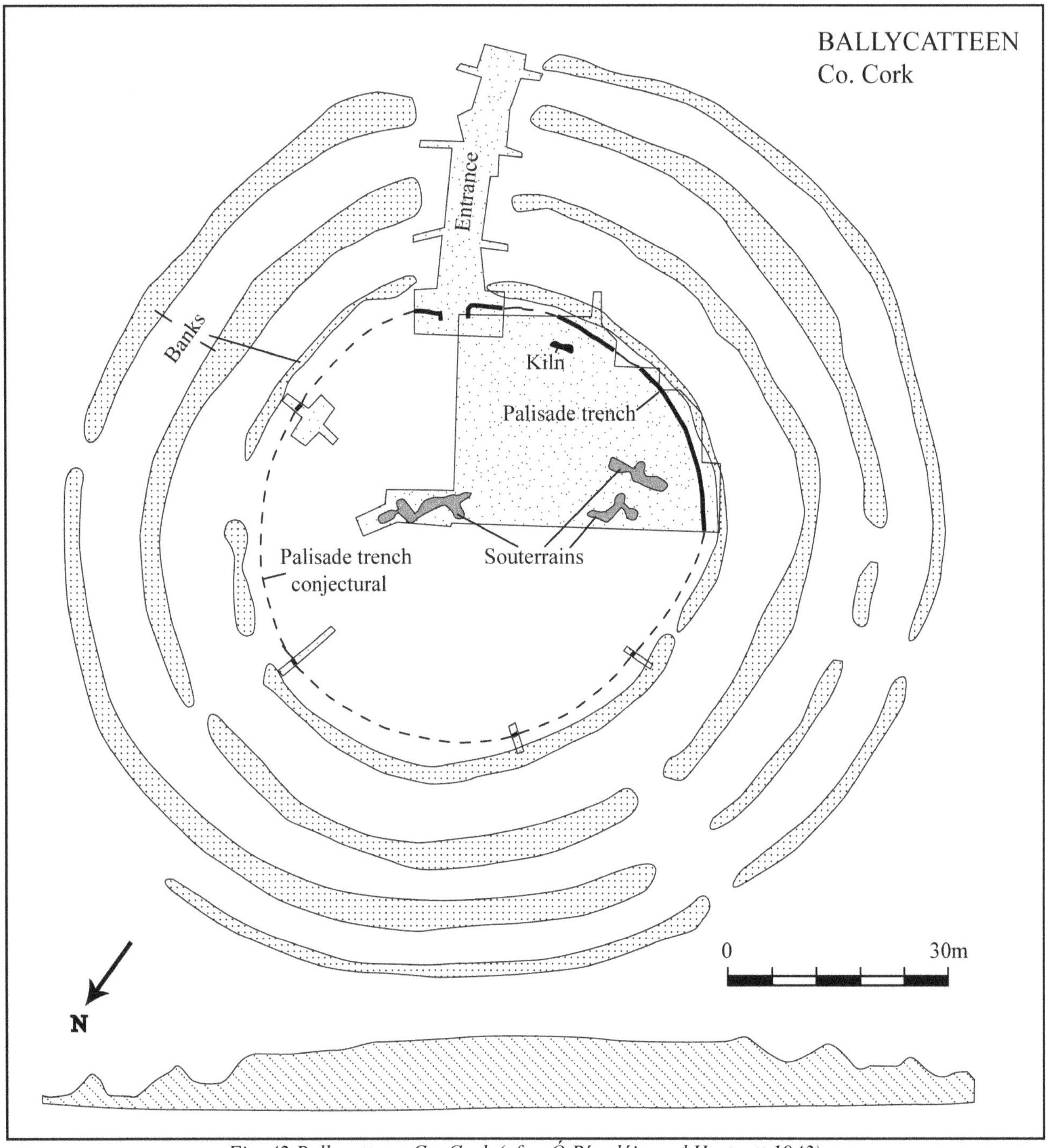

Fig. 42 Ballycatteen, Co. Cork (after Ó Ríordáin and Hartnett 1943).

It is difficult to identify such 'factory' sites where output is mostly destined for external trade. Perhaps Moynagh Lough could be viewed in this way, its large quantities of industrial evidence reflecting extensive non-ferrous metalworking which probably produced a site surplus.

That agriculture formed the basis of the economy on the excavated crannógs is clearly evidenced by the recovery of large quantities of animal bone. Approximately 6,623kg were recovered from Ballinderry II, much of which was used in the construction of the crannóg. Agriculture also formed the basis of the economy at Lagore where roughly 22,680kg of animal bone was found. All of the crannógs 'appear to have been essentially self-sufficient, obtaining most of [their] needs in the immediate locality' (Bradley 1991, 23). Artefacts such as quern-stones, ploughing equipment and sickles also provide evidence of agriculture on the relevant sites.

Copper-alloy and iron metallurgy is often found in association with other specialised craft activities in excavated Early Historic settlements. The latter include gold-working, the production of glass, enamel and millefiori, and the fabrication of objects using such materials as amber and jet. Where the locations of such activities can be identified, they are usually adjacent to, if not in the same place as, the copper-alloy working. Such organisation probably reflects the use of these materials in the fabrication of composite non-ferrous artefacts. Evidence of other quasi-industrial activities may also be inferred from a study of artefacts found in excavated ringforts and crannógs. These crafts include leatherworking, textile production and bone-working. The remains of leather shoes were found at Ballinderry II, Co. Offaly, textile production is evidenced by such finds as spindle-whorls and scraps of fabric at Lagore, while bone-working is evident on many of the sites in the form of combs, pins, motif-pieces and handles for various implements. Eogan says of the Knowth settlement; 'Mixed farming and to a lesser extent industry formed the economic basis of the settlement which appears to have been largely self-contained' (1977, 75). This can be said of most of the sites.

5.6. Trade and exchange

Trade and exchange played an important role in the economy. The exchange of goods was conducted on an inter-personal basis, used to form alliances and cement relations between families and groups. Trade, on the other hand, was impersonal, with no ties between the participants other than that of profit. Although the majority of large settlements in the Early Historic Period appear to have been more or less self-sufficient, there remained certain items that had to be acquired either through trade or exchange.

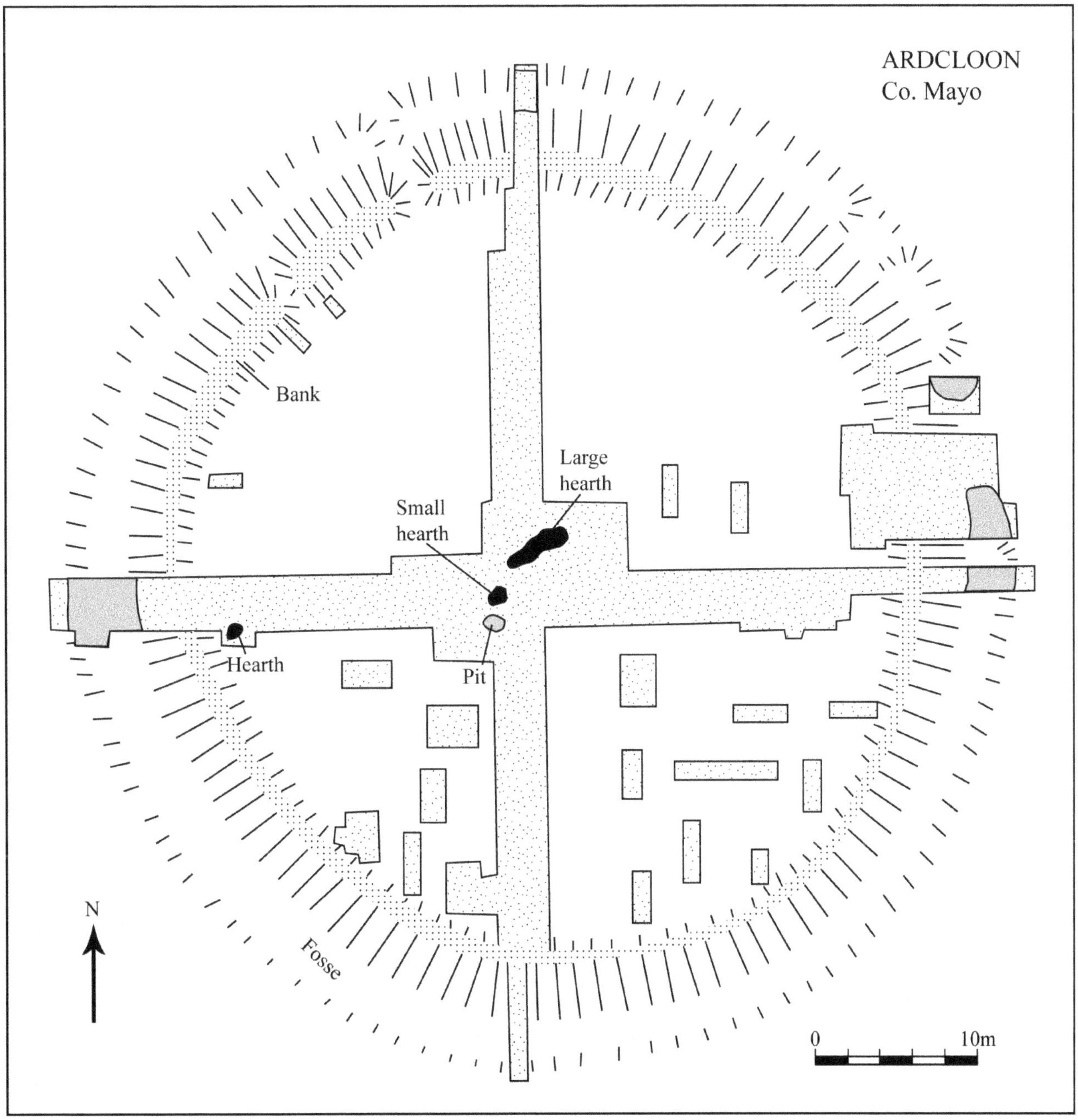

Fig. 43 Ardcloon, Co. Mayo (after Rynne 1956).

The exchange of gifts or tributes appears to have been central to the smooth functioning of early Irish society (Ó Cróinín 1995). The social hierarchy consisted of six main levels, the most important of which was the high-king. Below this came the king of three or four *tuatha* (petty kingdoms), the king of one *tuath*, the noble grades including warriors and craftsmen, the client or freeman who rented the king's land, and finally the serf who could not leave the land, and the slave, the property of the king. The exchange of gifts between these levels ensured their survival. Those of a higher status provided protection for those of inferior standing in return for their military and financial support (Edwards 1990, 8).

The acquisition of raw material was clearly of vital importance for non-ferrous metalworking, whether the source was within Ireland or abroad. It seems likely that native sources of copper ore were exploited in the Early Historic Period with local groups perhaps facilitating its distribution. The circulation of ore, smelted ingots and scrap would seem to have been controlled or monopolised by the most powerful in society. The average client or farmer would not have been entitled to a tribute or gift of such value (see chapter 2). The two lumps of copper ore found at the royal site of Lagore (see chapter 6) may have been mined in the locality, perhaps by a client of the kings of Lagore. The lead ore found in a small ringfort settlement at Ardcloon Co. Mayo may, on the other hand, have been mined by the inhabitants for use as an item of exchange or trade (Fig. 43).

The distribution of a particular artefact type, probably through reciprocal exchange, can be seen quite clearly in the zoomorphic penannular brooches of Kilbride-Jones' groups C5 and D. These were made at Clogher, Co. Tyrone, though examples have been found at Eaglish in Co. Armagh, Gransha in Co. Down, Navan rath in Co. Armagh and even one as far south as Killucan in Co. Westmeath (Kilbride-Jones 1980). These are dated to the fourth and fifth centuries, the Westmeath example perhaps reflecting the allegiance of the Airgialla (located at Clogher) to the powerful Uí Néill, newly arrived in the Meath area. Most of the brooches did not travel far, possibly the result of exchange between the king at Clogher and his subject kings/lords.

Two pieces of tin were found at Garranes, perhaps imported from either Cornwall or Brittany. Other imported items found on the relevant sites include amber, studs of which were found at Lagore, Kilpatrick and Cathedral Hill. Early imports during this period include a fragment of hack bronze from a Roman bracelet found at Clogher, and Roman glass found at Lagore, both of which were intended for recycling.

Sherds of E-ware form the largest body of evidence for importation on Early Historic settlement sites in Ireland. E-ware consists of domestic vessels dated to the sixth and seventh centuries AD. These were imported from the Continent, possibly from the Saintonge area of Aquitaine. These vessels may be associated with the importation of wine from the Continent, the larger vessels may even have contained wine. Of the relevant sites, E-ware is found at Clogher, Lagore, Garranes, Moynagh Lough and Lough Faughan.

The majority of goods imported from abroad may be seen to represent trade as opposed to gift-giving. It would seem unlikely that a large number of Continental kings or rulers felt the need to cement relations with Irish kings. While some, perhaps those with Irish connections, may have engaged in reciprocal tribute-exchange, merchants are likely to have facilitated the rest.

In the description of Columbanus' expulsion from Gaul in Adomnán's *Life of Columba*, the port of Nantes is described as a place frequented by Irish merchants. There, Columbanus is given gifts of wine, wheat and beer to take back to Ireland (Anderson 1961). Again, in the *Life of Ciarán of Clonmacnoise*, it is recorded that wine from Gaul was brought to the monastery by merchants (Plummer 1910, vol.1, 214). Such goods as salt, metal products and wine may have been imported in return for animal-hides or slaves. Unfortunately, the merchants mentioned in the early literature are not identified in the archaeological record so it is impossible to determine whether they were attached to specific communities/sites or whether they acted as independent 'middlemen'.

Before the growth of major urban centres, tribal or dynastic *oénach* (fairs) provided the setting for peaceful trade and exchange in Early Historic Ireland. The fairs of *Tailtiu* (Teltown), *Cruachu* (Rathcroghan) and Lynally, Co. Offaly are mentioned in *Irish Triads* no.35 which is dated to the ninth century (Meyer 1906, 5). However, by the end of the eighth century, monasteries appear to have taken over the function of the *oénach*. The first reference to a monastic *oénach* is recorded in the Annals of Ulster for the year 799 when a local king died after falling from his horse at a fair in Lusk, Co. Dublin. The use of many monasteries as 'storage depots' for the surrounding countryside made them ideal locations for markets.

A change from this system of trade and exchange at local *oénach* can be seen in the tenth century (Ó Cróinín 1995). With the growth of urbanism came the expansion of trade and the use of coinage. The Viking settlements with their easy access to international trade influenced and encouraged the growth of monastic towns as redistribution centres. Viking trade with the native Irish can be seen, for example, in the quantities of Viking silver found throughout the country, including the hoard from Carraig Aille II in Co. Limerick. While some of this silver may have been taken in raids, a good percentage was probably peacefully obtained through trade. By the tenth century provincial fairs had virtually ceased to exist (Ó Cróinín

1995). Both common and exotic goods could now be obtained quite easily via the Norse or monastic towns.

5.7. Settlement size

The status of a site is generally reflected in its size and the amount of effort invested in its construction (Figs. 47 and 48). Earthworks, however, cannot be compared with either cashels or crannógs on the basis of size. The status of all three types, though, can be compared using their artefact assemblages.

It is clear from excavated raths (Tables 4 and 5) that size is directly related to the wealth of the inhabitants. A clear hierarchy of sites is evident. Those raths with diameters measuring between 20m and 50m have produced evidence of agriculture and limited iron, textile and leather working. These include Ballywillwill I, Crossnacreevy, Rathbeg, Seacash, Coolcran and Killyliss. Ardcloon, with a diameter of 58m, produced evidence of agriculture in association with ironworking and limited non-ferrous remains. It is slightly larger than those sites first mentioned and accordingly provided evidence of a somewhat broader, wealthier economy. At the highest end of the scale are Garranes and Knowth (an earthwork though not strictly a rath) measuring 110m and 90m in diameter, respectively. These produced evidence of the most varied and wealthy economies.

NAME	COUNTY	TYPE	COMMENT	SIZE	REFERENCE
Dooey	Donegal	Sandhills	Interim report only	17 x 20m	Ó Ríordáin & Rynne 1961
Garryduff I	Cork	Ringfort	Fully excavated	20m diam.	O'Kelly 1963
Carraig Aille II	Limerick	Ringfort	Fully excavated	40m diam.	Ó Ríordáin 1949
Ballywillwill I	Down	Ringfort	Bivallate, limited ex.	31 x 26m	Waterman & Collins 1952
Rathbeg	Antrim	Ringfort	Univallate	34m diam.	Warhurst 1969
Seacash	Antrim	Ringfort	Univallate	35 x 28m	Lynn 1978
Crossnacreevy	Down	Ringfort	Univallate, limited ex.	36 x 31m	Harper 1973/4
Killyliss	Tyrone	Ringfort	Univallate, limited ex.	42m diam.	Ivens 1984
Coolcran	Fermanagh	Ringfort	Univallate, limited ex.	43m diam.	Williams 1985
Ardcloon	Mayo	Ringfort	Univallate	58m diam.	Rynne 1956
Garranes	Cork	Ringfort	Trivallate, limited ex.	110m diam.	Ó Ríordáin 1942
Moylarg	Antrim	Crannóg	Antiquarian ex.	27m diam.	Buick 1893
Lough Faughan	Down	Crannóg	Partially excavated	34m diam.	Collins 1955
Ballinderry II	Offaly	Crannóg	Fully excavated	35m diam.	Hencken 1942
Moynagh Lough	Meath	Crannóg	Research excavation	36 x 16m	Bradley 1991
Rathtinaun	Sligo	Crannóg	Report forthcoming	36 x 29m	Raftery forthcoming
Lagore	Meath	Crannóg	Fully excavated	41m diam.	Hencken 1950
Island McHugh	Tyrone	Island site	Poorly stratified	-	Davies 1950
Dalkey	Dublin	Island site	Limited excavations	0.25 ha	Liversage 1968
Clogher	Tyrone	Hillfort	Unpublished	-	Warner 1973
Knowth	Meath	Passage tomb	Fully excavated	90m diam.	Eogan 1977

Table 4 Site status.

NAME	Agricultural evidence	Iron-working	Copper-working	Lead-working	Silver/gold-working	Amber-working	Glass-working	Millefiori/ enamel-working	Textile-working	Leather-working	Bone-working	Finds of E-ware	Finds of Copper	Finds of Iron	F.Silver/gold
Dooey	S	S	S	L							L		P	P	
Garryduff I	S	EX	S			P	P		L		L	S	EX	S	L
Carraig Aille II	EX	S	S		S				L				S	S	
Ballywillwill I									L				L		
Rathbeg	S													L	
Seacash	S								L	L			L	L	
Crossnacreevy														L	
Killyliss	L									L				L	
Coolcran	L	EX												L	
Ardcloon	L	S		S									L	L	
Garranes	S	S	EX	L			S	S	L		L	S	EX	EX	
Moylarg		L	L	P					L		L		L	L	
Lough Faughan	EX	L	L			L	L		L		L	L	S	S	
Ballinderry II	EX	L	S			P	P		S		S		S	EX	
Moynagh Lough	S	EX	EX		S	L	S	L		L	S	S	L	S	
Rathtinaun	EX	L	S			L	P		S	L	L		S	S	
Lagore	EX	EX	EX	L	L	L	EX	L	EX	S	EX	L	EX	EX	

EX = Extensive evidence S = Substantial evidence L = Limited evidence P = Possible evidence

Table 5 Site economy.

Several problems are encountered in a survey of ringfort size and its relationship to status. Garryduff I and Moynagh Lough, for example, are small in size yet have produced evidence of a range of specialist activities. The probability that these sites are representative of larger settlements seems likely. An examination of size must take into account the clustering of sites such as the two cashels at Garryduff and also at Carraig Aille. Many of the more mundane domestic or agricultural activities may have taken place in nearby enclosures or perhaps in unenclosed areas, difficult to identify in the archaeological record. At Carraig Aille II, a small group of stone structures abutted the exterior of the cashel on the north side (Fig. 41). It is possible that wooden equivalents were employed at rath sites or on the lake-shores adjacent to crannógs, their lack of preservation or identification resulting in a biased view of settlement size.

Unfortunately, it is not possible to compare the crannógs in a similar manner as very few have been excavated. Those that have appear to be of a high-status, however differences between certain sites can be distinguished through an examination of archaeological evidence. It is clear that crannógs such as Lough Faughan and Moylarg did not operate at the same level as such sites as Lagore and Rathtinaun. They produce evidence of a similar range of activities, though not on the same scale. Perhaps this reflects the differing status and power of local and regional kings.

Little can be said of the two cashels other than that they both produced evidence of wealthy inhabitants. Taking into account all of the sites surveyed, non-ferrous metalworking is associated with the settlements of high status, whether the latter is indicated by the artefact assemblage or the enclosure size and form. No sites of an exclusively industrial nature can be identified, agriculture appears to play an important role in the economies of most, if not all of these settlements.

One of the largest problems encountered in a survey of Early Historic secular settlement is the quality of both excavations and their published reports. A large number of ringforts were excavated under rescue conditions where only a portion of the site could be investigated. As such, the resulting reports may provide biased impressions of the nature of the site. However, this is true of large and small enclosures alike, such as Coolcran and Garranes. Crannógs are difficult and expensive to excavate, the main reason for this being their waterlogged nature. This has resulted in the relatively small number of excavated crannógs. Also, several of the published examples were investigated by antiquarians whose reports cannot be compared with more recent examples. Monastic sites cannot be judged using the same principles, as spiritual concerns affected their size and form.

Garranes ringfort has been tenuously identified as *Rath Raithleann*, the ruling seat of a branch of the Eóganacht dynasty of Munster (Ó Ríordáin 1942). Whether or not this is a valid association is not relevant here as the archaeological evidence proves this to be a site of some importance, with probable royal links. On enclosure evidence alone (Fig. 39), the excavator judged the fort to have been a settlement of considerable importance. The links with *Rath Raithleann* would suggest a settlement that functioned as a tribal centre and which included domestic and industrial activities within its organisation.

While acknowledging that metalworkers had a relatively high status in society, as evidenced by the early literature (see chapter 2), their power and influence was not strong enough to allow them to personally control the sources and circulation of valuable metal. Neither were they of a high enough status to enable them to construct, possess or exclusively inhabit a site of such magnitude. It is much more likely that they were supplied with raw materials by wealthy patrons for whom they worked. The discovery of specialised craft activities like non-ferrous metalworking on high-status sites may reflect this patronage.

Another excavated fort in Co. Cork is that of Garryduff I which consists of a well-built stone rampart with rock-cut external fosse. Dated between 500 and 600 AD, the metalworkers who lived on this site had 'all the knowledge, artistic ability and technical competence to enable them to produce' (O'Kelly 1963, 120). It is of some importance to note, though, that they also had the ability to acquire the valuable raw materials required. When compared with Garranes, Garryduff emerges as the home of a much smaller community or maybe that of a wealthy family (Fig. 40). Perhaps the slightly later site of Garryduff reflects the general move from tribal to dynastic society which occurred around the sixth century AD.

Like Garranes, although somewhat more definite, the site of Lagore crannóg, Co. Meath has been identified as an Early Historic royal site. *Loch Gabair* was the seat of the kings of southern Brega many of whom also served as high-kings. The finds from the site substantiate the historical evidence regarding status (chapter 6).

A settlement which was neither ringfort, crannóg, nor monastic enclosure, but which seems to have been on a par with Lagore, is that of Knowth, Co. Meath. This site was built on top of the earlier passage grave mound, perhaps something of the original importance of the site remaining significant in the Early Historic period. The site has been identified as *Cnogba*, the seat of the kings of northern Brega, this link serving to emphasise and reinforce the importance of the archaeological evidence. According to the literary references the inhabitants of both this site and Lagore were branches of the Uí Néill dynasty, continually vying with each other for political supremacy.

The Early Historic settlement at Knowth was largely undefended and consisted of at least nine houses and two souterrains. The digging of two large ditches around the site is dated by the excavator to the pagan Iron Age (Eogan 1977, 70). However, these ditches were allowed to partially silt up in the Early Historic Period with no obvious effort made to re-fortify the site. In the southern part, an area of fine metalworking was identified including the remains of a fire, heating pans and clay crucibles. Traces of gold-working were also found, suggesting high status. The dates suggested by the early literature indicate that the kings of northern Brega were living here from the ninth to eleventh centuries AD. 'It is interesting to note that this is also the date that, in the main, one would also apply to the second phase of the settlement on archaeological grounds' (Eogan 1977, 75).

Whether or not there is a different status between the relevant sites is difficult to determine. The Garryduff I site is a cautionary example where settlement size does not necessarily correlate with the practice of specialist crafts. Sites such as Cathedral Hill and Lagore do appear to be more impressive than Reask or Lough Faughan (Collins 1955), while the differences between Garranes ringfort and enclosures such as Ardcloon are obvious.

5.8. Schools of metalworking

Comparisons have occasionally been drawn between individual items of fine metalwork from Early Historic Ireland. However, only one 'school' of metalworking producing a particular artefact type and associated with a specific workshop has been identified. This is the brooch 'factory' at Clogher, Co. Tyrone. Kilbride-Jones in his 1980 study associates two groups of zoomorphic penannular brooches with this site. Group C is dated to the fourth century when the site became the capital of the Airgialla, while group D is dated to the fifth century by finds of imported B-amphorae.

The seven brooches identified as belonging to group C are 'clearly the work of one individual' (Kilbride-Jones 1980, 63). The decoration on all seven is meticulous and extremely delicate, though perhaps the stronger unifying element was the inability of the smith to produce two identical terminals on any of these brooches. In all cases, an inconsistency between the two terminals is evident.

All but one of these brooches have sweated-on terminals, several of which were lost due to incompetent workmanship. The brooch rings were fabricated from a length of thick wire or rod, the terminals from beaten sheet metal. They were decorated with *champlevé* red enamel, chased and incised ornament. One example (Kilbride-Jones no. 128) reflects the introduction of Christianity into a pagan society. It is decorated with both the Maltese Cross and the pagan motif of spherical triangles with dots at their centres.

The final brooch from this group (Kilbride-Jones no. 130), whilst of the same basic design, differs from the others in its method of manufacture. It was cast and is crude by comparison with the other six, and may represent an experiment by the smith. Of the seven, two were found at Clogher, one in the bog of Clogher, two unprovenanced, one at Killucan in Co. Westmeath and one at Navan rath in Co. Armagh.

Group D consists of 13 brooches, one of which was found in the ditch outside the palisade surrounding the inner 'citadel' at Clogher. These brooches break with the tradition established by group C with the emphasis transferred from the eyes (which are omitted) to the ears and all the brooches were cast. This group includes a raw terminal casting (Kilbride-Jones no.131) from Clogher, probably discarded due to a casting flaw. The brooches of this group are decorated with millefiori, enamel, and one (Kilbride-Jones no.135) with dots of pink coral. Eight of the brooches are unprovenanced, one found at Clogher, two from Co. Armagh, one from Gransha in Co. Down, and one from Eaglish in Co. Armagh.

Also found at Clogher was a lead pin (Kilbride-Jones no. 150) that acted as a pattern for some of the pins of group D brooches, with no. 135 being the nearest bronze equivalent. This, in association with the hack-bronze Roman bracelet, crucibles and the raw terminal casting, points to Clogher as the place of manufacture of these two groups of brooches. Their difference from one another was most likely the result of a change in the resident chief or master smith. These brooches do after all span at least two centuries. The royal nature of this site, as capital of the Airgialla, would have supported such an industry.

Schools of metalworking, or workshops producing a range of fine objects over a lengthy period of time, probably existed throughout Early Historic Ireland. Similar artefact studies may enable the association of other major production centres both with each other and with particular styles of metalworking. They could also point to, as yet unidentified, places of manufacture or may indicate the scale of non-ferrous metalworking on a particular site or in a certain area.

5.9. British parallels

Evidence of non-ferrous metalworking has also been found on first millennium AD sites in Scotland, Wales, and England. As in Ireland, this evidence appears to be associated with sites of high status. The most well known of these are Dunadd and the Mote of Mark in Scotland and Dinas Powys in Wales.

Described as 'undoubtedly the most impressive site in Early Christian Wales' (Laing 1975, 109), Dinas Powys was not without Irish connections. Amongst the finds from this fortified hilltop was a lead pattern or model for a penannular brooch which stylistically belongs to the Irish series of such brooches. This 'along with the slight evidence for millefiori working on the same site, can perhaps suggest the presence of an itinerant Irish smith in south Wales' (Laing 1975, 98).

One lead die and a trace of millefiori, however, does not indicate the presence of an itinerant Irish metalworker. The period dating from the fourth to the sixth centuries AD saw substantial Irish settlement in Wales, the result of displacement of older tribal groupings by new dynastic families in Ireland (Ó Cróinín 1995). Definite evidence of the presence of Irish colonists in Wales is provided by the distribution of ogham stones in south-west Britain, dated to the fifth and sixth centuries AD. The community at Dinas Powys may have been of Irish descent or included in their number people of Irish origin. These, in turn, may have introduced the knowledge or practice of Irish metalworking. Alternatively, it is possible that in an area with strong Irish connections, the lead model and millefiori might have been imported.

The settlement at Dinas Powys saw the reoccupation of a hillfort in the sixth century AD. Located at the north end of a hill, the site consists of relatively insubstantial earthworks approximately 46m in diameter. In a courtyard area were found the remains of two buildings and a number of hearths, some of which were associated with metalworking. Finds included scrap metal and lidded clay crucibles. 'Bronze working seems to have been quite extensive: presumably the site was a stronghold of a local chieftain' (Laing 1975, 110; Alcock 1963). Other Welsh sites producing relevant evidence include Pant-y-Saer (Phillips 1934), Dinas Emrys (Alcock 1963) and Longbury Bank (Campbell and Lane 1993).

The site of Dunadd, Argyll, in Scotland is of a similar status, associated in the annals with Gabran, King of Dalriada, again of ultimately Irish origin (Craw 1929/30). The site consisted of stone built defences with rectilinear stone buildings inside. Amongst the finds were 'distinctively Irish objects' (Laing 1975, 76) such as slot-headed iron implements and a hand-pin. The metalworking evidence included hundreds of moulds for ingots, pins, rings and penannular brooches, lidded clay crucibles and carved stone motif pieces. The site is dated to the end of the fifth/beginning of the sixth century AD (Lane and Campbell 2000).

The most important site in south-west Scotland is the Mote of Mark, Kirkcudbright (Curle 1913/14). Located on the summit of a rocky outcrop, the site is delimited by a stone, earthen and timber rampart and an inner stone wall. Dated to the sixth and perhaps the seventh century AD, the Mote of Mark may have been an exclusively industrial site. No evidence of agriculture was found, however, ironworking (finds of the iron ore haematite and iron-smelting slag) and non-ferrous working was extensive.

Evidence of bronze working included a possible non-ferrous smelting furnace. This consisted of an ash-filled hollow measuring approximately 0.45x0.3x0.1m deep. No slag was found though associated finds included fragments of moulds and crucibles. Other relevant finds included tuyère fragments, two bronze ingots, scrap bronze, a bone model for a pin and the mould made from it. Gold and silver may also have been worked, a fragment of silver and a small coil of gold ribbon were recovered from the site (Swindells and Laing 1980, 121–128). Other Scottish sites producing relevant evidence include Dunollie (Alcock and Alcock 1987), Dundurn (Alcock *et al.* 1989) and Clatchard Craig (Close-Brooks 1986).

The evidence from Anglo Saxon England also appears to be largely associated with sites of high status, most notably the royal palace at Cheddar and the town of Southampton which was closely associated with the nearby royal and ecclesiastical Winchester (Wilson 1981). Dated to the tenth century, the Saxon palace of Cheddar in Somerset was the rural palace of the kings of Wessex and produced numerous clay crucible fragments. The evidence from the ninth-century town of Southampton consisted of hundreds of pits which bore signs of burning, clay mould fragments, a lead ingot and bronze 'drippings'. The thirty-hectare site produced evidence of iron, bronze, lead and silver working.

The range of site types which have yielded non-ferrous metalworking evidence in Ireland are paralleled in Britain. In Scotland, the dun at Kildonan Bay is an equivalent of the Irish cashel (Laing 1975). Here were found moulds, a crucible, whetstones, and bronze fragments, evidence very similar to the Irish remains. A 'high status' crannóg which produced much the same economic evidence as the Irish examples is that of Buston crannóg, Ayrshire (Laing 1975). The importance of this site is reinforced by the discovery of crucibles associated with gold-working. Again, the British evidence reflects the ecclesiastical association evident in Ireland. The monastic foundation of Glastonbury, for example, produced evidence of both glass and metalworking. Yet again power, wealth and status appear to be associated with non-ferrous metalworking (Wilson 1981).

5.10. Potential and problems

An examination of the non-ferrous metalworking evidence from Early Historic settlement sites in Ireland reveals much about social organisation between the fifth and twelfth centuries AD. On a local level, internal site layout is obviously influenced by the location of the industrial area. Such areas, it would appear, are always located as far away from the domestic area as possible, yet remaining inside the main settlement enclosure. The layout of a typically confined Early Historic settlement like a ringfort or crannóg, generally sees the domestic area in the middle of the enclosure, with the peripheral parts of the site utilised for industrial or craft activities. The latter can include corn-drying, glass-working, metalworking, or perhaps large-scale carpentry or bone-working. On larger sites, the domestic area need not be centrally located. It can be situated in a particular sector or quadrant e.g. Dunmisk, Co. Tyrone, possibly abutting the enclosing element.

As metalworking generally leaves behind substantial evidence, its specific location within a site is usually easier to identify than that of other activities such as textile production where much of the equipment and product was organic and has since decayed. Industrial zones can often be identified by locating the metalworking area, thus enabling the possible location of other craft activities. At Nendrum, for example, the bronze foundry and school were identified inside the middle rampart (Lawlor 1925). The remains of four other stone buildings were found in the same area, suggesting their possible use as workshops (Fig. 90).

The location of the industrial zone can also aid the identification of the domestic area. At Dunmisk, for example, the industrial areas were identified in the north-east and south-west leaving either the north-west or south-east as the domestic area (Fig. 37). This was further narrowed down by the discovery of a large cemetery and the remains of a possible timber church in the south-east sector. Though no direct evidence was found, the domestic area has been placed in the north-west of the site (Ivens 1989).

It should also be remembered that on some of the larger sites (Dunmisk and Knowth for example), the non-ferrous metalworking area is located more or less diametrically opposite the ironworking area. The discovery of one of these, then, may help locate the other if such exists on a site. As an indicator of high-status, non-ferrous metalworking can aid the establishment of a hierarchical pattern of settlement (see chapter 5). Sites where extensive relevant evidence is found can be regarded as important, high-status settlements. The size of the site and its artefact assemblage should help indicate its exact nature.

On a broader level, the practice of non-ferrous metalworking has the potential to aid the location of areas of wealth and prosperity. It may also reflect the availability of raw materials in a particular area. The average-sized ringfort at Ardcloon, Co. Mayo, for example, can hardly be viewed as high-status because of the discovery of lead ore on the site. This possibly reflects the availability of lead ore in the vicinity.

There are, however, many problems involved in studying the settlement contexts of non-ferrous metalworking. These problems concern the quantity and quality of the available information. The published excavation reports of several sites are lacking both in detail and scientific analyses. In many cases, the reports are biased by the percentage of the site actually excavated and perhaps by the expectations or opinions of the excavator.

5.11. Conclusion

A study of settlement evidence from Early Historic Ireland suggests that blacksmiths working in iron produced tools and weapons on a wide range of sites. Non-ferrous metalworkers, though, concentrated on finer items such as penannular brooches and decorative mounts and such work appears to have been a status-related activity.

All of the settlement types have produced a similar range of evidence with no apparent differences between religious and lay sites, either in terms of wealth or the range of items being manufactured. Moulds provide evidence of the objects manufactured on a site and generally represent the casting of pins, brooches and decorative mounts. One might expect the evidence from an ecclesiastical site to reflect religious items such as chalices and croziers. Most of these artefacts, however, are composite in nature, perhaps formed largely of sheet metal and then decorated with separate panels. Decorative panels were used to adorn both secular and religious items, making it impossible to distinguish between the two from the mould evidence alone.

On a national level, concentrations of known relevant sites occur in Munster, Meath, and Ulster (Fig. 32). This distribution may reflect the political power bases in Early Historic Ireland. The four royal sites of *Rath Raithleann*, *Cnogba*, *Loch Gabair* and Clogher are located in counties Cork, Meath and Tyrone. Ulster also boasted one of the most important monastic sites in the country, namely Cathedral Hill, Co. Armagh recorded in the annals as having been founded by St. Patrick.

Within the larger sites, specialisation according to the type of metal worked appears to have taken place. The metalworking area, whether specific to non-ferrous working or not, was invariably located on the edge of the settlement. Such positioning was a wise precaution considering the threat posed by fire to the habitation area. The location of non-ferrous metalworking areas inside ringforts and other settlement enclosures may be seen as a security measure designed to protect stock-piles of metal (see section 2.4. for early literary reference to theft of metal).

There is no substantial archaeological evidence of itinerant fine-metalworkers from Early Historic Ireland. If travelling smiths did exist, they practiced their trade within or adjacent to settlement sites. The vast quantities of relevant evidence found at many sites must reflect the activities of a resident metalworker(s) working under patronage.

The sites producing lesser amounts of relevant evidence may also have had permanent workshops or working areas though smiths need not have been exclusively or continuously employed in non-ferrous metalworking. They may also have been occupied with other craft activities or perhaps engaged in agricultural work at certain times of the year. If itinerant smiths did exist during the Early Historic Period, they have left no trace in the archaeological record.

Fine metal objects played an important role in the economy of Early Historic Ireland. They represented the wealth and importance of the owner, reinforcing divisions of class and status. The production of such items, then, is logically associated with a social elite. Non-ferrous metalworking was a carefully organised activity, a manifestation of power and social standing, closely controlled by the most powerful in society.

CHAPTER 6
LAGORE CRANNÓG
- A re-evaluation of the relevant evidence

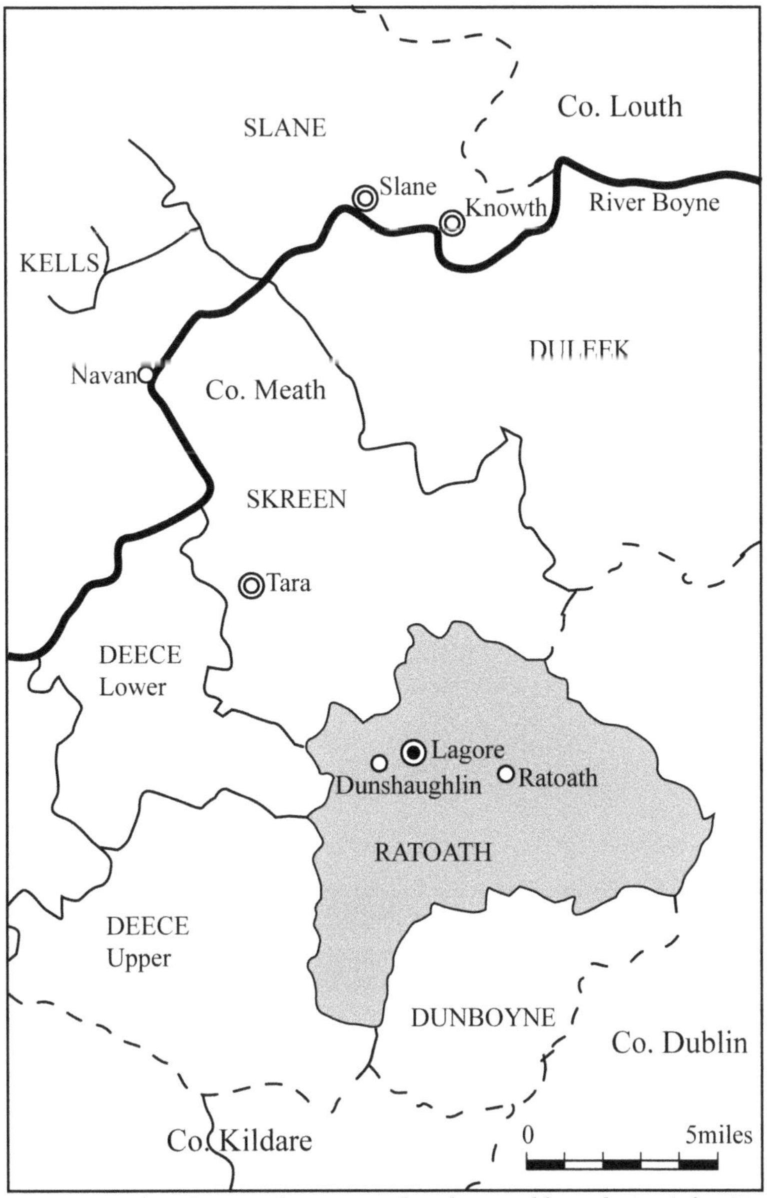

Fig. 44 District surrounding Lagore showing modern baronial boundaries (after Hencken 1950).

6.1. Introduction

The site of Lagore crannóg is located in Lagore Big townland in the barony of Ratoath, Co. Meath, approximately one mile north-east of Dunshaughlin village (O.S. 6" sheet 038, national grid reference N9862/5285, Fig. 44). Remains of the crannóg are situated at the eastern end of an old lake bed. Originally, this was a shallow, marshy lake, full of vegetation 'perhaps holding open water in the middle' (Hencken 1950, 35). Before excavation the site was marked by a low mound divided in two by a 'river' or artificial drainage channel (Fig. 45).

The crannóg was first noticed in 1839 when the drain was being deepened and large quantities of bone were discovered. Sir William Wilde visited the site and the following year read a paper to the Royal Irish Academy regarding his discoveries. He described the site as 'a circular mound, slightly raised above the surrounding plain, its highest central part being about eight feet above the margin, and the circumference of the mound measuring 520 feet' (Wilde 1840, 420).

The site was again brought to attention in 1848 and 1849 when the drain was being cleaned and the area subjected to turf-cutting.

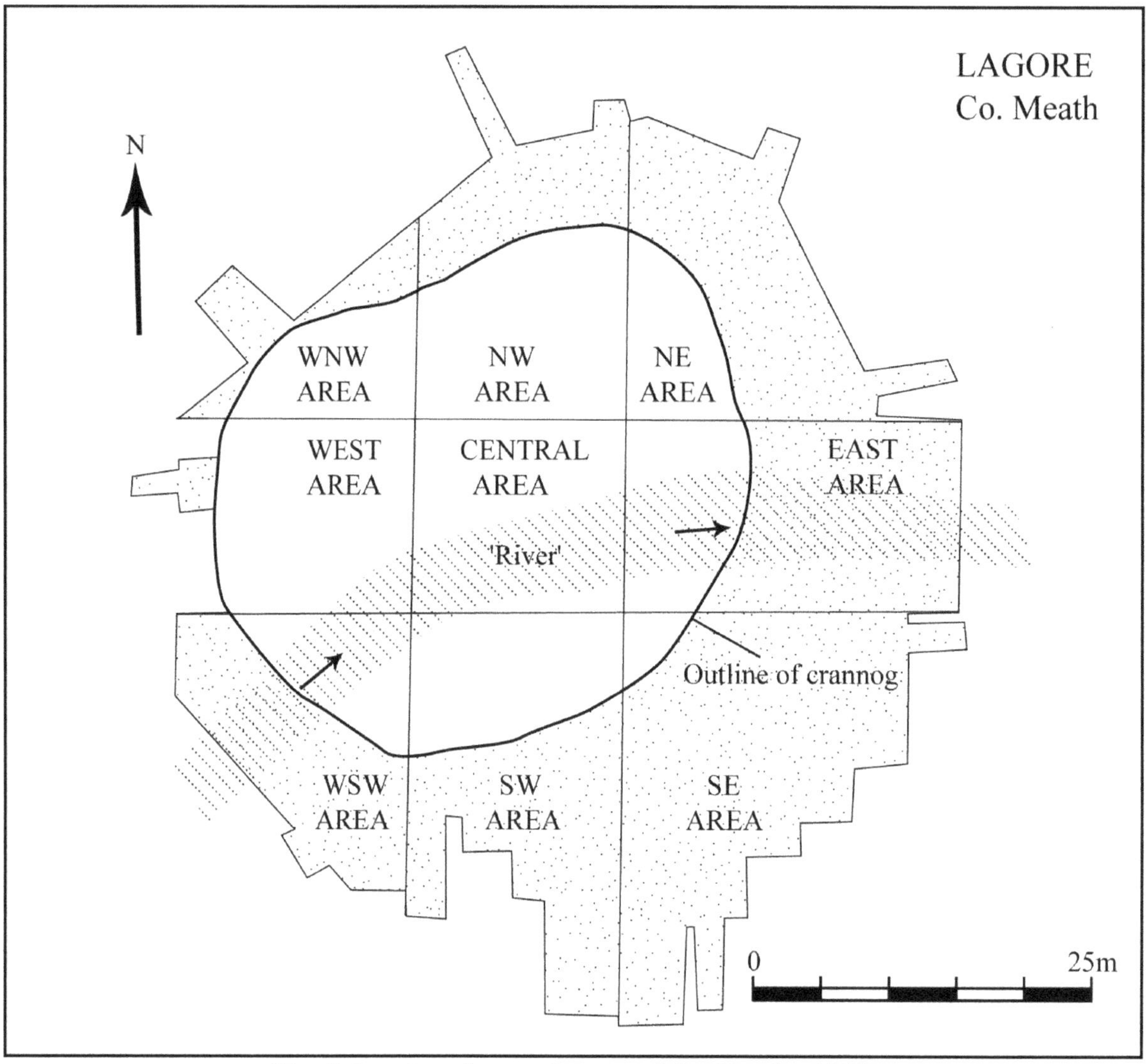

Fig. 45 Location of excavation areas (after Hencken 1950).

The antiquarian William Wakeman visited the site though he failed to publish a report of his findings until 1882, over thirty years after the event. This report was brief and dealt mainly with structures Wakeman identified as houses (see 6.4). The final description of the site was made by Wood-Martin in his book *The Lake-Dwellings of Ireland* in 1886. He gives a short list of previous finds from the site, though again, very little context information or structural detail is provided.

In 1934 and 1936 the Harvard archaeological expedition to Ireland, led by Hugh Hencken, excavated the site. In the century between its initial discovery and its excavation, the crannóg suffered much damage at the hands of turf-cutters and those searching for antiquities. A large quantity of bone was also removed from the site, to be used by the locals in the manufacture of fertilizer. The excavation, despite all previous disturbance, did produce some stratified areas and a large assemblage of finds. A 250-page excavation report was published in the *Proceedings of the Royal Irish Academy*, vol. 53c, in 1950.

6.2. Site chronology

Hencken divided the site into four main phases, beginning with period Ia which he dated to the seventh century AD. Finds from this level included sherds of E-ware, a horse-bit and a gold filigree ornament (Fig. 46). The excavator used this evidence to date the period to the seventh or possibly early eighth century AD. This date is supported by the more recent work of Richard Warner:

> The absence of Mediterranean pottery and penannular zoomorphic brooches, and the presence of E-ware at the lowest levels would seem to me to indicate very strongly that period Ia, and the start of occupation, does not pre-date the 7th century, and might even be as late as the 8th.
>
> (Warner 1985/86, 77)

Hencken also utilises historical data to support his chronology.

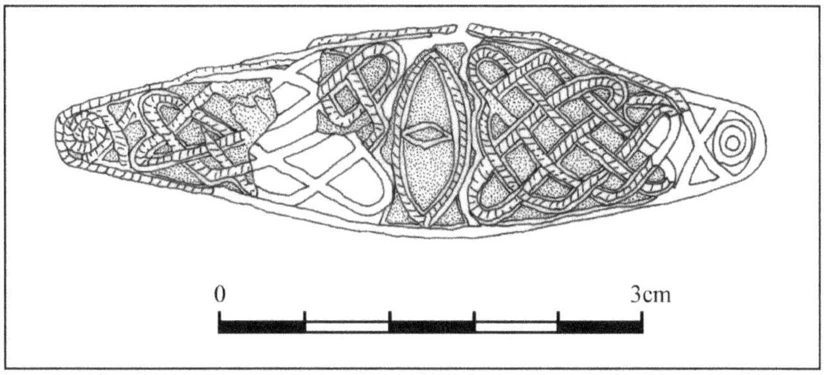

Fig. 46 Gold filigree panel from Lagore (after Hencken 1950).

It is recorded (see 6.3) that in the year 651 AD the high-king, Diarmait Ruanaid, was gathering his armies at Lagore in preparation to avenge the murder of his two nephews. Hencken felt that, in conjunction with the archaeological evidence, the literary references indicated construction not long before 651 and perhaps even by Diarmait himself.

Period Ia was represented by the body of the artificial island itself, identified by the excavator as the period of construction with all the material deposited at the one time. This interpretation, however, has been disputed by Lynn (1985/86). He identifies the bow-shaped timbers found at this level as the possible remains of a palisade, and the fragments of wattle walls and wicker mats as evidence of occupation. Hencken attributed these, and several spreads of ash, to 'the builders who might occasionally have wanted some temporary shelter while working on the crannóg' (1950, 42). It would make more sense, however, to build temporary structures on the nearby shore where they would not have to be destroyed and re-built repeatedly as new levels were added to the body of the crannóg.

The structure of the crannóg consisted of several layers of brushwood, each one associated with traces of wattle walls and wicker matting. They were most common on brushwood layer 3 in the Central area where the remains of a quite substantial wall were found. The posts around which the withies were woven were 0.08m in diameter.

> To sum up, it is proposed that the substructure of Lagore crannóg from which finds attributed to period Ia came was not a substructure but a series of alternating occupation and build-up of layers which contained a number of structures and a range of finds which would not be out of place in an important Early Christian period crannóg or ringfort.
>
> (Lynn 1985/86, 71)

Period Ib was designated the first occupation at Lagore by Hencken, though this now seems doubtful. It is dated to the seventh century by finds including the remains of a glass-stud manufacturing workshop (seventh/eighth century) and a single Romano-British sherd (fifth-seventh century). The period I crannóg was surrounded by a pile palisade while period II is marked by the building of a post palisade (Fig. 47). From the historical evidence (see 6.3) Hencken identified this as a refortification undertaken after the burning of the crannóg in 850 AD. Finds associated with this period did not provide definite dating evidence.

Period III is also marked by the construction of a palisade (of planks), possibly a final refortification following the historically recorded destruction of the settlement in 934 AD. The latest finds from the site include three or four Viking combs, a Viking spearhead and an English penny, dated 901–925. The archaeological evidence suggests that Lagore was abandoned at the end of the tenth or beginning of the eleventh century, and the last historical reference is dated 969 AD. A forthcoming C-14 programme on the animal bones from the site, by Queen's University Belfast, should aid the clarification of Lagore's chronology.

6.3. Historical references

(Based on Price's report in Hencken 1950, 18–34)

The history of Lagore is inextricably bound to that of the southern Uí Néill. The territory of the southern Uí Néill was divided in three. The western area was ruled by the Clann Cholmáin from Uisneach and later from the twin royal sites of Dun na Sgiath, a fortified earthwork on the shores of Lough Ennell, and Cró Inis, an off-shore crannóg, all of which are in Co. Westmeath. The eastern area was subdivided into the territory of northern Brega ruled by the Clann Uí Chonaing from Knowth (*Cnogba*), and southern Brega ruled by the Clann Chernaigh Sotail from Lagore (*Loch Gabair*).

Annalistic references appear to make no distinction between the terms 'King of *Deiscert* Brega' (southern Brega) and 'King of *Loch Gabair*'. This suggests that Lagore was the seat of the kings of *Deiscert* Brega. The lands around Lagore, the modern baronies of Skreen and Ratoath, were owned by the high-king Aéd Sláne who died in 604 AD.

THE SOUTHERN UÍ NÉILL

	AÉD SLANE (†604)	
COLMÁN MÓR (†557)		
SUIBHNE (†600)	CONGAL (†634)	DIARMAIT RUANAID (†664)
CONALL GUTHBIND (†635)	CONAING (†661)	CERNACH SOTAL (†663)
CLANN CHOLMÁIN (Midhe) Uisneach Dún na Sgiath Cró Inis	*CLANN UÍ CHONAING* (North Brega) Cnogba	*CLANN CHERNAIGH SOTAIL* (South Brega) Loch Gabair

Table 6 Based on information provided by Price in Hencken 1950, 18–34.

There is no mention of his having occupied Lagore, however his son, Diarmait Ruanaid (died 664 AD), is associated with the site of *Loch Gabair*. In preparation to avenge the murder of his two nephews, Diarmait gathered his armies at Lagore. The death of his nephews is recorded in the annals for the year 651 AD. As a story this is probably much embellished though it is clearly not a total work of fiction (Meyer 1894, 70).

The kings of Lagore/*Deiscert* Brega were all descended from Diarmait Ruanaid and were called Clann Chernaigh Sotail after Diarmait's son, Cernach Sotal (died 663 AD). The kings of *Tuaiscert* Brega (northern Brega) were descended from Diarmait's brother, Congal, and are referred to as Clann Uí Chonaing after Congal's son Conaing. These appear to have resided at *Cnogba* or Knowth, also in Co. Meath. Most of the high-kings were of the Clann Chernaigh Sotail until the eighth century when it was taken over by the Clann Cholmáin, descendants of Colmán Mór, Aéd Sláne's brother (see Tables 6 and 7).

HIGH-KINGS OF IRELAND
(from the Clann Chernaigh Sotail, associated with *Loch Gabair*)

Year of Death	Name
664	Diarmait Ruanaid
664	Blathmac, Diarmait's brother
671	Sechnasach, Blathmac's son
675	Cennfaelad, Blathmac's son
695	Finsnechta, Diarmait's nephew
723	Fogartach, Diarmait's great grandson

Table 7 Based on information provided by Price in Hencken 1950, 18-34.

The first specific mention of *Loch Gabair* in the annals is in 676 when the Leinstermen invaded Brega. The high-king, Finsnechta Fledach (who is not recorded as having resided on the crannóg), defeated them near Lagore. The site was probably the intended destination of the attackers, indicating its importance at the time whether or not the high-king was in residence there. In 785 the death of the king of *Loch Gabair*, Maelduin macFergusa, is recorded. He was a grandson of the high-king Fogartach and 'was probably a person of some importance in his day' (Price in Hencken 1950, 22). In 786 Fogartach macCumascaigh, the then king of Lagore, was killed by the high-king, Donnchadh, of the Clann Cholmáin (see Table 8).

KINGS OF LOCH GABAIR
(specifically mentioned as such in the annals)

Year of Death	Name
785	Maelduin macFergusa
786	Fogartach macCumascaigh
805	Cernach macFergussa
824	Oenghus macMaelduin
836	Coirpre macMaelduin
864	Tigernach macFocarta
868	Diarmait macEitirsceili
908	Maelograi macConghalaigh
969	Beollan macCiarmaic

Table 8 Based on information provided by Price in Hencken 1950, 18–34.

The deaths of the next three kings of Lagore are also recorded, Cernach macFergussa in 805, Oenghus macMaelduin in 824 and Coirpre macMaelduin in 836. Several references are made to the next king of Lagore, Tigernach son of Focarta. In 848 he defeated the Norsemen at an unspecified location. In 849, in association with the high-king Maelsechnaill macMaelraunaidh of the Clann Cholmáin, Tigernach plundered Dublin. In 850 Lagore was burnt by Cinaedh macConaing and his Norse allies, probably a counter-attack after the attack on Dublin. In 851 the two allies abducted and drowned Cinaedh. Tigernach died in 864.

In 868 the king of *Loch Gabair*, Diarmait macEitirsceili, was killed in battle and Maelograi macConghalaigh in 908. 934 sees the destruction of Lagore by Amlaibh ua h-Imair, probably Amlaibh son of Gothfrith (died 940). The last reference to *Loch Gabair* is in 969 when its king, Beollan macCiarmaic, died. By the eighth century the high-kings no longer resided at Lagore, the high-kingship being taken over by the Clann Cholmáin (with one exception; Congalach of *Cnogba*, Clann Uí Chonaing who died in 956 AD). The destruction of the crannóg by the Norse in 934 may have brought on the rapid decline of Lagore though, in reality, the decline started with the shifting of power to the Clann Cholmáin.

6.4. Structural evidence

The body of the crannóg is 3m thick at its deepest and consists of layers of brushwood, timbers, peat, clay and sand. The lowest level of the crannóg consisted of large quantities of animal bones. These were especially plentiful around the edges of the site possibly acting as a firm base into which the piles of the initial palisade could be driven. Where the ground was particularly soft (the West area) this dump of bones was enormous. Many human bones were also found which Hencken suggested were evidence of a massacre of the crannóg builders. However, if the lower levels were in fact occupational then the bones were possibly those of previous inhabitants or their slain enemies, perhaps even the Leinstermen defeated in 676 AD (see 6.3).

The bow-shaped timbers in the eastern part of the site, upon which rested a discarded dugout canoe, were also inserted at this level. Hencken surmised that these timbers were used to help support the foundations of the site. However, Lynn suggested the possibility that they once formed the earliest palisade enclosing the crannóg. On top of these, brushwood layers 4 and 4a were laid. Some parts were woven into a mat like surface with timbers included elsewhere. This layer was very thick with two substantial log platforms associated with it and was possibly the first habitation level at Lagore. The larger of the two platforms, that in the Central area, consisted largely of birch with some oak and was approximately 0.65m thick.

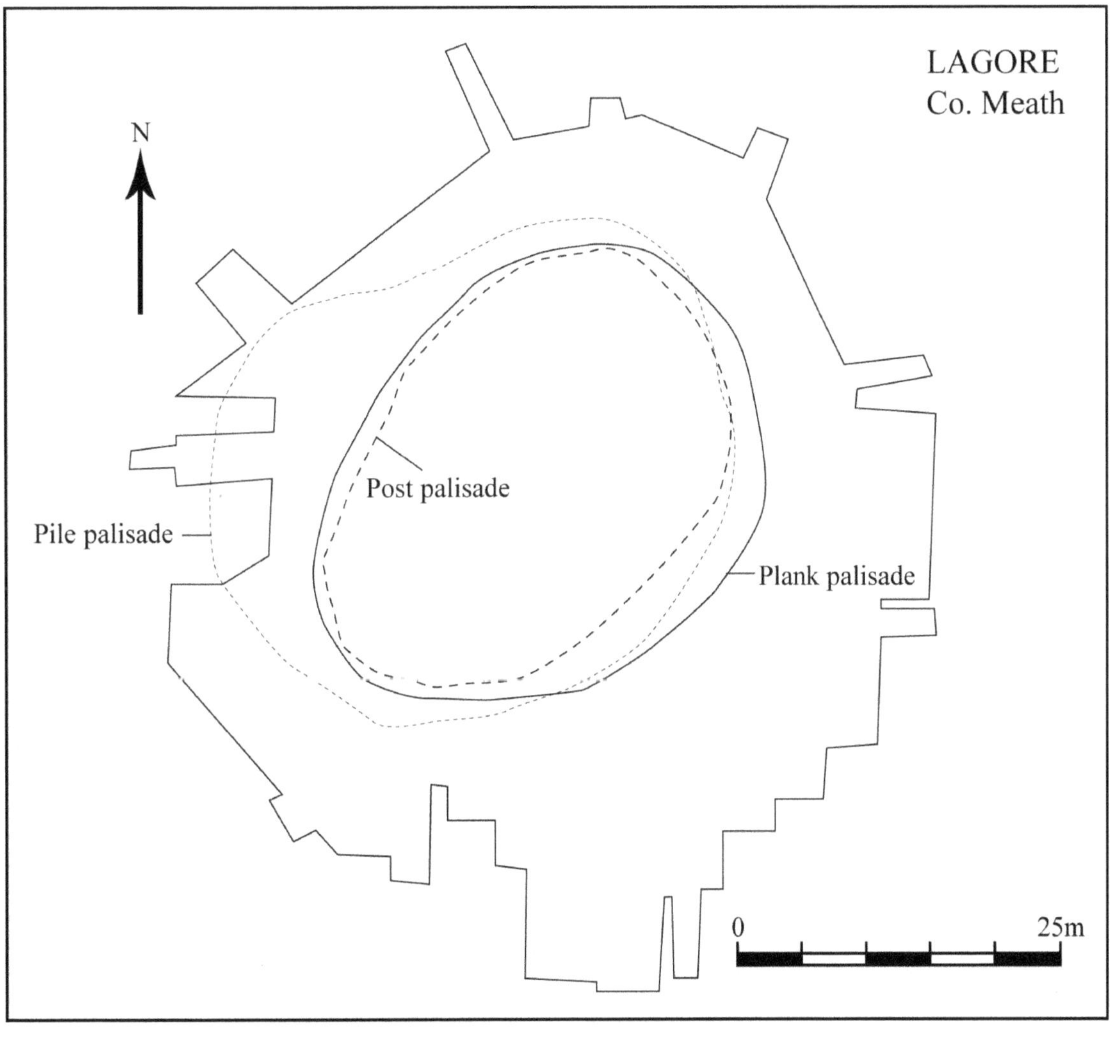

Fig. 47 Location of palisades (after Hencken 1950).

The body of the crannóg was then built up of alternating layers of brushwood and peat with timbers, wattle, and wicker matting at each brushwood level. These layers were thinner than the original foundation level and were, according to Hencken, 'never really continuous' (1950, 41). These he called layers 1, 2 and 3 with some other layers designated 1a, 2a or 3a, all of which he felt were contemporary. He states that the 'woven patches often looked like floors of huts but this proved illusory' (1950, 41). He does not, however, explain this or offer any reasons as to why they were illusory.

Of the palisades surrounding the site (see Fig. 47), the earliest was the pile palisade. It was also the largest measuring 41m in diameter. Hencken saw randomly placed piles outside the palisade as analogous of the stone *chevaux de frise* of land-based forts. Whether this is true or whether these were additional foundation supports is difficult to determine. The second palisade was constructed of roughly squared posts and enclosed an area approximately 36m in diameter. Part of the western edge of the original crannóg was abandoned at this time, probably due to instability and flooding. The final plank palisade surrounded an area 38m in diameter and consisted of heavy 'H'-sectioned oak posts placed at regular intervals of roughly 2m. Timber planks had then been slotted between these grooved posts.

The previously discussed wattle walls were perhaps the earliest hut-structures on the crannóg. In Hencken's own words, 'probably owing to previous digging' (1950, 54) he was unable to identify any structures in the remains he uncovered. He did find two large post-holes in the Central area which 'must have belonged to something substantial' (1950, 54). Unfortunately, previous interference with the site destroyed much of the stratigraphy in the centre of the crannóg making it impossible to determine the relationship between these post-holes and other features such as the substantial wattle wall and several hearths. Much burnt daub was recovered from the site though only two pieces were stratified (to periods I and II), suggesting the original presence of wattle and daub structures from at least two of the occupation phases.

Both Wilde and Wakeman included descriptions of wooden structures in their reports on the site. Wilde writes of compartments formed by *septa* of oak beams between two concentric circles of interspersed posts with connecting planks, 'divisions that intersected one another in different directions' (Wilde 1836–40). Wakeman goes into more detail regarding the structures he saw;

> a foundation formed of four roughly-squared planks of oak, each about 12 feet in length (so arranged as to enclose a quadrangle), the ends of which were carefully fitted together. From the angles of this square rose four posts, also of oak, to the height of about 9 feet. In these grooves were cut, into which roughly-split planks of oak had been slipped, so as to form the sides of the house. The irregularities between the boards were tightly caulked with moss; a low and narrow opening in one of the sides had evidently served as an entrance. There were no traces of window or chimney.
>
> (Wakeman 1879/1882)

Though it is difficult to reconcile one account with the other, both appear to describe the plank palisade or, at least, the method of construction it employed. Wilde's compartments may be explained by two posts in the North-East area. These posts bear slots for planks on their inner faces as well as on each side. Planks inserted in such slots may have formed small internal chambers or divisions. Wilde also mentioned a double circle of planks though no evidence of it was found in Hencken's excavation.

Wakeman refers to a series of rectangular structures inside the plank palisade, perhaps a distortion or different interpretation of what Wilde saw. Hencken suggests that a series of compartments along the inside of the palisade may be 'analogous to the little stone dwellings ..... built against the inside of the inner wall at the stone fort in Clare called Cahercommaun and to similar evidences of occupation found inside the walls of Ballycatteen and Garranes' (1950, 46). Unfortunately, neither antiquarian report included illustrations or exact details so their information cannot be relied upon for a realistic representation of the site and its possible structures as they appeared in the nineteenth century.

Hencken briefly notes an arrangement of piles forming two rectangular areas (or perhaps a larger subdivided one) outside the pile palisade in the East and South-East areas. These are dismissed without a purpose being suggested (see ground plan, Fig. 77). That the body of the crannóg extended beyond the palisades is evident in the brushwood layers and large quantities of finds discovered outside the wooden enclosures. It would have been illogical to construct a palisade right on the edge of the crannóg, risking its collapse if the foundations subsided into the lake. At Ballinderry II, Hencken found a hearth between the palisade and the water's edge, confirming that such areas were being utilised (Fig. 35). Therefore, it is possible that the piles were the remains of one or two timber structures, protruding slightly from the line of the palisade.

Fig. 48 Pyramidal crucibles from Lagore.

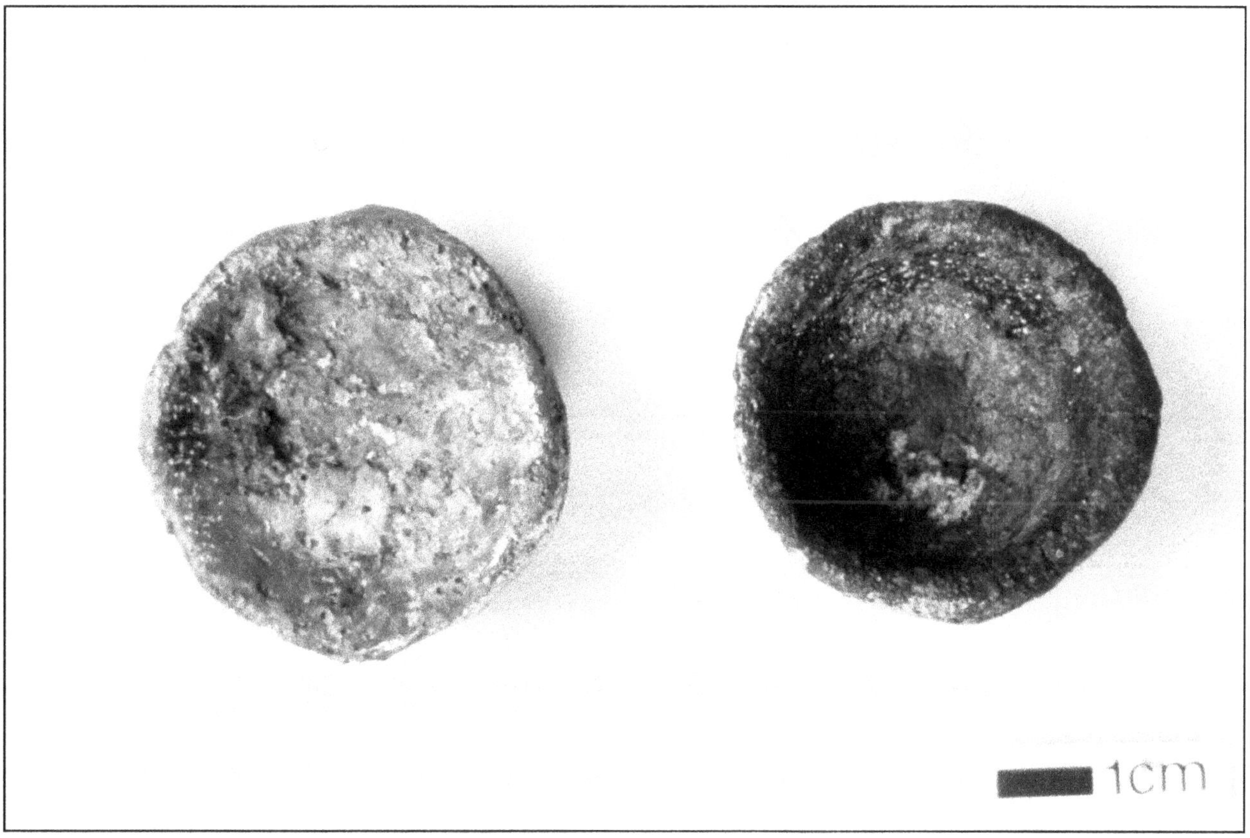

Fig. 49 Interior of flat-bottomed crucibles from Lagore.

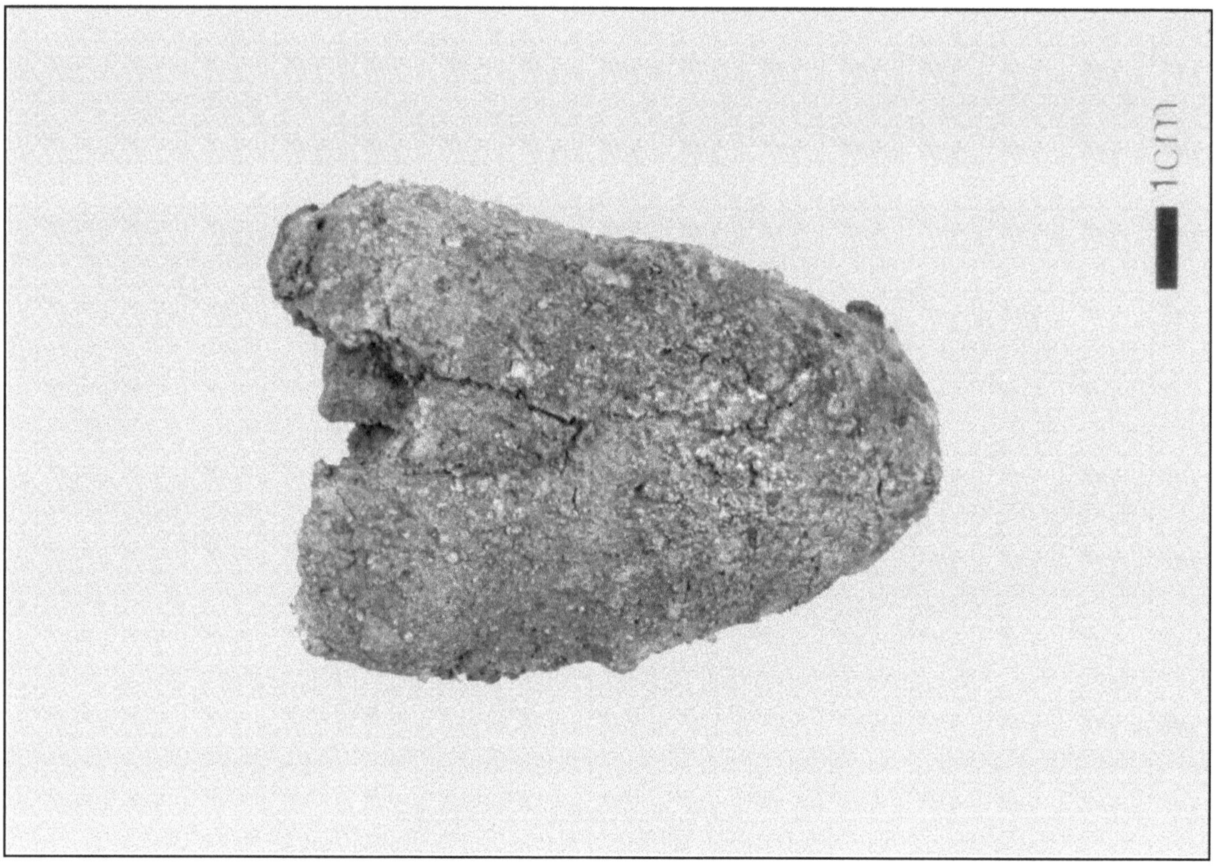

Fig. 50 Bag-shaped crucible from Lagore.

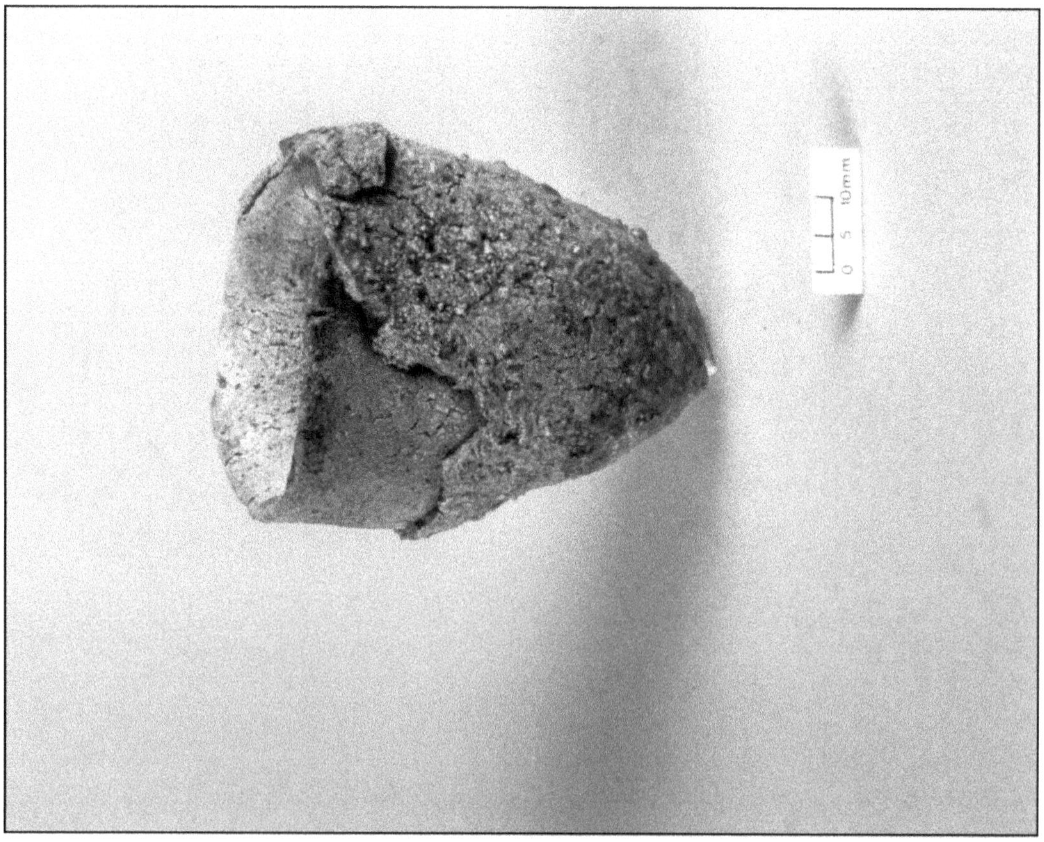

Fig. 51 Slag coated crucible from Lagore.

6.5. Evidence of non-ferrous metalworking

The vast majority of relevant artefactual evidence from Lagore consists of crucibles. There is a total of 279 crucibles and fragments thereof from the site, now stored in the National Museum of Ireland, as opposed to the 263 noted by Hencken in his report. Practically all of these bear various residues and accretions. These are a result of, and reflect, the metalworking processes in which the vessels were used. As previously discussed in chapters 3 and 4, both the pyramidal (Fig. 48) and bag-shaped (Figs. 50 and 51) crucibles bear these residues externally and the flat-bottomed vessels (Fig. 49) internally.

There are three main types of residue visible on the Lagore crucibles. It seems that many of the vessels were heated by placing them in the smelting furnace. This resulted in the adherence of grey porous slag from the furnace to the exterior of the crucible. Such slag is evident on a large number of both the complete and fragmentary pyramidal and bag-shaped crucibles. It is clear, from some of the bag-shaped crucibles in particular, that vessels were heated by inserting them into the top of a furnace.

No.41 (in Appendix 3) is heavily encrusted with dark grey porous slag. This encrustation partly overflows the rim on one side, indicating the almost complete submersion of this crucible in the furnace. Such 'slag lids' would also explain the missing rim sherds on many of the vessels. In order to retrieve the contents of the crucible the 'lid' had to be broken off, breaking away some rim sherds also.

Vitreous material is visible on nearly all the used crucibles and sherds (Fig. 53). This vitreous material is generally red in colour though it occasionally varies from green to black. Maby identified this red glaze as a 'fused silica', its colour depending on 'the type of flame employed (whether oxidizing or reducing)' (in Hencken 1950, 238). This glaze has previously been interpreted as smelting slag or the result of spillage. It is, however, 'due to the heat-induced migration of silicates from within the ceramic fabric' (Warner in Youngs 1989, 186).

Occasionally, metallic residues adhere to the interior of the pyramidal and bag-shaped crucibles. Maby also analysed these and found that they consisted mainly of copper sulphide with some iron sulphide. Such residues would be derived from the use of chalcopyrite ore. One of the crucibles examined by Moss was recorded as having come from the 'black bog of Dunshaughlin' and it contained metal made up of 85.05% copper and 11.60% tin (1927, 184). This vessel, at least, was used in the manufacture or melting of bronze.

> The sum of the evidence then leads to the conclusion that the pyramidal and bag-shaped crucibles were used for melting copper (or bronze) smelted from natural copper ore consisting of copper and iron pyrites.
>
> (Hencken 1950, 239)

The function of flat-bottomed vessels as stands for pyramidal and bag-shaped crucibles has been discussed in chapter 4. Two of the flat-bottomed vessels from Lagore bear roughly circular impressions in the vitreous material in their interiors (nos. 65 and 69; see Fig. 49). It would seem likely that these were caused by the base of another vessel. In practical terms pyramidal crucibles sit very well and are quite stable in an appropriately sized flat-bottomed vessel (see Fig. 26). The large flat-bottomed fragment, catalogue no. 67, bears no internal or external residues which would argue against it having been in direct contact with a source of heat. This fragment is possibly that of a heating tray, suggested by its comparatively large size and clean appearance. Another crucible from Lagore is fitted with legs to enable it to stand independent of any support (Fig. 54).

A number of clay fragments identified as, and stored with, the crucible remains may in fact have been part of a furnace lining (nos. 280–285). They differ in appearance from the crucible sherds, both in size and colour. Their thickness is generally greater than that of the crucibles and their fabric is orange in colour. This orange colour is the result of intense exposure to heat, giving the fragments the appearance of having been fired. Some are rim sherds and all bear vitreous material and/or slag on one face. It is also possible, of course, that these were crucibles exposed to heat for a greater length of time than other vessels. A crucible of this thickness (*c.*10–20mm) would have had to be heated for a long time in order to melt its contents which would explain its fired appearance.

Four clay fragments were recovered during the excavation, three of which are possible tuyère fragments, the other definitely (Fig. 55). The definite example (no. 288) is hemispherical in shape with a small amount of residue visible around its opening, which has a diameter of 0.01m. Of the other three fragments (nos. 286, 287 and 289), two have a fired appearance while the shape of all three suggests that they may have formed part of a clay tuyère. The interior curvature of the three is quite small, though large enough to have held the end-piece of a bellows or blow-pipe. Several small pieces of slag (nos. 290–301) were recovered along with the crucibles. Some of these have curved interiors with traces of clay visible. This would suggest that they were dislodged from crucible exteriors. They are not mentioned by Hencken in his report.

Six stone (nos. 307–312), five bone (nos. 302–306) and one wooden (no. 313, Fig. 57) motif-piece are known from the site, some of which are Old Finds (i.e. pre-excavation).

Fig. 52 Pyramidal and flat-bottomed vessels from Lagore.

Fig. 53 Heavily encrusted crucible from Lagore.

Fig. 54 Polypod crucible from Lagore.

Fig. 55 Fired ceramic and tuyère fragment from Lagore.

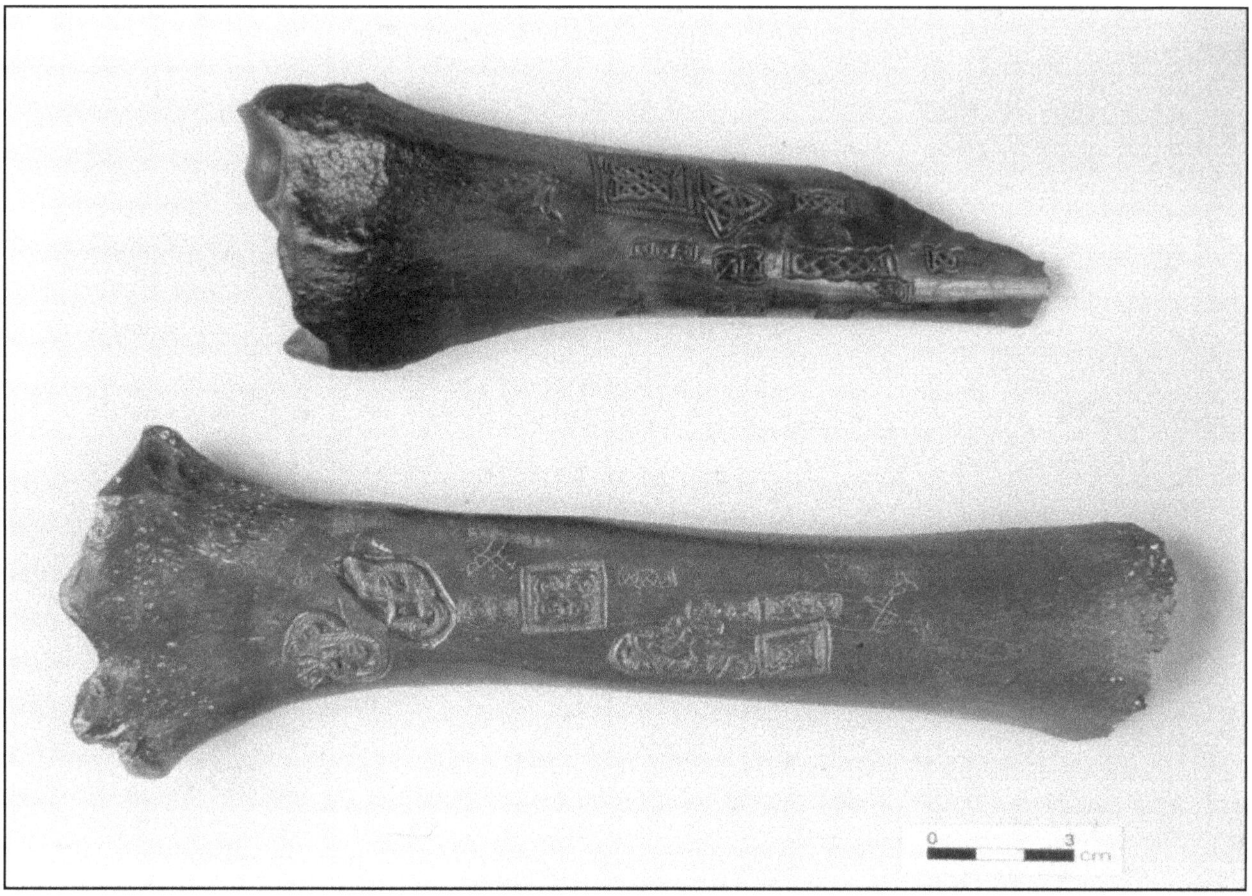

Fig. 56 Bone motif-pieces from Lagore.

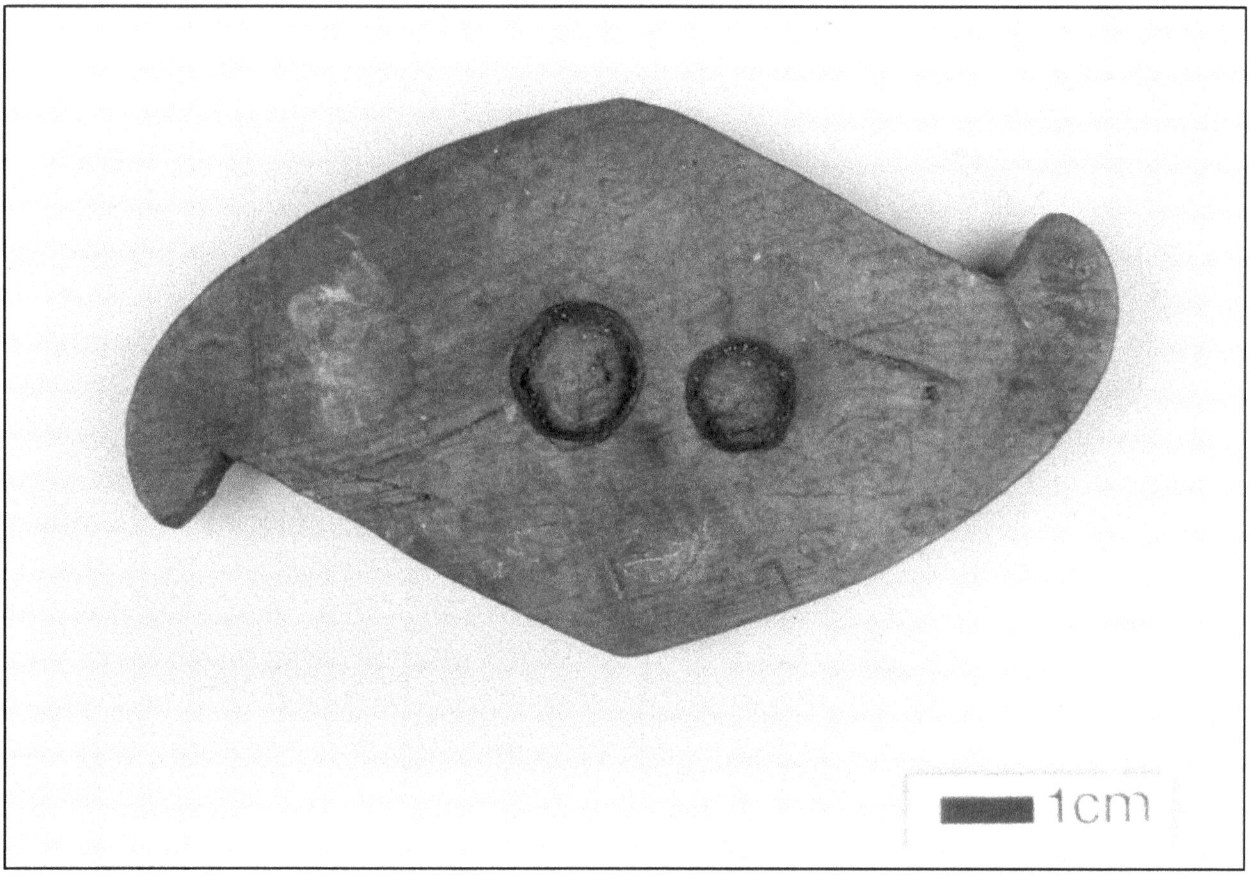

Fig. 57 Wooden motif-piece/mould from Lagore.

Henry suggested that the finer bone pieces (Fig. 56) may have been used to make lost-wax models in the production of various of the finished bronze artefacts from the site (in Hencken 1950, 183). However, 'in general it can be established that the pattern cutting on the pieces is less careful than that on the decorated copper alloy artefacts found on the site which all are of higher precision and more controlled workmanship than are the motif-pieces' (Ó Meadhra 1987, 68).

However, just because none of the recovered finished pieces of fine metalwork directly relate to the motif-pieces does not mean that they weren't used for such purposes. Ó Meadhra does point out that one old find from the site, a bronze penannular brooch, can be linked to at least one of the motif-pieces. The left-hand terminal of the brooch bears a circle within which is a duplex with interlaced crossing points (see Fig. 58). An identical motif, the design being unusual because of the interlaced crossing points, can also be found on catalogue no. 304 (Fig. 165) while a rough etching of it is visible on catalogue no. 306 (Fig. 166). Ó Meadhra suggests that 'the motif-pieces might have had some role to play in the manufacture of the brooch and that this took place at Lagore' (1987, 70).

Fig. 58 Bronze penannular brooch (after Hencken 1950).

Fig. 59 Clay mould fragment from Lagore.

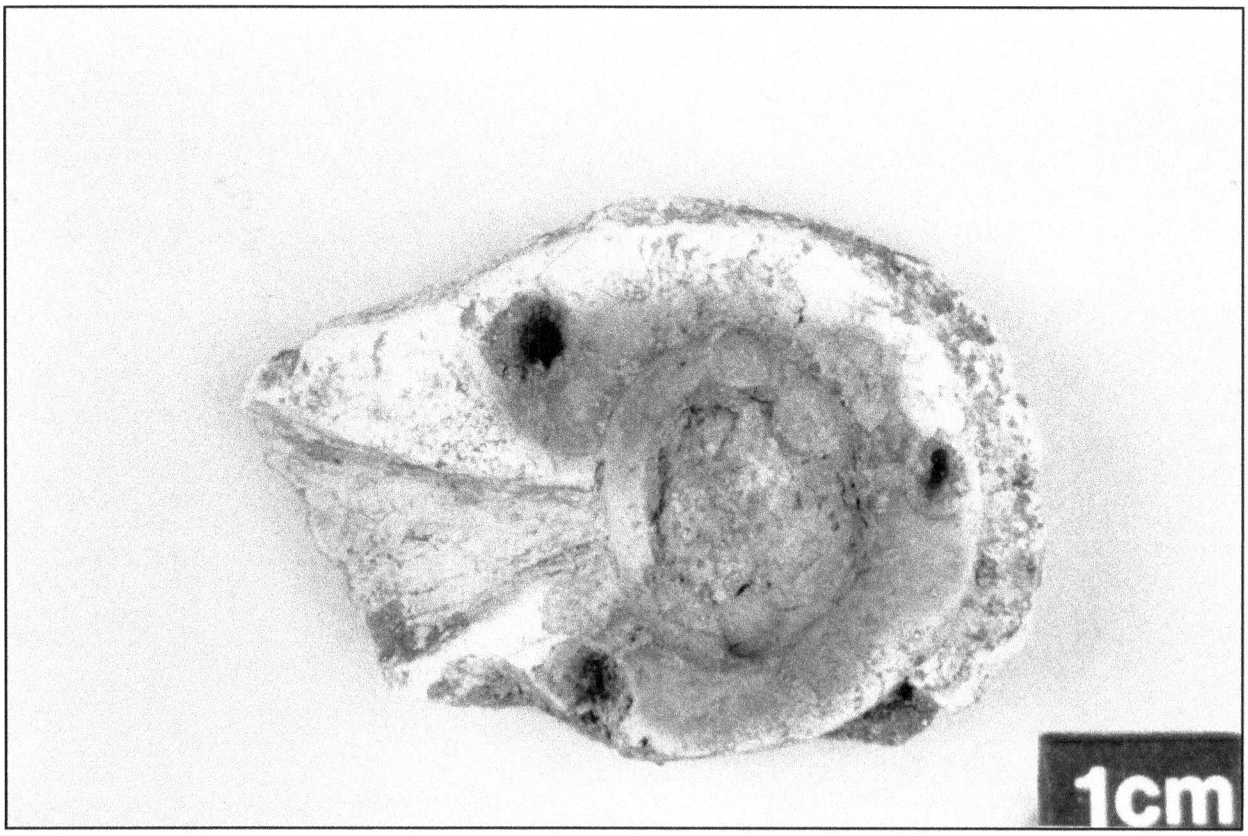

Fig. 60 Clay mould valve from Lagore.

Fig. 61 Stone moulds from Lagore.

Fig. 62 Stone ingot mould from Lagore.

The work on the stone motif-pieces appears less experienced than that of their bone counterparts though this could be a result of the greater difficulty involved in the precision carving or etching of stone. The wooden artefact differs from the others in that its overall shape has been worked as well as the working of its surface. Also, the incisions on this piece appear as thoughtless scratches as opposed to attempts at definite, planned motifs.

A small number of moulds were found at Lagore, including two of clay and seven of stone. Again, some of these were Old Finds. One of the clay moulds (no. 315, Fig. 60) was used for casting the ring of a ring-headed pin. Several such pins were found on the site though none that were made in this particular mould. The matrix and surrounding areas of the mould are blackened. This may have been caused by the hot molten metal or perhaps reflects the use of soot to prevent the molten metal sticking to the mould. The other clay mould (no. 314) was used to cast a large ring, possibly that of a brooch such as Hencken's no. 492 (see Fig. 59).

Six of the seven stone moulds (nos. 316–322, Figs. 61–62) were used for casting bar-ingots. These ingots were 0.04-0.1m in length and 0.01m thick. No. 316 bears one ingot matrix and a circular matrix. Both of these are now rust coloured with some traces of blackening also visible, probably residues from casting. Another stone mould (no. 321) bears two such circular depressions. These may represent the matrices for a form of metal ingot, perhaps of differing value or metal to the bar-shaped ingots.

The final stone mould (no. 322) bears two grooves running along its length and may have been used to cast/form pin shanks. Four circular depressions, two on either side, exist on the wooden (yew) mould/motif-piece. One of these matrices is clearly that for a small brooch. It is likely that these matrices were used for casting wax or lead models. Whether or not any of these moulds had a second valve is uncertain. None were found though some of the moulds do have relatively flat upper surfaces which may indicate their use with a flat covering stone or second valve.

Various pieces of bronze recovered during the excavation are identified as either waste or scrap intended for re-use. These include two pieces, spilled while molten (nos. 338 and 339, Fig. 63), twelve small fragments of sheet bronze cut from larger pieces (340 and 341), six lengths of wire (348–353) and five metal rods (343–347) (Fig. 64). Also included is a lump of bronze, circular in shape (342), which Hencken suggested 'probably came from a flat-bottomed crucible' (1950, 86), despite his earlier suggestion that such vessels may have been used as stands. This piece of bronze is most likely a metal ingot, cast in a matrix such as that of the stone mould no. 316. A similar bronze ingot was found at Nendrum (Lawlor 1925). Finally, the Lagore finds also included a piece of scrap silver, a fragment cut from a Viking bracelet (354).

In his excavation report, Hencken completely omits a quite important find. A small lead model for the ring of a ringed pin was recovered (355, Fig. 65). Its presence may indicate the use of a 'lost-lead' casting process at Lagore. Alternatively, it may have been an enduring, not very valuable, master pattern from which bivalve clay moulds were made.

The final group of artefacts relevant to non-ferrous metalworking are a range of metal and wooden tools. These tools include an iron punch with intact striking platform (323), a badly corroded iron wedge probably used as a chisel (324), and a small (54mm long) socketed iron implement that originally had three small prongs, one of which has since broken off (325) (Fig. 67). The function of such tools has already been discussed in chapters 3 and 4. The punch was most likely hand-held and struck with a hammer or mallet. The pronged tool originally had an organic handle, while there is no indication of whether or not the wedge/chisel was hafted.

Three iron gouges were also found, the largest of which was socketed (326) (Fig. 68). The other two are similar, though they have looped ends possibly to enable their suspension. They differ from each other in that the loop of one (328) is bent backwards to form the handle while the other is bent forwards (327). Four iron awls were identified (329–332, Fig. 69), two of which were corroded though all four taper to a point. Amongst the Old Finds were four pairs of iron shears (333–336, Fig. 70). The blades of the largest of these (0.22m long) have stepped bases joining them to the stems. The exterior flattened surface of the spring is decorated with simple incised lines in a criss-cross pattern. The blades of all but one (334) of the shears overlap at the base. While three of the examples have simple single-looped springs, one has a double-loop (336). Such a spring can also be seen on a pair of shears from Cathedral Hill (Fig. 81).

A small bronze tweezers from the site consists of a thin, flat strip of bronze bent in two (337, Fig. 66). Such an implement could have been used for such delicate work as the manipulation of metal wire in the creation of filigree. Several other relevant items are mentioned by Hencken in his report. Unfortunately, these could not be located in the National Museum of Ireland. They include a large socketed iron chisel (Fig. 30, D), a small hammer made completely of iron (Fig. 30, C), a large iron awl (0.22m long) and a smaller example (67mm long), two fragments of iron shears, and three slotted and pointed iron implements. One of the latter was an Old Find and bears a strong resemblance to the iron punch in both shape and the presence of a striking platform.

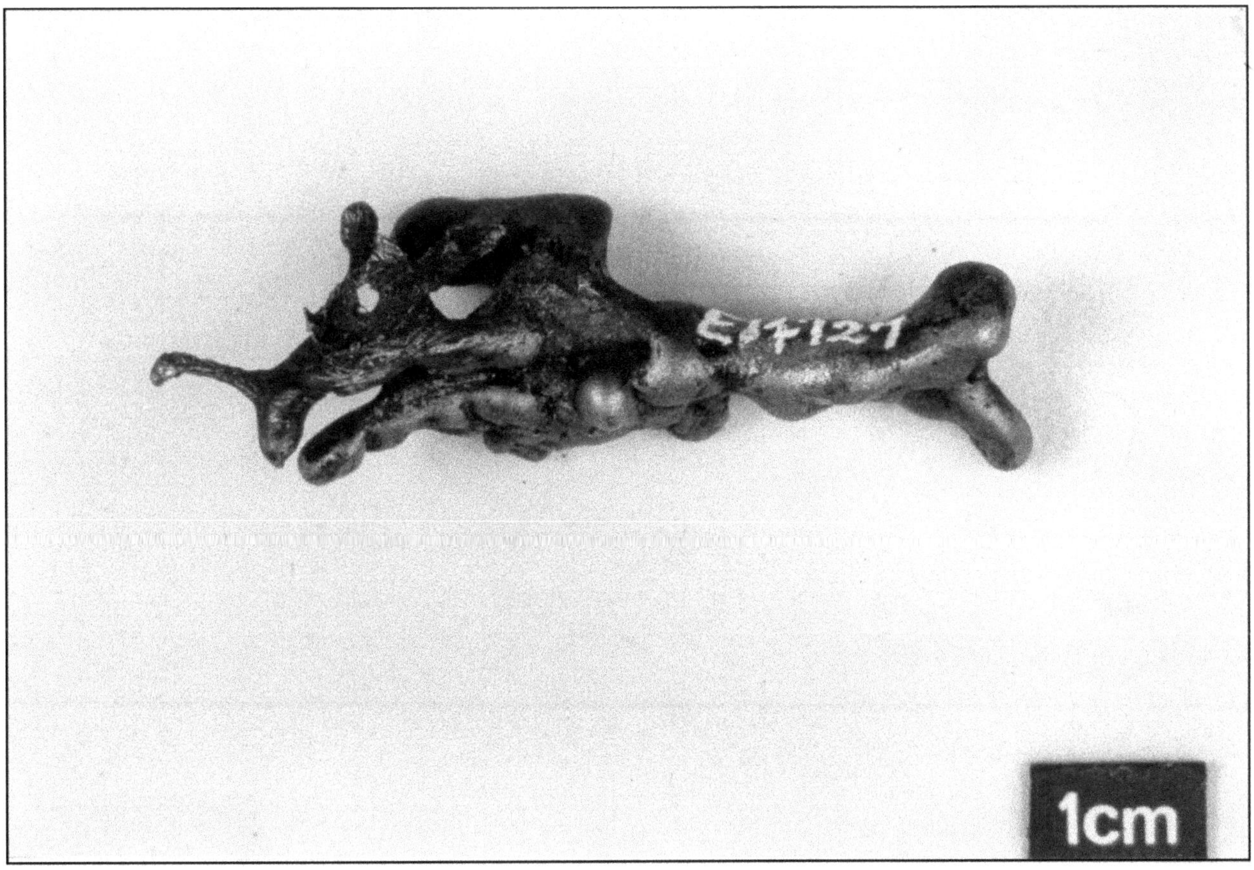

Fig. 63 Spilled bronze from Lagore.

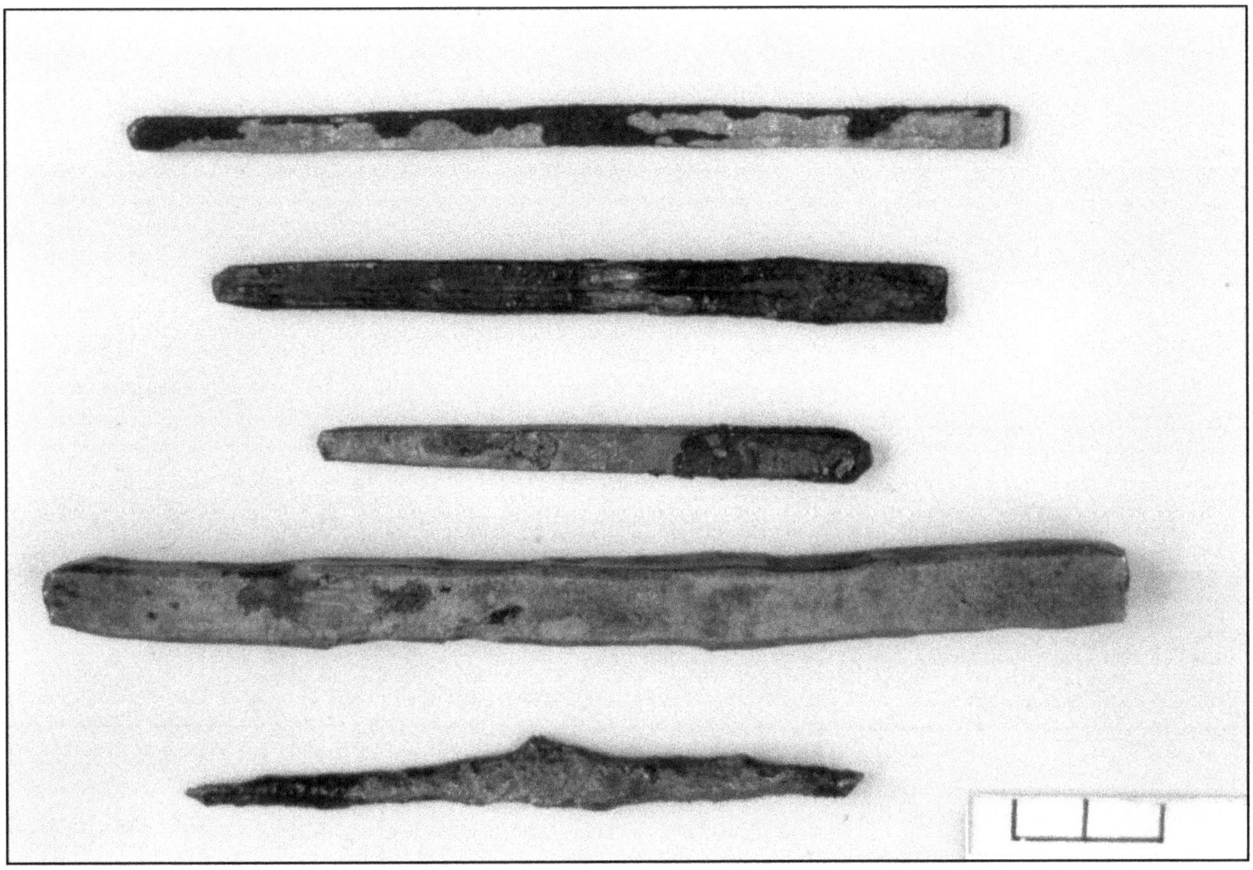

Fig. 64 Bronze rods from Lagore (scale: 5mm divisions).

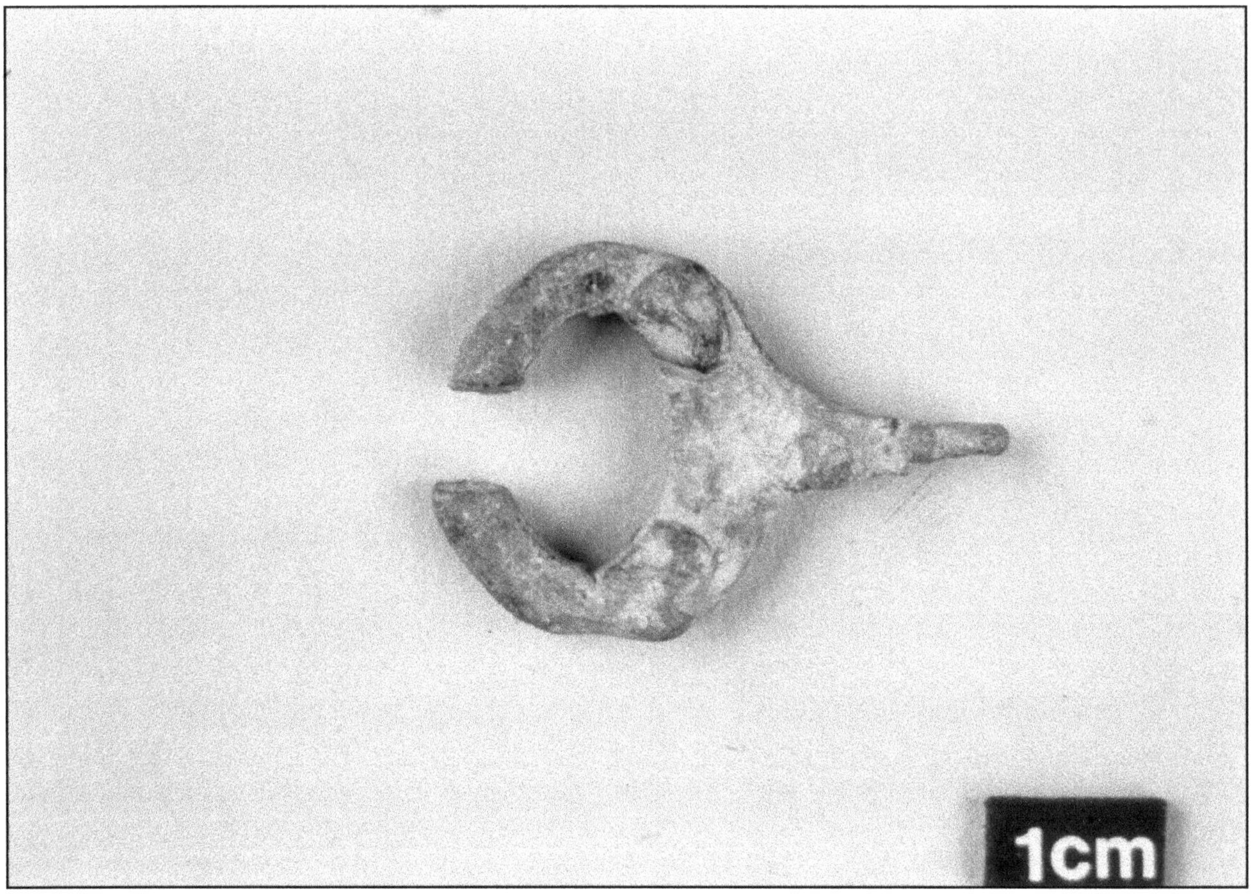

Fig. 65 Lead model from Lagore.

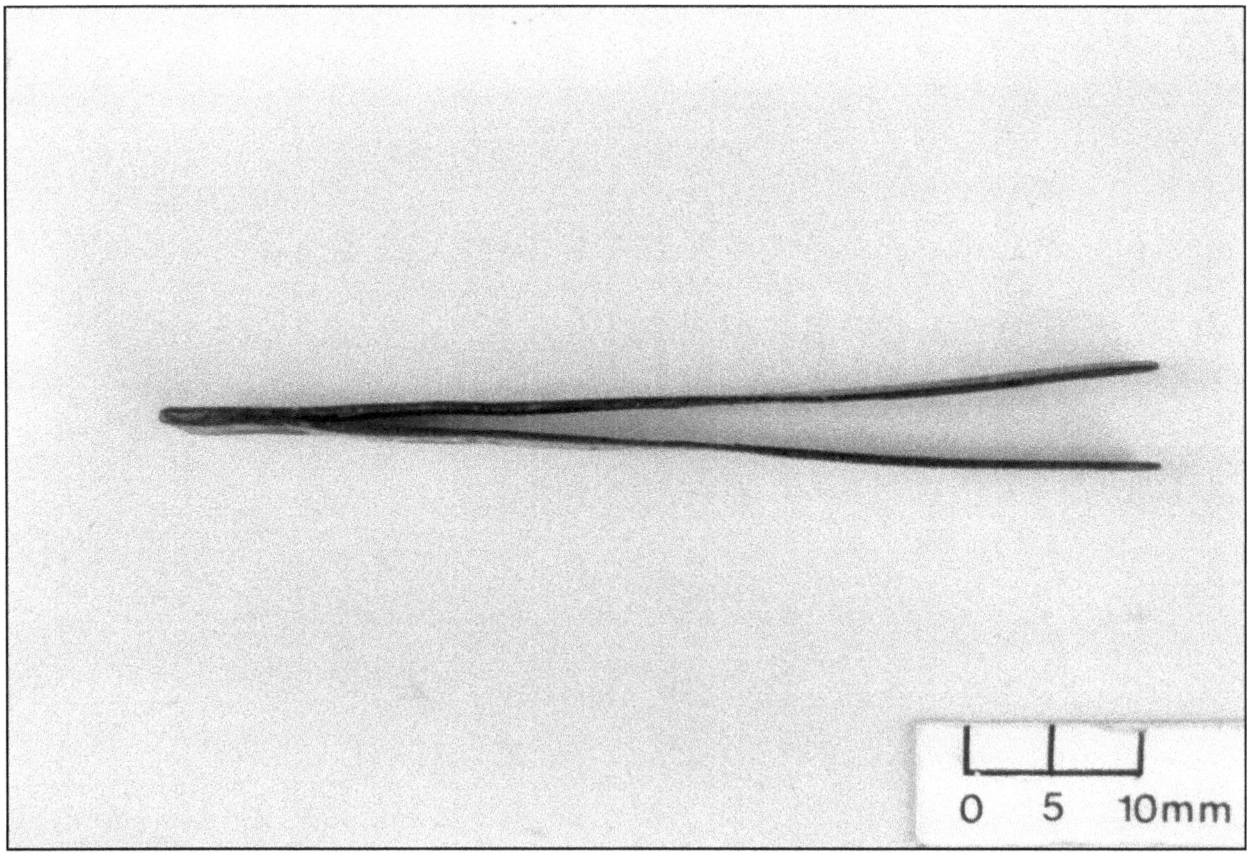

Fig. 66 Bronze tweezers from Lagore.

This possibly functioned as a punch, though held in place with a piece of wood or bone inserted in the slot.

Two wooden mallets were recovered, both made of oak (Hencken 1950, 163). These could have been used in conjunction with the punches, socketed gouge or chisel. While no direct evidence of a wooden lathe was found, the presence of such at Lagore can be inferred from the remains of several wooden vessels. Five lathe-turned bowls of alder, poplar and hazel were found, as were three pieces of alder roughed out to form deep bowls. One of these had a hole for attachment to a lathe in the middle of its rounded side (Hencken 1950, 156).

Also of significance in terms of metalworking was the discovery of two lumps of copper ore (Hencken 1950, 240). No other Early Historic settlement site has produced raw copper ore. Both were found in the East area, one stratified to period III, the other found unstratified outside the palisade. Maby examined the ore and identified it as copper pyrites, a copper sulphide. Copper pyrites was exploited in Co. Meath in recent centuries, at Beauparc and Brownstown, twelve and eight miles away (respectively), however there is no known evidence of mining in the Early Historic period.

The only other find from the site relevant to smelting was a chalky powder found in several of the hearths on the eastern side of the crannóg. This was identified by Maby as calcium carbonate and may have been used as a flux in the smelting process. This chalky powder was provided in the form of a chalk-mud from the lake-bed and was also found in one piece of ironworking slag. With regard to its function, Maby wrote

> It is not likely to have been used as a flux in smelting, as one might have thought, for in that case it would have occurred in more than one piece of slag, and, if it had passed through the smelting process, it would probably have been turned into calcium oxide (lime) by the heat.
>
> (in Hencken 1950, 232)

No furnaces were identified by Hencken and none of the numerous hearths are mentioned as having been employed in non-ferrous working, though it would seem probable that some were.

Hencken gives very little detail on the hearths, in fact he fails to mention some of those marked on his ground-plan. Some of the hearths had chalk-mud floors, that mud having probably come from the lake-bed. Layers of ash and charcoal were found on top of these floors. Several hearths on the eastern part of the crannóg contained enough traces of iron slag to suggest their use in ironworking.

6.6. Location of relevant evidence within the site

In terms of non-ferrous metalworking, there is a total of approximately 376 (including those unavailable for cataloguing) relevant finds from Lagore crannóg, 279 of which are crucible remains. In his site report, Hencken attributes 8 crucibles to period Ia, 58 to Ib, 8 to II, and 9 to III. There is, however, a discrepancy between these figures and those recorded in the National Museum of Ireland registers. Here, 8 crucibles are attributed to period Ia, 14 to Ib, 2 to II, and 2 to III. Of the other relevant artefacts, 3 are dated to period Ia, 7 to Ib, 2 to II, and 2 to III (see Table 9).

In all three cases the largest quantity of relevant finds came from period Ib (Table 9). The number of stratified artefacts, though, is quite low in relation to the total number of relevant finds. Depending on which record of the crucibles is taken, either 83 or 32 examples of the total 279 are stratified, while 14 of the 97 other finds are dated to one of Hencken's periods. These figures may be too low to attempt a serious chronological analysis.

Hencken was convinced that Ib was the chief period of non-ferrous metalworking at Lagore. Dated to the seventh century, this was a period when Lagore was a powerful and wealthy place, the seat of the high-kings. The status-related activity of non-ferrous metalworking would have thrived in such a setting. Whether or not the situation was markedly different either before or after is uncertain. Non-ferrous working appears to have been carried out in all periods, though perhaps to a lesser extent as the occupation at Lagore declined in status.

In examining the distribution of the relevant artefacts a definite eastern bias becomes apparent.

PERIOD	HENCKEN REPORT CRUCIBLES	N.M.I. REGISTER CRUCIBLES	N.M.I. & HENCKEN OTHER NON-FERR.
Ia	8	8	3
Ib	58	14	7
II	8	2	2
III	9	2	2

Table 9 Chronological distribution of non-ferrous metalworking evidence.

Fig. 67 Iron tools from Lagore.

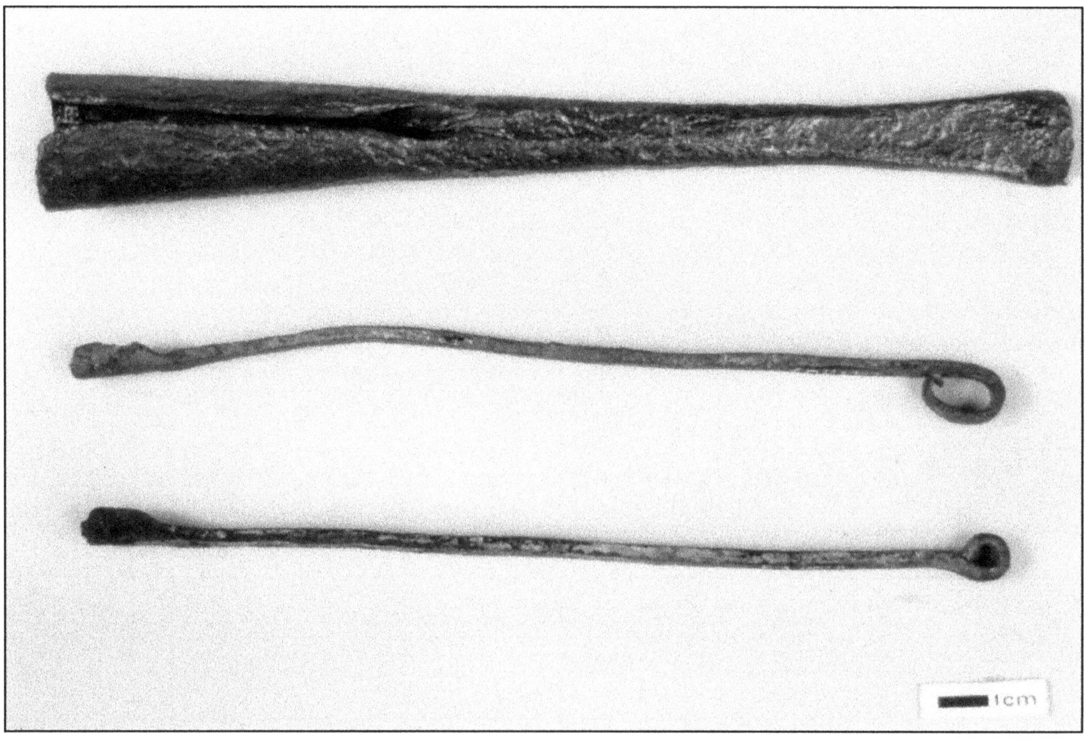

Fig. 68 Iron gouges from Lagore.

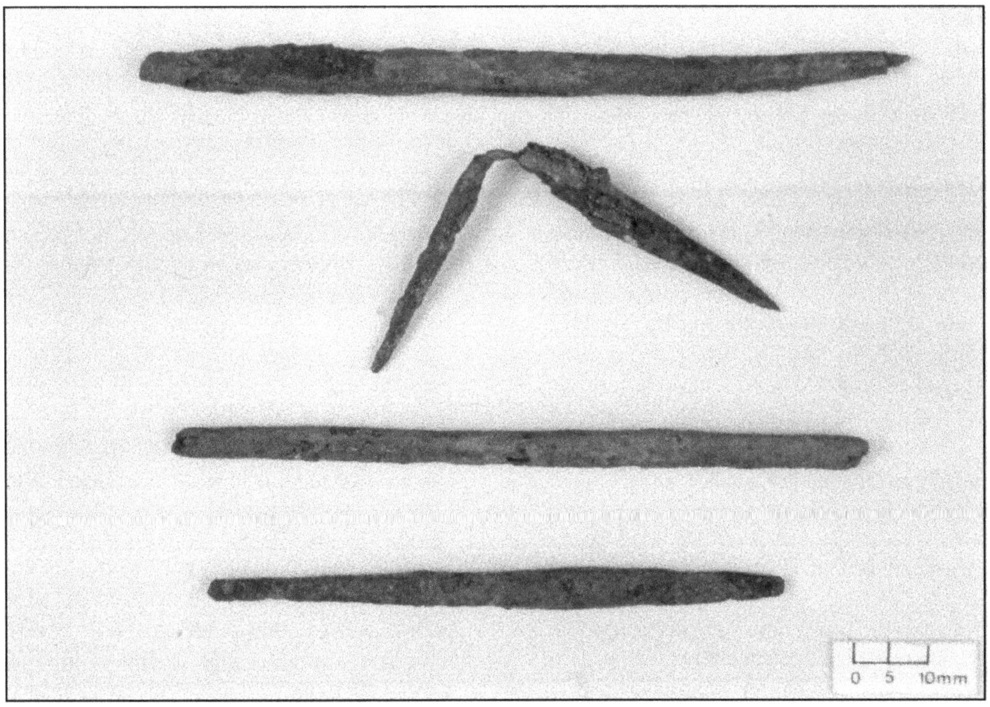

Fig. 69 Awls from Lagore.

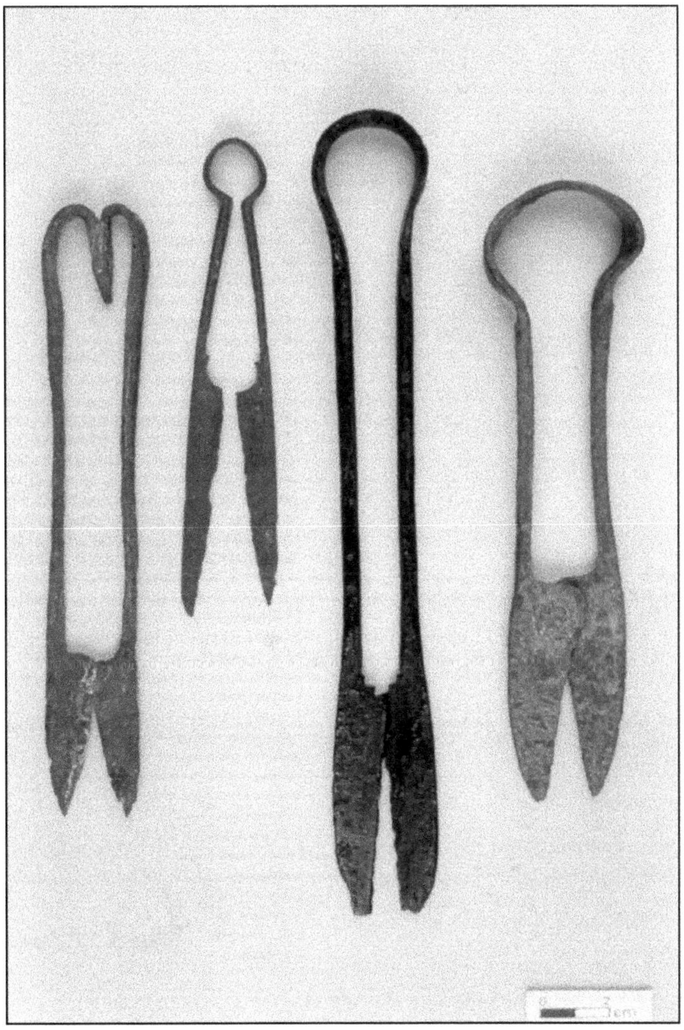

Fig. 70 Iron shears from Lagore.

223 of the total number were found in the East, North-East and South-East areas which would suggest this part of the site as the location of the non-ferrous metalworking (Fig. 71). This is also the area identified as the site of ironworking and glass-working, clearly an industrial zone.

Hencken didn't interpret any of the features from this area as having been specifically involved in non-ferrous metalworking. Considering the finds, though, it seems likely that at least some of the hearths were utilised in the fabrication of fine metal objects. There is also the possibility that the remains of the rectangular structure may represent a workshop of some nature, its offsetting from the line of the pile palisade perhaps reflecting the general pattern of distancing metal workshops from the domestic area.

Any interpretation of site structure at Lagore is compromised by the fact that large areas of the West and North-West sectors were not excavated, due either to flooding or damage by previous interference. The overall view of site organisation, then, is not a completely balanced one. Fragments of wicker, wattle and daub were found in the Central sector which may have been the domestic area.

Relevant artefacts are not confined to the three eastern sectors, they are found all over the site. Smaller concentrations of crucibles, for example, may represent refuse dumps, disturbance, small workshops of differing periods, or possibly a production/recycling area. Other artefacts such as motif-pieces could have been worked on any part of the site while metal tools may have been employed in other activities such as carpentry or leatherworking. Such artefacts are not tied to the metalworking area though they may, at some time, have been used there.

The apparent lack of relevant structural evidence is possibly due to the practicing of both ferrous and non-ferrous metalworking in the same area. The two finds of copper ore suggest that smelting was taking place at Lagore and such a process must have produced waste. Copper smelting slag, however, may appear very similar to iron slag, large quantities of which are recorded from the site. There is also the possibility that the non-ferrous working went into decline after the eighth century, while ironworking continued on much the same scale as always. This continuing work may very well have destroyed or re-used any non-ferrous workshops.

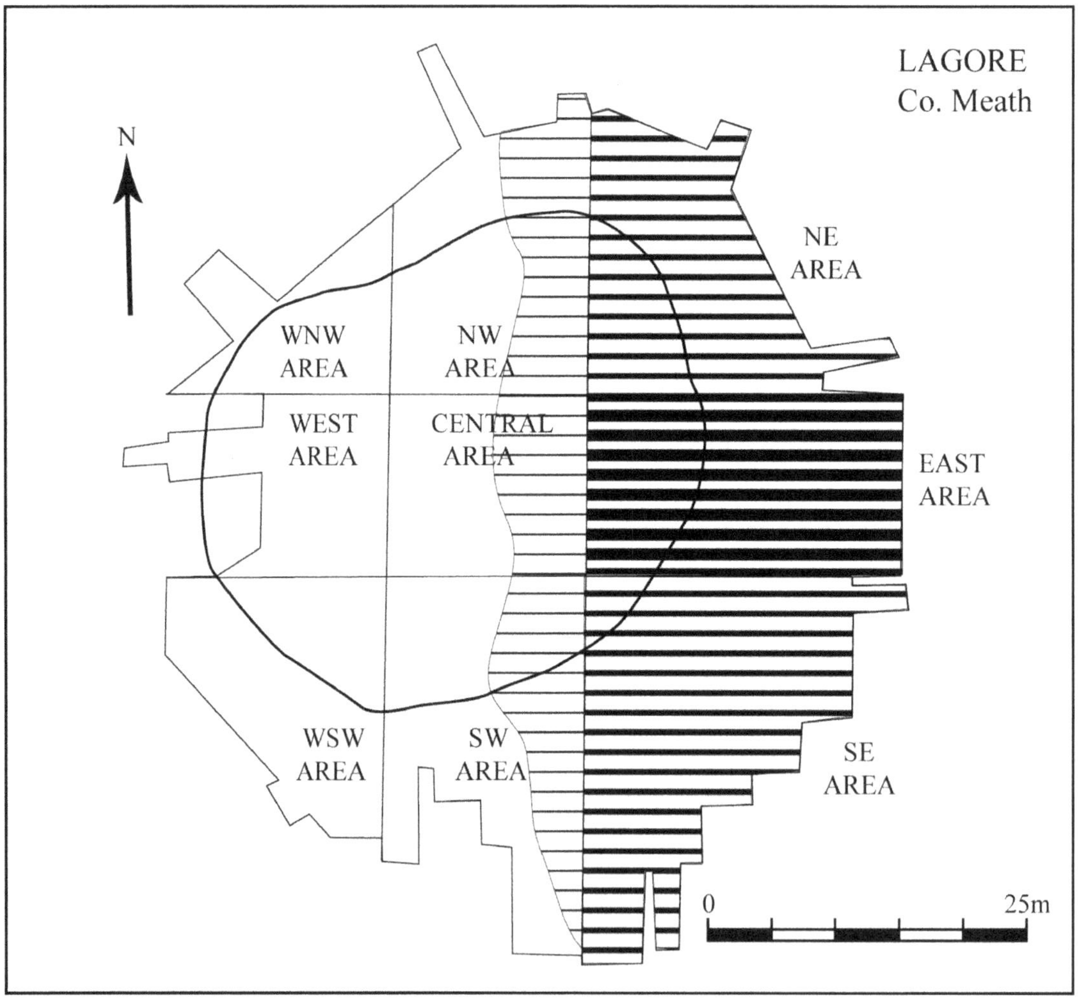

Fig. 71 Main concentrations of relevant evidence.

A

PERIOD	OX SKULLS	PIG SKULLS	SHEEP SKULLS
Ia	54	11	23
Ib	103	21	22
II	52	38	22
III	11	2	
Disturbed areas outside palisade	368	101	86

B

PERIOD	OX BONES	PIG BONES	SHEEP BONES
Ia	84.10%	8.20%	7.70%
Ib	84.40%	9.40%	5.80%
II	75.40%	13.40%	11.20%
III	72%	15%	13%
Disturbed areas outside palisade	80%	12%	8%

Table 10 Chronological distribution of A. animal skulls and B. animal bones (after Stelfox in Hencken 1950).

6.7. Site economy

Agriculture

Lagore's status as a royal settlement is supported by the economic evidence from the site. As with all Early Historic settlements, the main-stay of its economy was agriculture. 22,680kg of animal bones were recovered from the site. Of these, cattle were by far the most numerous with smaller quantities of sheep and pigs (see Table 10). There appears to be a slight rise in the percentage of sheep in periods II and III, perhaps reflecting a growth in the wool trade. Horses, cats, dogs and fowl were also kept. Hunting for sport may be represented by the small quantities of red deer, fox and hare bones. The inhabitants of Lagore, or their clients, also appear to have been expert fowlers with wild birds providing *c.*80% of the total number of bird bones from the site. It must be remembered that many of these bones may have been the results of tributes to the Lagore kings. No fish bones were identified which led Hencken to surmise that the inhabitants of Lagore were not fishermen. This seems highly unlikely considering the lacustrine nature of the site. It is more probable that the small size of the bones caused them either to decay or be missed during excavation.

Several agricultural tools were found at Lagore, including a heavy iron ploughshare and coulter for a pair of oxen. Also recovered were iron sickles and fragments of rotary querns. Amongst the building materials of the crannóg was a quantity of wheat straw. This evidence suggests that both cereal cultivation and cattle farming were practiced on the surrounding lands.

Ironworking

Non-ferrous metalworking was not the only industry or craft undertaken at Lagore. On-site blacksmiths most likely produced the range of iron implements and weapons recovered from the crannóg. The weapons produced may have included several, if not all, of the finds of swords, spearheads, spear ferrules and shield bosses on this site (Fig. 72).

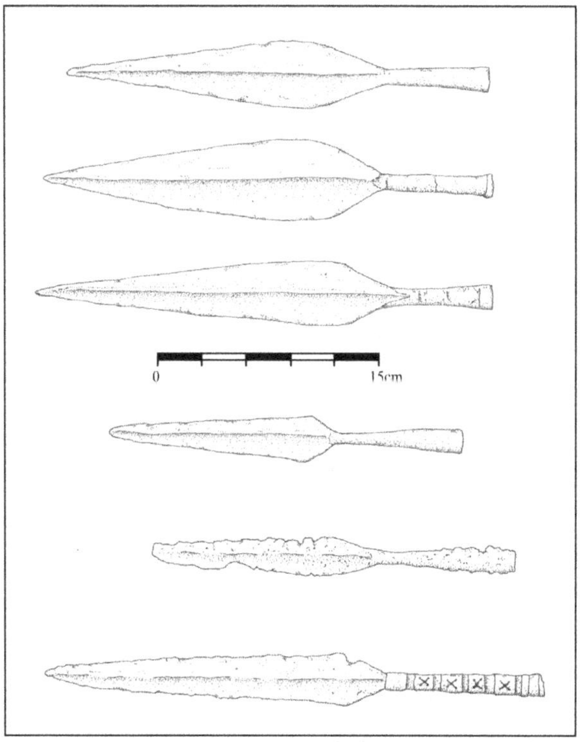

Fig. 72 Iron spearheads from Lagore (after Hencken 1950).

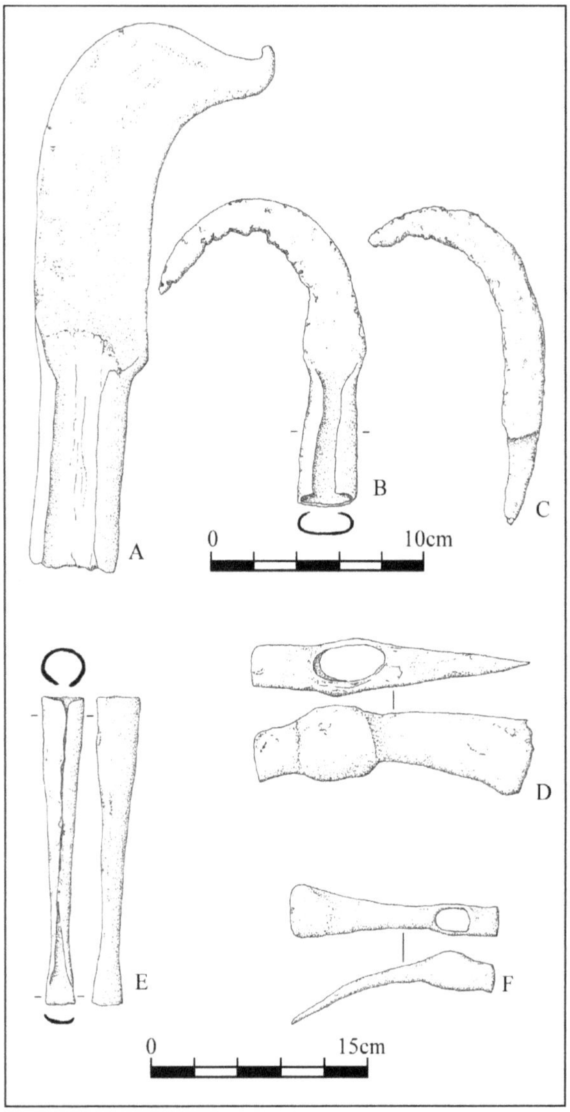

Fig. 73 Iron implements from Lagore; A-C bill-hooks, D axe-head, E gouge, F adze-head (after Hencken 1950).

The tools or implements from the site include sickles, bill-hooks, plough coulter, saws, gouges, shears, knives, nails and bucket handles (Fig. 73). Non-ferrous items possibly produced on the site include penannular brooches, belt buckles, pins, finger-rings and tanged studs (Figs. 74 and 75).

There is a clear division evident between the product ranges of the two activities in terms of function. Ironworking produced the necessary implements for domestic life and the weapons essential to the survival of any king, while non-ferrous working provided fine ornamental items. A slight overlap does occur, however. A crude iron scabbard chape and a piece of iron decorated with red and yellow enamel were recovered from the site as were two sheet-metal bronze bowls, bronze needles and tweezers. The scale of either output is impossible to determine from the fragmentary archaeological remains.

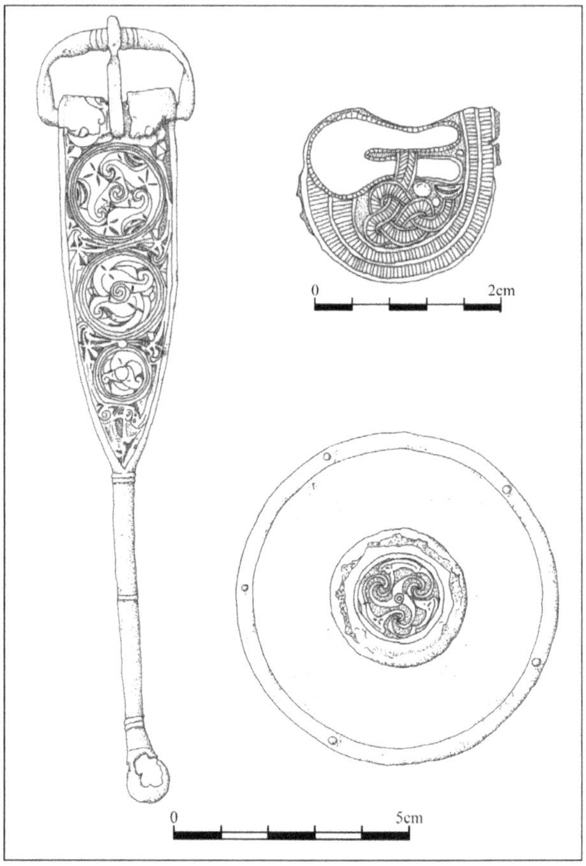

Fig. 74 Bronze artefacts from Lagore (after Hencken 1950).

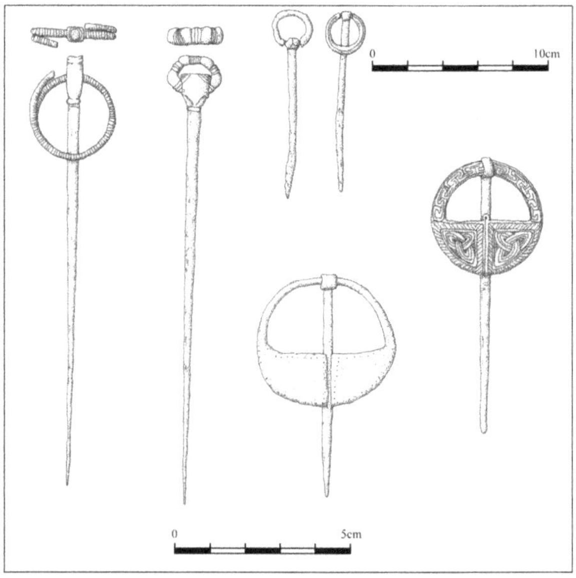

Fig. 75 Bronze pins and brooches from Lagore (after Hencken 1950).

Evidence of ironworking consisted of 'ironworking floors' and relatively large quantities of slag. As with non-ferrous metalworking, no furnace was found. Three different types of iron slag were identified by Maby (in Hencken 1950, 230–233). The largest pieces of slag recovered were furnace bottoms, plano-convex lumps that formed at the bottom of a furnace.

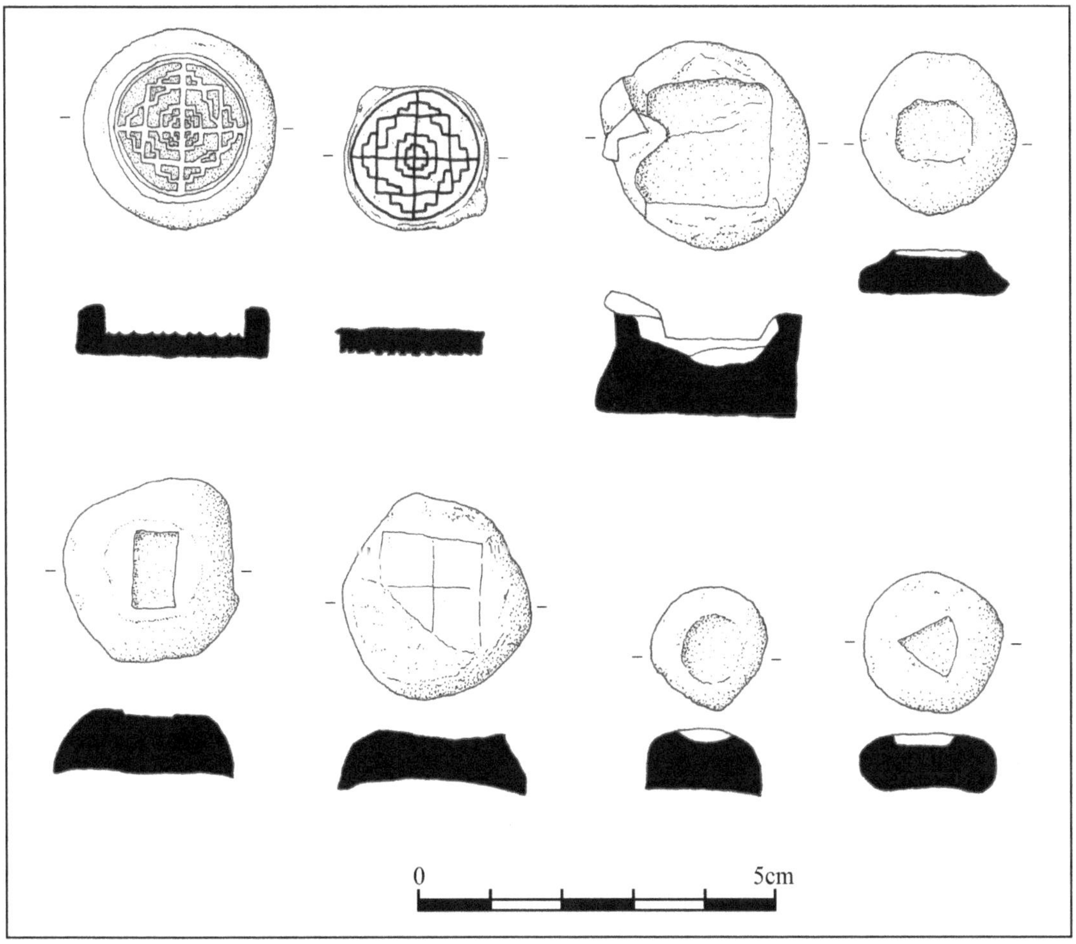

Fig. 76 Moulds and glass stud from Lagore (after Hencken 1950).

The size of the Lagore examples ranged from 0.12–0.19m in diameter. They were composed of magnetic iron, iron oxide and sulphide, mixed with sand and wood charcoal.

Smaller lumps of coarser slag were also found. These contained small stones, charcoal, fused silica, gritty sand, iron sulphide and a little magnetic iron. The third type of slag consisted of a vitreous substance with a high content of free magnetic iron, fused silica and traces of charcoal. Maby identified it as 'the overflowings of iron slag from furnaces or in some cases sand, etc., that had been close enough to a furnace to be fused by the heat' (in Hencken 1950, 233).

Hencken's 'ironworking floors' or trodden hearths consisted of layers of a chalky substance, ash, charcoal and sufficient amounts of iron oxide and free magnetic iron to associate them with ironworking. The largest of these, hearth A, was located in the East area and dated to period III. It was composed of calcium carbonate, wood charcoal, a little clay, a substantial amount of iron oxide and a little fine sand. Several smaller, though similar, hearths were also located on the eastern side of the site.

This eastern bias is supported by the slag distribution. Of the 34 furnace bottoms found, 6 came from the North-East area, 5 from the South-East, and 19 from the East. 276 lumps of coarse slag were found, 47 in the North-East, 86 in the South-East, and 67 in the East. Again, 20 pieces of vitreous substance were recovered in the South-East and 65 in the East. It would seem clear, then, that ironworking took place in the same area as non-ferrous working.

Hencken identified Period II as the chief seat of ironworking as all of the stratified furnace bottoms (number not mentioned), 7 of the 14 stratified pieces of coarse slag, and 19 of the 36 stratified pieces of vitreous substance came from Period II. The large ironworking floor or hearth A, however, is dated to Period III and, as with the non-ferrous evidence, perhaps the number of stratified pieces is too low in comparison to the total number to accurately determine a definite chronological pattern.

The blacksmith was an essential member of the retinue of any king in his capacity as weapon-producer and, on the larger sites at least, was probably a permanent member of the community. This can definitely be said of the Lagore smith(s) considering the site's unsettled history.

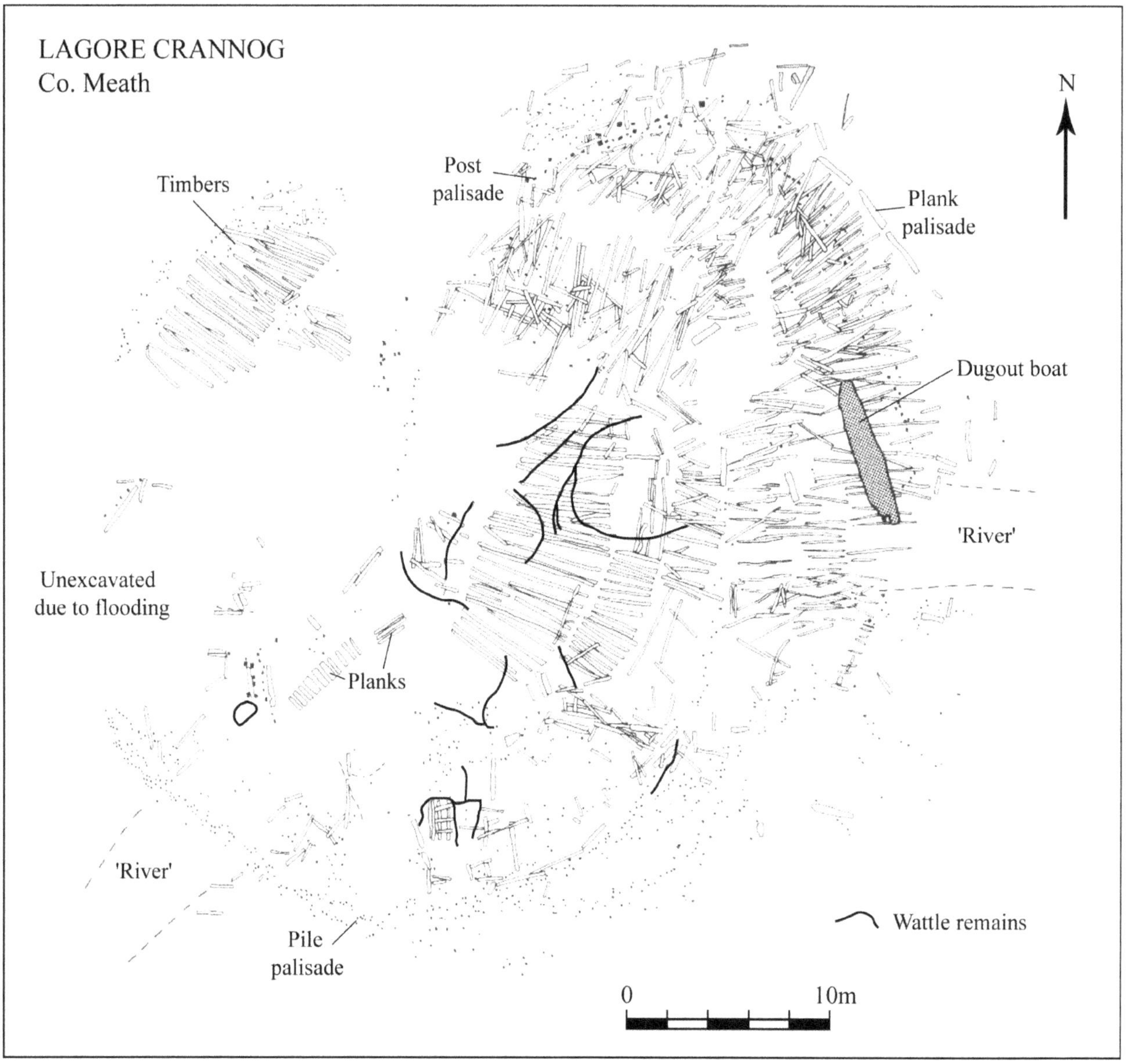

Fig. 77 Lagore ground plan (after Hencken 1950).

Craft activity

Often associated with metalworking are the crafts of glass-, amber-, millefiori- and enamel-working. Fragments of imported Roman glass intended for recycling, moulds for studs (one with a stud still *in situ*, see Fig. 76), and a number of beads and bracelets were all found. The moulds indicate the manufacture of glass studs on site, though there is no such evidence regarding either the beads or bracelets. However, both were probably made in Ireland if not at Lagore itself. Hencken located glass-working at Lagore within the Eastern area of the site in period Ib also. Three small rods of millefiori were found on the site, as was a tiny lathe-turned amber stud. The latter could be indicative of small-scale amber-working or may have been lost from a complete object such as a penannular brooch. No evidence of on-site enamel-working was discovered.

Other crafts included woodworking (domestic vessels, tools), stone-working (querns, moulds motif-pieces), leatherworking (shoes, scraps of leather, wooden shoe last), bone-working (combs, pins, motif-pieces), and spinning and weaving (textile fragments, wooden spindles, spindle-whorls, fleece and animal hair). Of all the crafts carried out at Lagore, non-ferrous metalworking and the glass/millefiori working are the most obviously status-related. The rest produced objects necessary for daily life while the latter two produced luxury items.

Trade and exchange

Very little pottery was found on the site. Those few sherds that were, were Continental imports. These consisted of fragments of E-ware, probably imported from France. Wooden vessels appear to have fulfilled most domestic needs. Other imports would have included the tin required in the production of bronze.

This may have been obtained through another Irish site acting as a 'middleman' or directly from sources in, most probably, Cornwall or Brittany.

The economy of a royal settlement was also supported by the receipt of tributes or rents from lesser kings and wealthy/strong farmers (Ó Cróinín 1995, chapters 3–5). The tributes due to a high-king would most likely be paid in the form of slaves, cattle or metal products. The latter could have included ore, ingots, scrap or finished artefacts. The more common crafts such as bone-working not only produced necessary items but also highly decorative objects including pins and combs. All crafts at Lagore reflected its status though none more so than the non-ferrous metal-working. Basic subsistence was provided by agriculture which, possibly in conjunction with royal tributes, allowed for the luxury of high-status crafts. Non-ferrous metal-working was a manifestation of the wealth and power of the inhabitants and its role in the economy was one of prestige not subsistence.

6.8. Summary

Despite the absence of a coherent stratigraphy and ground plan, Lagore crannóg is a good case-study for research into non-ferrous metalworking in Early Historic Ireland. As a status-related activity, this metalworking is best examined at a 'royal' site, with Lagore recorded as one of the most important settlements of the Early Historic period.

Hencken's idea that the East area in period Ib was the chief seat of non-ferrous metalworking must now be considered more as a strong possibility than as fact. The number of stratified relevant artefacts is too low to conclusively prove his theory of dating. His identification of the eastern side of the site as the metalworking area, however, does appear to be correct. The large quantities of relevant material found here also suggests the presence of a permanent or residential non-ferrous smith or *cerd* on the crannóg. Itinerant craftsmen would not leave behind such extensive evidence.

The high status of the site also made it a focus of political attention. Hencken wrote that 'possibly the increase of ironworking and the seeming decline of bronze-working after the first occupation may represent a shift from the arts of peace to those of war and was occasioned by the Norse invasions' (1950, 8). This, however, was not the case. There was no distinct increase in ironworking after the first occupation, though there was an apparent decline in non-ferrous working. No increase in weapon production was required to meet the Norse threat, as the Vikings did not, after all, introduce warfare to Ireland. The history of Lagore before their arrival was just as troubled or unsettled as after their coming. The settlement at Lagore was never a very peaceful one, its proximity to Viking Dublin making some measure of contact inevitable. It was native Irish politics that determined the hostile nature of that contact. The constant feuding between the Clann Chernaigh Sotail of Lagore and the Clann Uí Chonaing of Knowth prevented both families having peaceful relations with the Norse of Dublin. When the Clann Uí Chonaing allied themselves with the Norse, the Clann Chernaigh Sotail automatically became their opponents.

These hostilities ensured the almost total insularity of the fine metalwork at Lagore. Only one example of contact with the nearby Viking settlement was recovered amongst the relevant evidence, namely a fragment cut from a silver bracelet. This was most likely stolen in a raid on Dublin or from a site such as Knowth, allied with the Norse. There is no evidence of trade with the Vikings, probably reflecting a stubborn opposition and refusal to interact with the allies of the Clann Uí Chonaing. Perhaps this is why the non-ferrous metalworking at Lagore went into decline. With Viking Dublin came the establishment of large-scale, non-ferrous workshops (Ó Ríordáin 1971). It may have been to the advantage of the native Irish living in Leinster to show their allegiance to Norse Dublin, as opposed to the native powers at Lagore. Trade and tributes which once supplied the non-ferrous metalworking on the crannóg may have been diverted to Dublin (or its allies), leaving Lagore in relative isolation.

The state of the non-ferrous metalworking at Lagore was a reflection of the contemporary social position of its inhabitants. It was not vital to the site's existence though it did indicate how wealthy the economy was. This appears to have been the case throughout the island, with the status of a site evident in its capability to produce items of very fine metalwork.

CHAPTER 7
CONCLUSION

7.1. Research problems

A relatively large gap exists in our knowledge of non-ferrous metalworking technology from Early Historic Ireland in the area of metal production. The recent discovery of copper smelting furnaces from this period within the mine site at Ross Island, Killarney, does point to contemporary mining at that location. The analysis of the Ross Island furnaces confirms they were using ore (chalcopyrite) from this mine, whether this was mined or scavenged from older spoil heaps, it still constitutes a mine source for copper in Early Historic Ireland. However, the technology employed in ore extraction remains uncertain. We do not know, for example, if the Bronze Age technique of fire-setting was still in use. Smelting evidence is also fragmentary and its association with non-ferrous metalworking tenuous.

The lack of definite archaeological evidence for primary smelting may, on many settlement sites, be due to the extent and quality of archaeological excavation. Antiquarian investigations sought out artefacts and these were detailed in their reports, though structural remains were very often ignored, e.g. Buick's 'excavation' of Moylarg crannóg in 1893. Later reports also tend to lack excavation detail, though not to the same extent.

We may include here the sites excavated by the Harvard archaeological expedition to Ireland in the 1930s and 40s. The lack of stratigraphic detail in the Lagore and Ballinderry reports, for example, is understandable considering the enormous quantities of information and artefacts those sites produced. The failure to record individual find spots for the large number of artefacts discovered probably reflects the short two or three seasons allowed for excavation.

Recent reports, such as Ivens work at Tullylish and Dunmisk, are generally more detailed. However, the use of scientific analyses on non-ferrous metalworking evidence is still much underused, usually confined to the examination of some crucible residues. In this regard, the findings of Moss in 1927 have not been expanded or developed. The identification of slag and furnace/hearth remains would also benefit from scientific analyses perhaps, in some cases, helping distinguish between ferrous and non-ferrous evidence.

Several published reports, such as that of Garranes, are biased by the limited extent of investigation/excavation carried out. Large sites are very expensive and time-consuming to investigate fully, while others are excavated under rescue circumstances or are limited by modern developments on, or adjacent to, the site. The small number of excavated crannógs is probably due to the difficulty and expense involved in such work. Problems which have to be overcome include flooding of the site and the preservation of organic finds.

Sites including Dooey, Clonmacnoise, Rathtinaun and Moynagh Lough have not yet been fully published, although the latter two are forthcoming. Interim reports or short articles are available on all but Rathtinaun. Unfortunately, little is known of eleventh and twelfth century sites except for those earlier settlements which survived into that period, while recent discoveries such as the Scrahane complex in Killarney were investigated under less than ideal circumstances.

The many problems encountered in examining the literary references have been discussed in some detail in chapter 2. These problems include the lack of technical knowledge of the author, editor or translator, the propagandistic use of written documents, and also the later, often undated, addition of glosses to many texts.

7.2. Future research

Despite the aforementioned problems, it is possible to construct some picture of non-ferrous metalworking activity in Early Historic Ireland. A correlation of technological, artefactual, contextual and literary evidence helps fill the gaps in any one area of study, or at least suggest answers to the questions posed. This study has attempted to do this with regard to fine metal production, however it is also applicable to other aspects of life in the period. Other crafts could be explored in much the same way, eventually arriving at a more complete understanding of Early Historic economy and settlement hierarchy. This has recently been addressed by the author in a doctoral study (Comber 2000).

Another area of potential future research lies in an examination of the affect of outside influences on all aspects of Early Historic life, and the routes which facilitated the spread of new ideas. A study of the non-ferrous metalworking evidence from Viking towns would be of particular relevance to this work. A comparison of such remains with the native evidence could provide some insights into the true relationship of the Norse settlers with their Irish neighbours.

Detailed scientific analysis of a range of finished objects of fine metalwork from Early Historic Ireland might enable the identification of assemblages manufactured by an individual smith or workshop. Such work clearly has much potential to provide information on the scale of industrial output, and on trade or exchange in the Early Historic Period. A study of this nature, however, would require detailed scientific knowledge and access to both the relevant equipment and a range of artefacts.

7.3. Non-ferrous metalworking in Early Historic Ireland: summary

The manufacture of a fine metal object in Early Historic Ireland entailed a relatively lengthy production process. The first step was the acquisition of raw materials in the form of ore, smelted ingots or cakes of metal, scrap or hack-metal. Direct exploitation of ore required the physical involvement of a community and it seems unlikely that a high-status or royal group would themselves participate in such work. They may have used slaves or subject groups to recover the ore for them, perhaps receiving it as a gift or tribute, or through trade. The use of ready-smelted ingots or cakes of metal reflects some degree of importation onto a site, via either tribute or trade. The recycling of older objects, however, possibly utilised stock-piles gathered over a period of time at a particular site. The archaeological record is surprisingly silent with regard to the organisation of mining or the circulation of raw metal in Early Historic Ireland.

Processing this raw material required either smelting and/or melting and alloying. If ore was being directly exploited, it had to be smelted to produce a molten metal which was then refined to remove impurities. On the other hand, if ready-smelted metal was imported onto a site, then re-melting and perhaps alloying were undertaken. The third step in the process involved the fabrication of the object by casting or hammering and joining. The majority of penannular brooches, for example, were initially formed through casting, though some such as Kilbride-Jones' group C5 (1980) were composed of beaten sheet-metal terminals sweated onto a hammered and shaped metal bar or rod. Once the basic form of an object was achieved, it was then decorated, the smith having a variety of different techniques at his disposal.

In an ecclesiastical context, metal objects had a religious significance in addition to being an emblem of wealth and status, perhaps remaining with the community in which it was produced or being used by that community as a tribute or in trade. However, the latter transactions were only entered into with those of high status. As seen in chapter 2, those of inferior status were not permitted to either possess or wear fine decorative items of bronze, silver or gold.

The processing of raw ore and/or metal required the use of furnaces and hearths. Unfortunately, the archaeological remains of such are often quite difficult to identify and interpret. The location of such structures on settlement sites, however, would not have been far from the area where fabrication and decoration took place. They may have been involved in such techniques as casting, gilding, the application of enamel or filigree. The metalworking area, marked by the aforementioned elements and a range of relevant artefacts, was generally situated on the periphery of the settlement. This distanced the domestic area from noise, dirt and air pollution.

The processing, fabrication and decorating of fine metal objects was undertaken by a smith, or *cerd* as he is sometimes referred to in the literary sources (see chapter 2). This smith may have been a permanent resident on a site or could have been an itinerant metalworker, travelling from patron to patron. The possibility also exists that some smiths may have been female. The literary evidence also points to a certain hierarchy or structure within the trade, a situation where master craftsmen trained apprentices as smiths. In a workshop setting there may have been a smith or master-smith with two or three apprentices or students working under his direction.

Sites producing evidence of non-ferrous metalworking had economies based on agriculture, while other activities central to the basic subsistence of the site included stone-working, iron-working, carpentry, leather and textile production. In addition to non-ferrous metalworking, crafts such as glass, enamel, millefiori, amber, lignite, bone and antler working produced luxury items.

7.4. Conclusions

Technology of non-ferrous metals

The possible sources of metal, already discussed, were probably all utilised in Early Historic Ireland and it seems likely that native mining satisfied most, if not all, copper requirements. Irish copper mines were exploited in the Bronze Age and were still known of in the Early Historic Period, even in Britain if Nennius is to be believed. It hardly seems likely that copper was smelted at Ross Island without the adjacent ore source being exploited. Native sources may also have provided gold during this period, though recycling and importation may have been equally as important. Native lead ore was probably exploited, however both tin and silver were most likely imported. Tin is found in Ireland though only in small quantities, while Roman (e.g. Balline and Ballinrees hoards, Edwards 1990) and Viking (e.g. Carraig Aille II hoard) sources appear to have supplied much of the silver utilised in Ireland.

It is impossible to compare many aspects of non-ferrous metalworking technology in the Early Historic Period with those of the preceding Iron and Bronze Ages. Much of the evidence from the late prehistoric period is either fragmentary or totally absent. It is possible to trace development in the techniques of fabrication and decoration from the three periods, though the Iron Age evidence is limited to finished artefacts. The non-ferrous technology of the Early Historic Period is seen by some as 'the development of applications by skilled and sensitive craftsmen, drawing on long-established technology. A new design of artefact, a new combination of decorative elements, sometimes tends to give rise to ideas of a new technology, when in fact it is an innovative application of an old technology' (Scott 1990, 148).

It is true that such techniques as sheet-metal production, clay-mould casting and *repoussé* were in use since the Bronze Age, and enamel and trichinopoly since the Iron Age. However, the Early Historic Period does see the introduction of some important new decorative techniques, as opposed to a mere 'combination of existing decorative elements'. *Pressblech, kerbschnitt* and filigree, for example, are all Early Historic innovations, whether or not they were introduced or influenced from abroad. Stitching, the method of holding decorative panels in place, is unique to Early Historic Ireland. While the fundamentals of non-ferrous metalworking technology were established in the Bronze Age and developed in the Iron Age, they were expanded upon and flourished in the Early Historic Period.

Outside influences always played a role in technological and artistic development in Ireland including the period in question. In the fifth and sixth centuries AD Irish fine metalwork was influenced by previous pagan Celtic work, such as the spirals and spherical triangles on zoomorphic penannular brooch terminals, and Late Roman ideas from Britain such as the use of millefiori as decoration. In the sixth and seventh centuries Christianity introduced many motifs to Ireland, the most basic of which was the simple cross. Anglo-Saxon influences are also evident in the use of *kerbschnitt* and in some aspects of filigree work, the latter also reflecting a possible Frankish influence. The biggest influence in later centuries was Viking, both in terms of available quantities of silver and the introduction of new art styles such as Urnes and Ringerike.

As a patron of the arts, the church provided one of the most important routes of influence in the Early Historic Period. Non-ferrous metalworking was an integral part of the economy of many large ecclesiastical foundations and smiths must have been among the numbers of people that constantly travelled between them. Pilgrimages to the Continent, via Britain, would have exposed a smith to new designs and new ideas which he doubtless experimented with when he returned home. Alternatively, or in addition to this, a metalworker may have been trained on one site and later moved to another.

The dissemination of ideas within Ireland would also have been facilitated by the movement of smiths. Both the clergy and smiths (and, indeed clerical smiths) were free to travel between *tuatha* with very few restrictions placed on them. Secular apprentice-smiths were usually, according to the literature, placed in fosterage with master craftsmen or smiths far from their homes, despite the fact that their fathers were generally metalworkers also (Kelly 1988). This allowed the apprentice to learn techniques unknown to his father or in his native area.

In later centuries, native metalworkers were influenced via the new urban centres, established by the Vikings in the ninth century AD. These towns provided easy access to international trade and ideas. Despite all of these influences on Irish metalwork, the native smith never blindly borrowed or adopted a new or foreign idea, he constantly adapted techniques and designs to suit his own purpose.

Gaps in the artefact record

It is evident from an examination of the processes involved and the recovered artefact assemblages that some items are missing from the tool-kit of the Early Historic *cerd.* Tools used in mining are unknown and will remain so until a relevant mine is identified and excavated. Tools such as hammers and picks were probably employed in this work. The majority of stone and iron hammers found on contemporary settlement sites were probably used in wood or metal working.

Of some items only a small number of examples are known. Surprisingly, the metalworking anvil is one of these. The example from Garryduff, in addition to being unique, is also very small. Its working surface is a mere 32mm square and was obviously intended for very fine, delicate work. The use of large stones as anvils may explain the strange lack of metal examples. Another object of which there is only one known example is the *pressblech* die, the example being from Rathtinaun in Co. Sligo. The suggestion that the finer bone motif-pieces may have fulfilled this function is tenuous, as the motifs evident on metal objects are of far better quality than those of the motif-pieces.

The small numbers of such tools as files, chisels and punches probably reflects the state of preservation of iron objects on most sites. Corrosion has rendered many iron items unidentifiable and several of the tools required in fine metalworking, due to their small size, have probably survived as formless lumps.

Three pieces of equipment are completely absent from the recovered tool-kit though their existence is suggested by various finished artefacts. For the initial production of metal in a furnace a bellows or blow-pipe was required. That no such items survive probably reflects their construction from organic components such as animal skin and wood. There may be similar reasons for the lack of lathes and trichinopoly frames.

Lathes must have been relatively common, being used in both wooden and metal vessel-spinning. The pole-lathe was manufactured of wood and may have since decayed or become unrecognisable in the archaeological record. Trichinopoly frames, however, were probably not commonplace as the use of the technique on finished items is limited to a small number of extremely fine objects such as the Ardagh Chalice and Tara Brooch. It is possible that they were made of wood, or perhaps metal examples that have yet to be recognised. Apart from these few gaps, the Early Historic metalworking tool-kit appears reasonably complete.

Site contexts

Within a site, it seems likely that metalworking took place beneath an open or flimsy shelter structure yet it must be admitted that there is very little structural evidence on which to reconstruct a metal workshop of this period. On sites where the non-ferrous working was carried out on a smaller scale, an unenclosed outdoor area may have been utilised. The metalworking area is generally removed from the domestic area though usually remains within the enclosure. It may, on some sites, have been located just outside the enclosure. Semi-industrial craft activities were often grouped together in the peripheral areas of a site, the most common exception being the location of ironworking at some distance from the rest. This may reflect the involvement of different individuals or groups in ironworking and the 'lighter' crafts of carpentry, textile production, non-ferrous metalworking, glass and bone working.

In an examination of relevant site distribution, concentrations are evident in Ulster, Munster and Meath (Fig. 32). This may reflect, to some extent, an excavation bias, however it is more likely an indication of the dominant political power bases of Early Historic Ireland. The church was very strong in Ulster as seen in Armagh's constant claim to primacy over the whole country. Several large monastic foundations have been excavated in the North, producing abundant evidence of non-ferrous metalworking. These sites include Tullylish, Dunmisk, Movilla Abbey and Cathedral Hill. The Eóganachta rulers of Munster, both tribal and dynastic, were originally based in Cork/Kerry, while the Uí Néill of Meath were powerful throughout most of the relevant period.

These political regions all had a common economic base that centred on agriculture. Successful farming provided a stable economic base necessary for any group or family with political ambitions. Only the wealthy could afford to engage in battles and attempt to establish themselves as local or regional kings. The fertile lands of Ulster, Munster and Meath were exploited for such purposes. Once established, the settlements of 'new' kings or rulers could then participate in the luxury activities, such as non-ferrous metalworking, glass production etc.

Site dating

The ever-changing political background of constant battles and feuding is reflected in the relevant sites. In the late fifth century AD large tribal groupings still existed in Ireland and the two most powerful of these are represented by high-status relevant sites. In Ulster, the Airgialla were dominant and they ruled from Clogher in Co. Tyrone, while in Munster, one branch of the Eóganachta ruled from Garranes.

Also founded in the fifth century was the ecclesiastical site on Cathedral Hill, Co. Armagh reflecting the introduction and future importance of the church. The only relevant establishments in the sixth century were monastic sites. These included Kilpatrick, Movilla Abbey and Tullylish. Though only three in number, these may represent the growing influence of the church.

The seventh century saw the establishment of the majority of relevant sites. Unlike the tribal centres of Garranes and Clogher, the new sites were smaller, though clearly of high-status. The smaller size of the sites may be the result of the replacing of tribal groups with dynastic families. Garranes was no longer in use, while the power of the Airgialla at Clogher was much diminished. Families such as the Uí Néill in Meath and individual branches of the Eóganachta in Munster now ruled from sites such as Lagore and Garryduff. The fragmentation of older tribes into feuding family branches gave rise to the increased number of high-status sites as individual family groups triumphed over their neighbours. Examples of such sites may include Raheennamadra, Ballinderry II, Moynagh Lough and Rathtinaun.

Only the most powerful of these families survived into the eleventh and twelfth centuries when local kings lost much of their power. The arrival of influential provincial leaders from the four provinces of Ulster, Connaught, Munster and Leinster and the Anglo-Normans from Britain caused much change in Irish society and politics. Carraig Aille II, Lagore and Knowth continued to be inhabited in this period though these centuries did see their decline and eventual abandonment. Knowth, for example, was taken over by the Anglo-Normans in the twelfth century.

The smith

Evidence regarding the role and position of individuals and groups engaged in non-ferrous metalworking is largely confined to the literary sources. In the literature, the *cerd* is portrayed as a specialist working in bronze and/or precious metals. Unlike ironworking, which can occasionally be spatially distinct from non-ferrous working, it is impossible to suggest the existence of an individual goldsmith or silversmith from the archaeological evidence. There is also the possibility that some *cerds* may have been female. If a smith did not have a male heir to succeed him, he could instead have a daughter trained in his profession (section 2.2.).

Status-wise, the smith was equal to that of the lowest grade of noble and as such enjoyed certain rights and privileges, one of the most important of which was freedom of movement. The smith could travel from region to region without losing status, though only allowed to practice metalworking in a new area when given permission to do so by that region's mastersmith. This would facilitate a travelling or itinerant smith, however whether this applied to non-ferrous workers as well as blacksmiths is not specifically stated.

There is also evidence to suggest a more permanent role for the metalworker in society. The demand for a fine metalworker was limited to high-status sites, whereas the blacksmith could be employed in the average ringfort, making and repairing domestic tools. However, larger sites probably had a resident blacksmith engaged in both tool and weapon production. The large quantities of non-ferrous evidence left behind on some sites appears to indicate the activities of a residential *cerd*, a permanent member of the community. It is impossible to say for certain whether or not all non-ferrous metalworkers held such a place in society.

It is also difficult to determine the importance of patronage in Early Historic non-ferrous metalworking. We know from literary references (section 2.6.) that the system was practiced in Ireland though its extent is uncertain. It would seem probable that the royal sites e.g. Lagore, and larger monastic centres e.g. Cathedral Hill, employed the most skilled craftsmen while others e.g. Reask, may have had to rely on the talents within their own community. It would appear that metalworkers operating under patronage did so within their employer's enclosure (secular or ecclesiastical), though whether or not they were permanently attached to a single patron or site remains a matter of conjecture.

Lagore crannóg, Co. Meath

As the seat of a branch of the Southern Uí Néill, Lagore crannóg epitomises a high status Early Historic settlement from Ireland and, as such, is useful in the examination of non-ferrous metalworking on a specific site. The economy of the site was broad and varied, reflecting the wealth of the inhabitants. Agriculture formed the basis of an economy that also included the working of stone, leather, bone, antler, iron, non-ferrous metals, leather, textiles, glass and millefiori.

The two finds of copper sulphide ore recovered from the site may have been mined in the locality, obtained through trade (including importation from abroad) or tribute. That ore was then possibly smelted on the site, though no actual furnace remains have been definitely identified. All metalworking took place on the eastern side of the crannóg, where fabrication and decoration were perhaps undertaken in the possible workshop structure. Objects may have been produced for internal site use or may have been used for gift exchange or as items of trade.

The smith at Lagore appears to have been a permanent member of the community judging from the quantities of relevant evidence recovered, as would be expected from a site of such status. The smith may also have employed apprentices, however this is impossible to determine from the archaeological record. It is likely, though, that the smith would have had some help, if only in the operation of bellows for the smelting furnace. The multiplicity of skills evident in a master work such as the Ardagh Chalice may indicate a team approach to the fabrication of the most ornate objects of this period.

Non-ferrous metalworking was an economic luxury at Lagore, practiced on the eastern edge of the site. There is a possibility that the smith worked within a workshop with an adjacent smelting furnace and/or metalworking hearth. Such facilities, however, may not have existed during all phases of occupation at Lagore, perhaps only during the period when it was at its most powerful, between the seventh and ninth centuries AD.

The Early Historic Period in Ireland 'saw the emergence of literacy and the production of some of the finest works of art on vellum, in stone and in non-ferrous metals produced anywhere, and at any time, in the Old World ... a time of heroes (mythical, pseudo-historical and historical), of saints, scholars, and of jurists' (Scott 1990, 99). As such, it facilitates the examination of fine, non-ferrous, metal production against a contrasting background of both political tension and upheaval, and scholarly and artistic innovation.

Appendix I

Relevant extracts from the early Irish laws

Blai Ord Indeoin

The exemption of the sledge [hammer] and anvil
... the person who plies the sledge [hammer, sledge-hammer] on the anvil is exempt from penalty for injuries arising from the work he is engaged on.
... he is exempt, though the anvil break [the hammer], or the hammer break the anvil, or though an idler has been thrust between them without his knowledge or his having seen him.
He is exempt
... in the first [glancing of the hammer off the anvil]
... if he has no knowledge of any defect [in the implements he is using]; and he pays
... compensation for injury to fellow workers
... whether he has seen them or not
... and for injury to beasts which he has not seen; but if he has seen the beasts, if (the fine) is full compensation.

If it be the second [glancing of the hammer off the anvil] ...
If it be the third slipping ...
If it be the fourth slipping ...

If it be the head of the [sledge-] hammer that has slipped [from its haft] ...
If it be from [the striker's] hand it (the sledge-hammer) slipped ...
If two sledge-hammers while being wielded have come in collision, and if injury to the persons engaged [in the forging operation] only has resulted ...
If it be injury to someone else ...

If it be the anvil which has slipped off the block ...
and if (this happened) in consequence of bad fixing [of the anvil to the block] ...
if it occurred in consequence of bad striking ...
If it (the slipping off) be in consequence of bad fixing and bad striking ...
The injuries from the sparks of the iron and of the hearth [forge fire] ...
The injuries done by the iron in carrying it from the fire to the anvil and from the anvil to the fire are to be paid for by the smith.
If the anvil has been [damaged] and if it was in consequence of bad striking ...
but if it was in consequence of the iron having been badly held ...
and if it was in consequence of bad holding and bad striking ...
If it be the sparks from the [forge fire] that have caused the injury ...
the bellows-blower alone is to pay for it ..., unless the smith has urged him to blow strongly ...

If it be the iron that has caused the injury when being struck, it is to be considered whether it was in consequence of bad holding or of bad striking ...
if it was in consequence of bad holding the smith alone is to pay ...
if in consequence of bad striking the striking party alone is to pay ...
If it be the sparks of the iron that have caused the injury ...
and if there has been an urging on [to blow up the fire vigorously] ...
If the smith is aware (of a hazard) and the apprentices do not know, the culpability is his alone ...

(Translation from Scott 1983 of eighth century tract, *C.I.H.* I, 267.2ff, III, 936.1ff and 1078.25ff.
Repetitious definitions of compensation are omitted while words within brackets are from the original translation and those within square brackets are Scott's comments or alternative versions).

Bla Miand Midchlais

The exemption of ore is a mine
The exemption of minerals is a mine.
He who digs ore from a mine is exempt from liability, provided that the ore was not privately owned, or if it was, provided permission has been given by the owner.
Exemption exists provided there was not excessive destruction, for if there was, it is to be treated as an unlawful excavation.
The exemption extends to idlers and unprofitable workers, provided that the work is being carried out with proper entitlement; there is one third restitution for all damage caused while moving it [the ore] up or down, this being a reduction from half honour-price to one third restitution.
[One third of restitution for collaborators, for every profitable person, and for every animal: *C.I.H.* III, 940.15]

If the ore is in private ownership and the owner has not given permission, there is a fine of five séts for it and the restoration of the ore in the form in which it is found, whether it be in ingots or semi-products or as finished artefacts.
And there is full liability for all damage caused in transporting it, as well as for the mine [itself] until the owner has acknowledged [its repossession] some time after it has been restored to its former state, and he has the option of not restoring it.

In this case full payment is due as for the first slipping of the sledge-hammer without fore-knowledge of a fault.

(Translation of *C.I.H.* I, 270.15f and III, 940.15f by Prof. MacEoin in Scott 1990, 178)

Dinnra Clasaige

A miner's *dírna*, that is a measure used by those who dig the mine: it is equal to two ounces, that is he harvests the copper ore. An artisan's *dírna* of red copper has six ounces and is worth a scruple. If it is white copper it is worth two scruples. The ounce of white copper is worth a penny and the ounce of red copper is worth a half penny, and the ounce of copper is equal [in value] to an ounce of silver and the red copper is equal in value to tin, and the silver penny (or perhaps 'the pennyweight of silver'), and eight grains of wheat are equal to a silver penny (again perhaps 'a pennyweight of silver') and four *dírna* of lead for a *dírna* of white copper, for it is of lead that the solder is made.

The [value of a] manufactured article is divided into three: one third for the material, one third for food and for the maintenance of the man by whom it is made, and one third for the smith etc. And whatever the value of the base (i.e. 'basic material', possibly) art adds twice as much to it. And the smith never fashions a 'white vessel' (i.e. 'of precious metal'), as the cerd (craftsman) does, but makes only a 'black vessel. The *'ól Pátraic'* of the *saer* (handworker) is the same [in value] as the *ól medach* (*'mead ól'*?) of the craftsman. The *cuad* ('bowl') of the craftsman is equal to the *follerb* (must mean a larger vessel of some kind) of the handworker, and their meisrín is also the same.

(Translation by Prof. MacEoin of a glossary dated approximately 1100 AD, *C.I.H.* V. 1571, 18–30, pers. comm.)

APPENDIX II

CATALOGUE OF RELEVANT SITES AND THEIR FINDS

See chapters 4 and 5 for detailed discussions of artefacts, structures and sites.

RELEVANT SITES

MONASTIC SITES

1. Ballyvourney, Co. Cork.
2. Cathedral Hill, Co. Armagh.
3. Cathedral Hill/Scotch Street, Co. Armagh
4. Clonmacnoise, Co. Offaly.
5. Dunmisk, Co. Tyrone.
6. Kilpatrick, Co. Westmeath.
7. Kiltiernan, Co. Galway.
8. Movilla Abbey, Co. Down.
9. Nendrum, Co. Down.
10. Reask, Co. Kerry.
11. Tullylish, Co. Down.

RINGFORTS

12. Ardcloon, Co. Mayo.
13. Ballycatteen, Co. Cork.
14. Carraig Aille II, Co. Limerick.
15. Castleskreen, Co. Down.
16. Cathair Fionnúrach, Co. Kerry.
17. Feltrim Hill, Co. Dublin.
18. Garranes, Co. Cork.
19. Garryduff, Co. Cork.
20. Gransha, Co. Down.
21. Lisdoo, Co. Fermanagh.
22. Lisduggan, Co. Cork.
23. Marlinstown, Co. Westmeath.
24. Raheennamadra, Co. Limerick.
25. Rathgurreen, Co. Galway.
26. Rathmullan, Co. Down.
27. Sluggary, Co. Limerick.

CRANNÓGS

28. Ballinderry II, Co. Offaly.
29. Ballydoolough, Co. Fermanagh.
30. Ballyjamesduff, Co. Cavan.
31. Ballymena, Co. Antrim.
32. Clea Lakes, Co. Down.
33. Clones, Co. Monaghan.
34. Cloneygonnell, Co. Cavan.
35. Corraneary, Co. Cavan.
36. Craigywarren, Co. Antrim.
37. Drumkeery Lough, Co. Cavan.
38. Lagore, Co. Meath.
39. Lough Faughan, Co. Down.
40. Lough Mourne, Co. Antrim.
41. Loughrea, Co. Galway.
42. Loughtarmin, Co. Antrim.
43. Moylarg, Co. Antrim.
44. Moynagh Lough, Co. Meath.
45. Randalstown, Co. Antrim.
46. Rathtinaun, Co. Sligo.
47. The Miracles, Co. Fermanagh.

MISCELLANEOUS SETTLEMENT SITES

48. Clogher, Co. Tyrone.
49. Dalkey Island, Co. Dublin.
50. Dooey, Co. Donegal.
51. Island MacHugh, Co. Tyrone.
52. Knowth, Co. Meath.

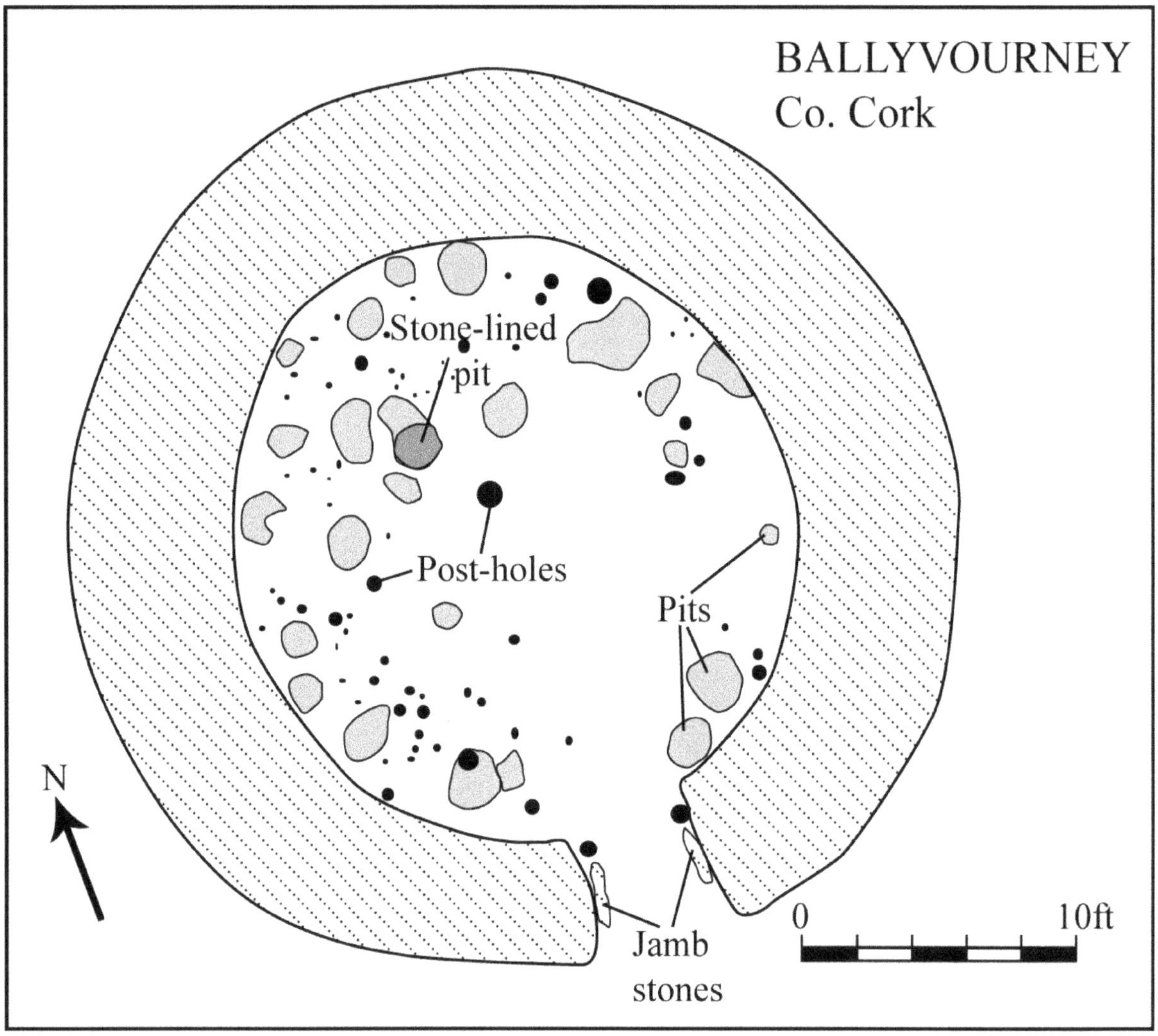

Fig. 78 Plan of St. Gobnet's House (after O'Kelly 1952).

1. ST. GOBNET'S HOUSE, BALLYVOURNEY, Co. CORK.

Townland Glebe *O.S. 6" map* O58

Fig. 78 *Nat. grid* 11971/07681

Site summary

In 1951, while clearing an area of ground before the erection of a statue, workmen uncovered a small crucible and a drain-like structure filled with a black deposit. The ensuing excavation revealed the remains of a circular stone house with one entrance, marked by two vertical jambs. Within the structure was found a series of hearths and pits, some of which were associated with iron-working. Two distinct layers of use were identified, both of which had associated pits and hearths. No dates were recovered from the excavation, however the site is known as an important Early Historic site associated with St. Gobnet.

Metalworking features

No features specifically associated with non-ferrous metalworking were uncovered. In the interior of the hut, along its walls, were found a series of hearths and pits, some of which contained large lumps of iron slag and furnace 'bottoms'. It is obvious that ironworking was undertaken in this structure while the only indicator of non-ferrous working was the discovery of two short lengths of bronze wire, a clay crucible and a fragment of another.

Relevant artefacts

1 complete bag-shaped clay crucible, 1 clay crucible fragment, 17 whetstones, 2 lengths of bronze wire.

Comment

The evidence of non-ferrous metalworking from the site may seem negligible, however it must be remembered that this hut is probably only a small part of the original site. Further investigation in the area may reveal more extensive monastic remains. The structure was completely excavated and is reminiscent of Hut G at Reask which also produced quantities of iron slag and several crucible fragments.

Reference

O'Kelly, M.J. 1952 'St. Gobnet's House, Ballyvourney, Co. Cork.' *Journal of the Cork Archaeological and Historical Society* 57, 18–40.

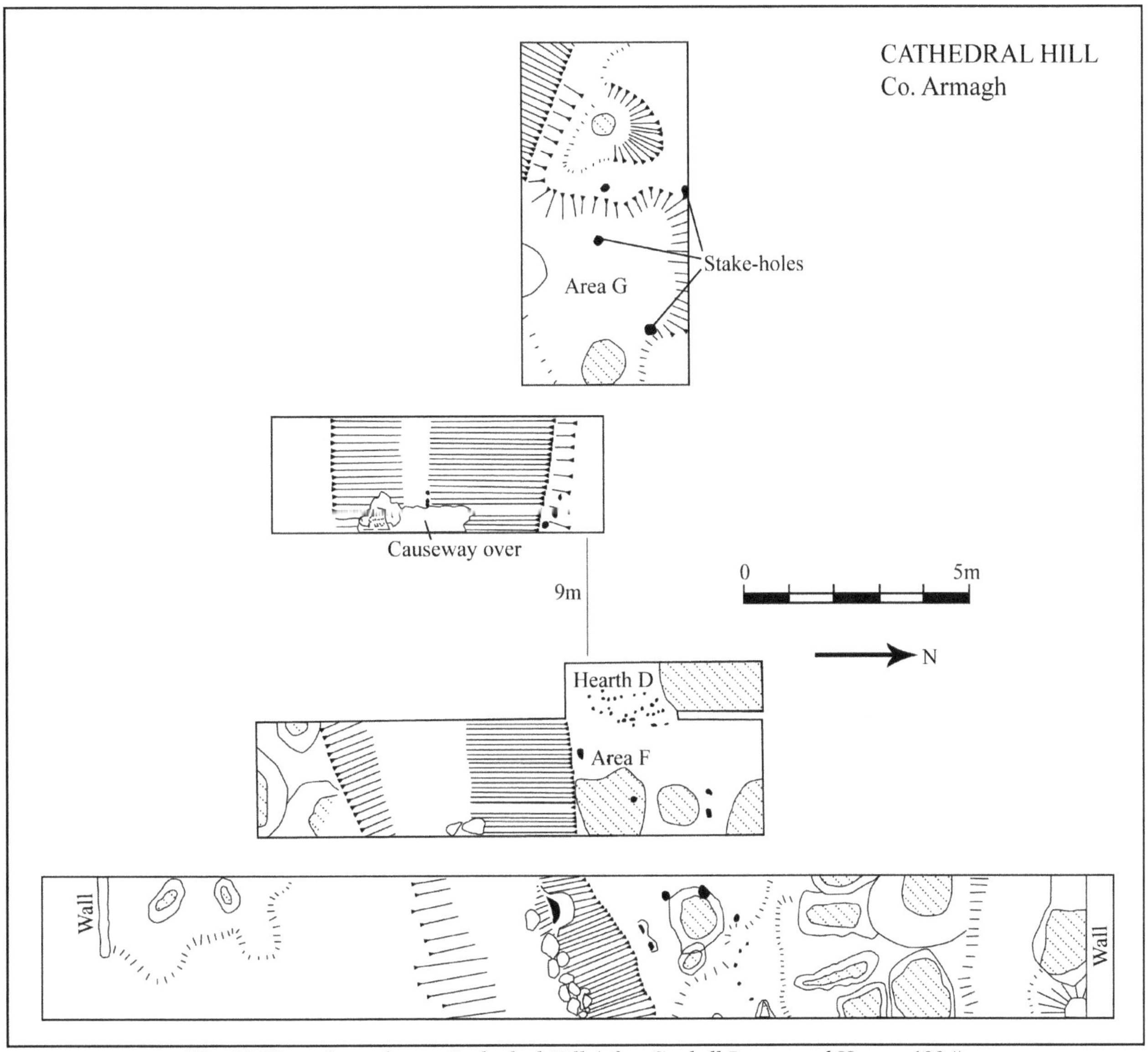

Fig. 79 Plan of trenches at Cathedral Hill (after Gaskell Brown and Harper 1984).

2. CATHEDRAL HILL, Co. ARMAGH.

Townland Parkmore or Demesne *O.S. 6"map* 012
Figs. 79–81 *Nat. grid* H 8764/4477

Site summary

This monastic site in Armagh town encloses the summit of Cathedral Hill and was originally delimited by a substantial ditch and outer bank, approximately 50m in diameter. Excavations undertaken on the south side of Cathedral Hill uncovered the bank and ditch. Armagh was supposedly founded by Patrick in either 444AD (Annals of Ulster) or 459AD (Annals of the Four Masters). It was well established by the second half of the fifth century and continued as a major ecclesiastical centre into the ninth century. Three main phases of activity were identified. Phase I saw the construction of the bank and ditch enclosure, the bank supported by a rough stone revetment. A radiocarbon date of 1660±80BP was obtained from the remains of three stakes driven into the ditch bottom. In phase II the bank was shoved back into the ditch to fill it. This phase is roughly dated to the Early Historic period. Activity in the ditch hollow marks phase III, dated 1430±85 BP.

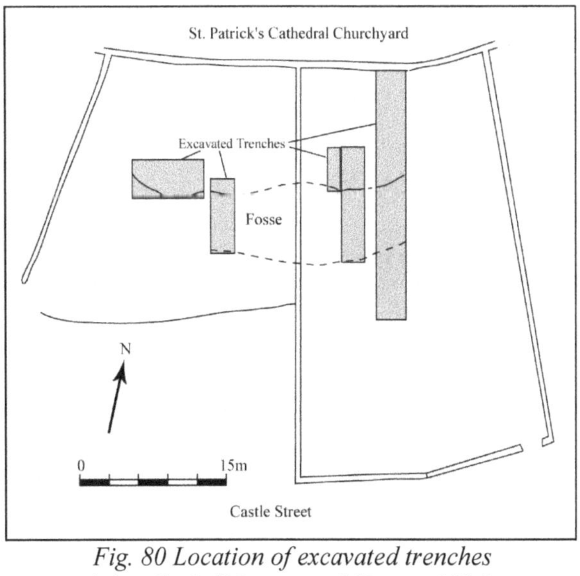

Fig. 80 Location of excavated trenches (after Gaskell Brown and Harper 1984).

Metalworking features

Although there are relevant finds from all three phases, the most substantial evidence comes from the ditch hollow in phase III. Here, the remains of a probable workshop structure were found. It was marked by four large post-holes forming a rectangular plan, with a high concentration of relevant finds and a nearby metalworking hearth (hearth D). This structure had been affected by fire twice as evidenced by two layers of burnt material in the interior. Also located on the site was a possible source of the clay used to manufacture the crucibles and moulds. This source (areas G and F) started as an excavation for clay, later the workshop/shelter was constructed upon it. Metalworking on the site is dated from the late fifth century to the ninth century AD, continuing throughout all three phases.

Relevant artefacts

52 pyramidal crucibles, 9 flat-bottomed crucibles, 4 large, fabric-reinforced crucibles, 3 other crucibles, 1 possibly Roman tuyère, 2 tubular tuyères, 8 circular tuyères, 1 key-pattern ornamented clay mould, 1 heart and stars ornamented clay mould, 1 pin clay mould, 2 ring clay moulds, 1 book-clasp clay mould, 45 clay mould fragments, 7 stone ingot/lamp moulds, 3 other stone moulds, 1 whetstone, 8 stone motif-pieces, 1 lignite motif-piece, 1 bone motif-piece, 1 bronze awl, 12+ fragments of scrap sheet bronze, 11+ lengths of bronze wire, 1 bronze chain, 1 iron shears, 1 iron awl.

Comment

It was not possible to conduct more extensive excavations at this site due to modern Church use. The excavated area comprises a very small part of the original monastic enclosure. However, it does appear to have been a non-ferrous metalworking area. The large quantity of relevant evidence points to it having been an important activity within the site. Due to the limited nature of the investigations it is impossible to examine the non-ferrous working in relation to other activities on the site.

Reference

Gaskell Brown, C. and Harper, A.E.T. 1984 'Excavations on Cathedral Hill, Armagh 1968.' *Ulster Journal of Archaeology* 47, 109–161.

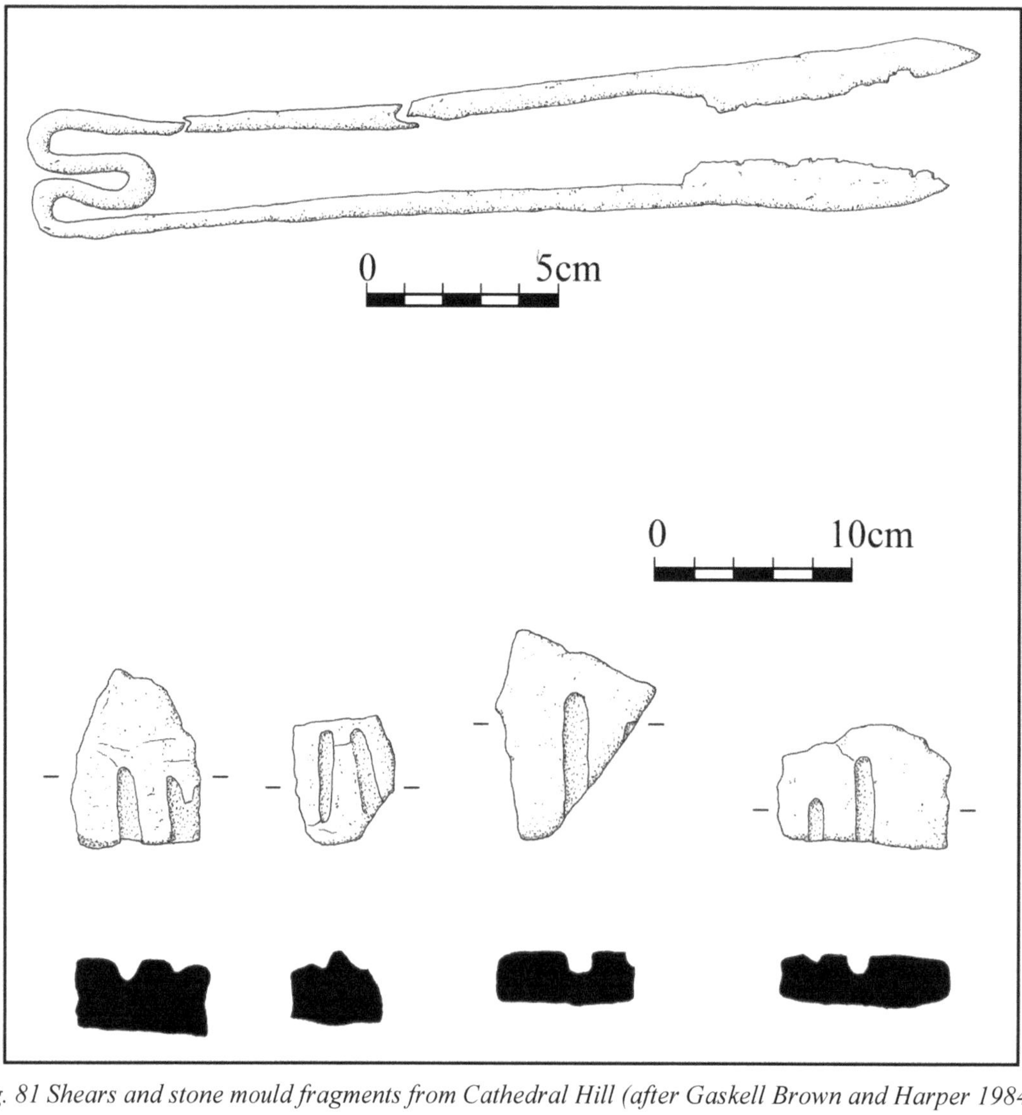

Fig. 81 Shears and stone mould fragments from Cathedral Hill (after Gaskell Brown and Harper 1984).

Fig. 82 Central enclosure at Clonmacnoise (photo; Dept. of Archaeology, NUI, Galway).

3. SCOTCH STREET, Co. ARMAGH.

Townland Parkmore or Demesne *O.S. 6" map* 012
Nat. grid H 8764/4477

Site summary

The excavation at nos. 46–48 Scotch Street uncovered part of an Early Historic cemetery. The numerous burials excavated were overlain by a later, though still Early Historic, occupation layer. This layer was too fragmentary to provide much information though it did produce the artefacts listed below. The burials date from the fifth to seventh century AD, while the occupation is later in date, probably seventh/eighth century. The site falls within the Cathedral Hill monastic enclosure.

Metalworking features

None discovered.

Relevant artefacts

1 clay crucible fragment possibly of the pyramidal type, a length of bronze wire, an iron plate.

Reference

Lynn, C.J. 1988 'Excavations at 46–48 Scotch Street, Armagh, 1979–80.' *Ulster Journal of Archaeology* 51, 69–84.

4. CLONMACNOISE, Co. OFFALY.

Townland Clonmacnoise *O.S.6"*005
Fig. 82 *Nat. grid* 20111/23043

Site summary

Limited excavations were undertaken in the new graveyard at Clonmacnoise after the discovery there in 1990 of Offaly's first recorded ogham stone. Work revealed an area of domestic and industrial activity located on the eastern edge of the monastic enclosure. The earliest evidence of occupation consists of a series of stake holes, the probable remains of post and wattle structures. Stone foundations of a later circular house with an internal clay floor and domestic hearth were found in association with evidence of iron-working, non-ferrous working, antler-working and corn-drying. More limited excavations 200m south-west of the Cathedral, though still within the monastic enclosure, also produced relevant artefactual evidence.

Metalworking features

None discovered.

Relevant artefacts

Clay crucibles, fragment of a bivalve clay mould, clay tuyère fragments, bone motif-piece, bronze tweezers, iron gouges, iron punches, iron chisels, fragment of a gold-working crucible, copper-smelting slag, whetstone fragment.

Comment

The large number of relevant finds from the new graveyard would suggest a metalworking area here, however the lack of any related features may indicate proximity to a workshop. The excavated site was clearly only a small part of a larger industrial zone and early ecclesiastical site.

Reference

King, H. 1992 'Excavations at Clonmacnoise.' *Archaeology Ireland* 6, 3, 12–14.
Ó Floinn, R. and King, H. 1994 'Archaeological investigations at St. Ciarán's National School.' In H. King (ed.), 119–132.

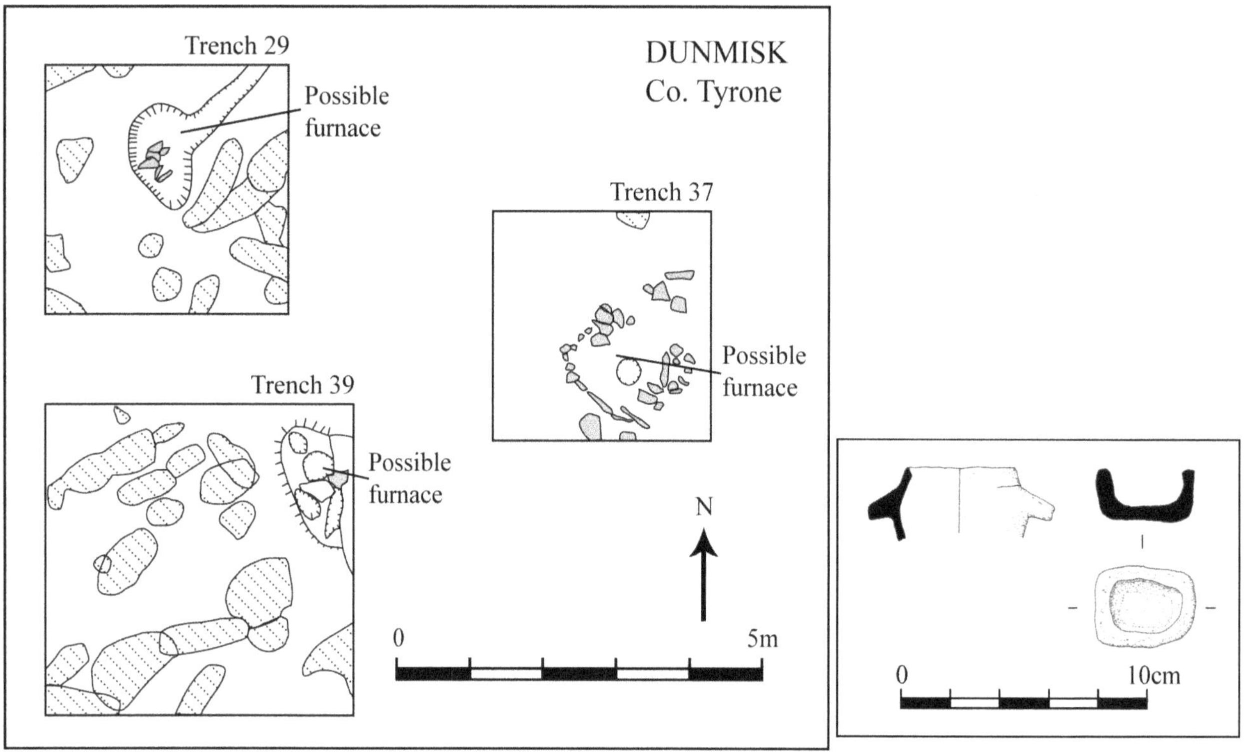

Fig. 83 Possible furnaces and crucible from Dunmisk (after Ivens 1989).

5. DUNMISK, Co. TYRONE.

Townland Dunmisk — *O.S. 6"* 036

Figs. 37 and 83 — *Nat. grid* H 6279/7074

Site summary

The name Dunmisk 'fort' is misleading as the site is ecclesiastical in nature. The 'dun' element of the name probably led to the misunderstanding. The site is located on a prominent, steep-sided glacial hill. This hill has a flat top and its surface covers an area of approximately 1800 square metres. Trial investigations were initially undertaken in 1984 (when the hill was threatened by a gravel quarrying scheme), the rewarding results of which prompted extensive excavations in 1985 and 86. The monastic site appears to have been divided into quadrants relating to use. Non-ferrous and glass-working took place in the north-east quadrant, iron-working in the south-west quadrant, while it seems that the north-west quadrant was a domestic area and the south-east a religious area. The religious area is marked by an extensive burial ground and the probable remains of a timber church.

Metalworking features

Ironworking was carried out in trench 2 in the south-west corner of the site. The area is marked by a working hollow and a series of metalworking hearths. On the opposite side of the site, non-ferrous evidence was uncovered in trenches 28–31, 36–39, and 42. No workshop remains were identified though a series of possible metalworking hearths and furnaces were discovered. Three possible furnaces were located in trenches 29, 37, and 39. In trench 29, the 'furnace' consisted of a deep pit with flat bottom and steep sides. This merged into a flat-bottomed, vertical sided channel. Both of these were cut into the natural gravel. Despite its probable use, no signs of burning were visible.

The possible furnace in trench 37 was a stone rectangular structure, 1.2m x 1.5m. The rectangular area was enclosed on three sides and contained a dense spread of charcoal. The north side of the structure was left open though a line of three large stone-packed postholes may have closed it in some way. Considering its location in the non-ferrous metalworking area it seems likely that it was used in a metallurgical process. The 'furnace' in trench 39 consisted of a pit cut into an artificial mound of friable, sandy, light brown earth. It contained large quantities of charcoal and calcined bone (possible fluxing agent) and was sealed with burnt clay and burnt stones. A series of later pits/furnaces were cut into the original and also contained black ash and charcoal-rich deposits. Also in trench 39 was a hearth sequence, the earliest of which was stone-lined. The latest hearth produced a collection of glass rods which suggests its use in glass working.

Relevant artefacts

145 clay crucible fragments, 9 clay tuyère fragments, 258 clay mould fragments, 1 possible clay mould core, plentiful copper-smelting slag.

Reference

Ivens, R.J. 1989 'Dunmisk Fort, Carrickmore, Co. Tyrone: excavations 1984–1986.' *Ulster Journal of Archaeology* 52, 17–110.

6. KILPATRICK, Co. WESTMEATH.

Townland Corbetstown *O.S. 6"* 020
Fig. 84 *Nat. grid* 05752/05548

Site summary

The monastic site of Kilpatrick survives today as a disused graveyard and fragmentary ruins of a medieval church within a stone wall enclosure. Aerial photography identified a large circular enclosure, of which the existing graveyard and ruins comprise the north-east quadrant. Research excavations carried out over a six year period revealed an enclosure measuring 80-90m in diameter and delimited by an earthen bank and ditch. The earliest activity at this site dates from the sixth–seventh centuries AD, the site being re-used in the tenth–eleventh centuries and from the fifteenth to the late seventeenth century. The southern part of the site was most excavated and appears to have been an industrial area.

Metalworking features

No features specifically associated with non-ferrous metalworking were identified. However, a forging area, two furnaces and a possible workshop structure were located in the southern part of the site. The forging area and furnace A were used in ironworking as indicated by the discovery of iron slag and iron spalls.

Furnace B produced no finds of iron slag. It consisted of bowl-shaped, clay-lined structure with associated charcoal and burnt matter. All three were located close together adjacent to a semi-circular foundation trench. The 'open' side of this structure was marked by a fine line of postholes, perhaps partially enclosing it. The excavator, Leo Swan, identifies this as a possible open-sided workshop. All of the relevant finds were discovered in its vicinity.

Relevant artefacts

10–15 clay crucible fragments, scrap fragments of bronze and tin sheet, a possible bronze tweezers, 3 tiny bronze points possibly compass points, clay mould fragment, iron gouge, iron chisel, 2 iron punches.

Comment

The artefacts listed above suggest that small-scale non-ferrous working was practiced on site. A small-scale operation would explain the lack of specifically non-ferrous metalworking features. It seems probable that the furnaces and forging area were employed in both ferrous and non-ferrous working.

Reference

Swan, L. 1994/95 'Excavations at Kilpatrick, Killucan, Co. Westmeath.' *Ríocht na Midhe* 9,1, 1–21.

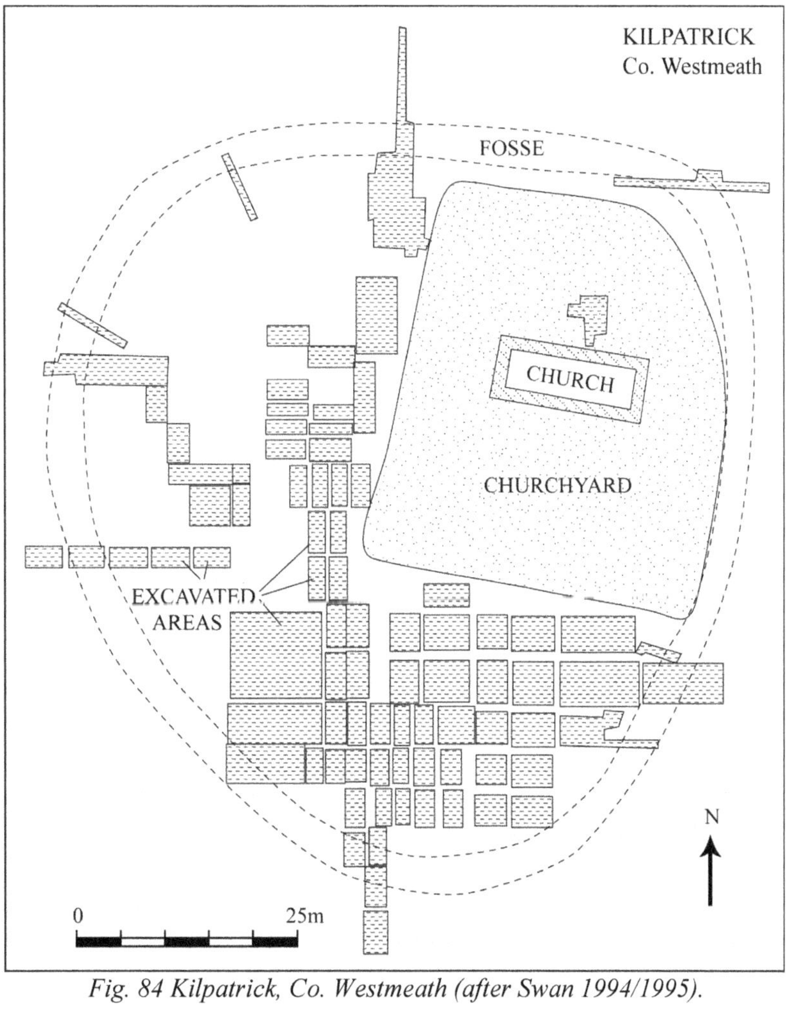

Fig. 84 Kilpatrick, Co. Westmeath (after Swan 1994/1995).

Fig. 85 Aerial view of Kiltiernan (photo; Dept. of Archaeology, NUI, Galway).

7. KILTIERNAN, Co. GALWAY.

Townland Kiltiernan East *O.S. 6" map* 103

Figs. 85–86 *Nat. grid* 14372/21559

Site summary

The monastic site of Kiltiernan consists of a large circular enclosure approximately 120m in diameter, covering 1.4 hectares. The site is enclosed by a drystone wall, 2.5m maximum width and 1m maximum surviving height. The interior is divided into fifteen or sixteen sub-enclosures by radial and other walls. Just south-east of the centre are located the ruins of an eleventh century (with later additions) church and multi-period cemetery. Up to ten houses are visible within the main enclosure, and others in peripheral sub-enclosures to the south, east and west. A souterrain was identified, though not excavated, to the west of the church. Limited excavations were undertaken by M.V. Duignan in the early 1950s, investigating the church are and three houses adjacent to the enclosure entrance (on east side). Artefacts recovered suggest that much of the recorded activity took place in the Early Historic and Medieval periods.

Metalworking features

None identified.

Relevant artefacts

4 crucible fragments, tear-shaped bronze slag, 4 hones, 1 pin-sharpener, 1 whetstone, 4 hammer-stones.

Comment

Three of the four crucible fragments were found in House III, the other in House I. A very small percentage of the overall site was excavated, producing a very biased impression of activity at Kiltiernan. In addition, medieval, post-medieval and modern activity has disturbed the earlier layers, further confusing an already difficult stratigraphy. It is not possible to draw reliable conclusions from the minimal evidence of non-ferrous metalworking found here.

Reference

Waddell, J. and Clyne, M. 1995 'M.V. Duignan's excavations at Kiltiernan, Co. Galway, 1950–1953.' *Journal of the Galway Archaeological and Historical Society* 47, 149–204.

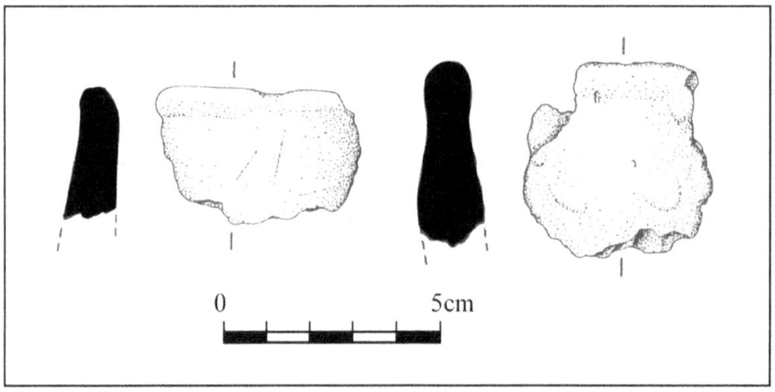

Fig. 86 Two crucible sherds from Kiltiernan (Waddell and Clyne 1995).

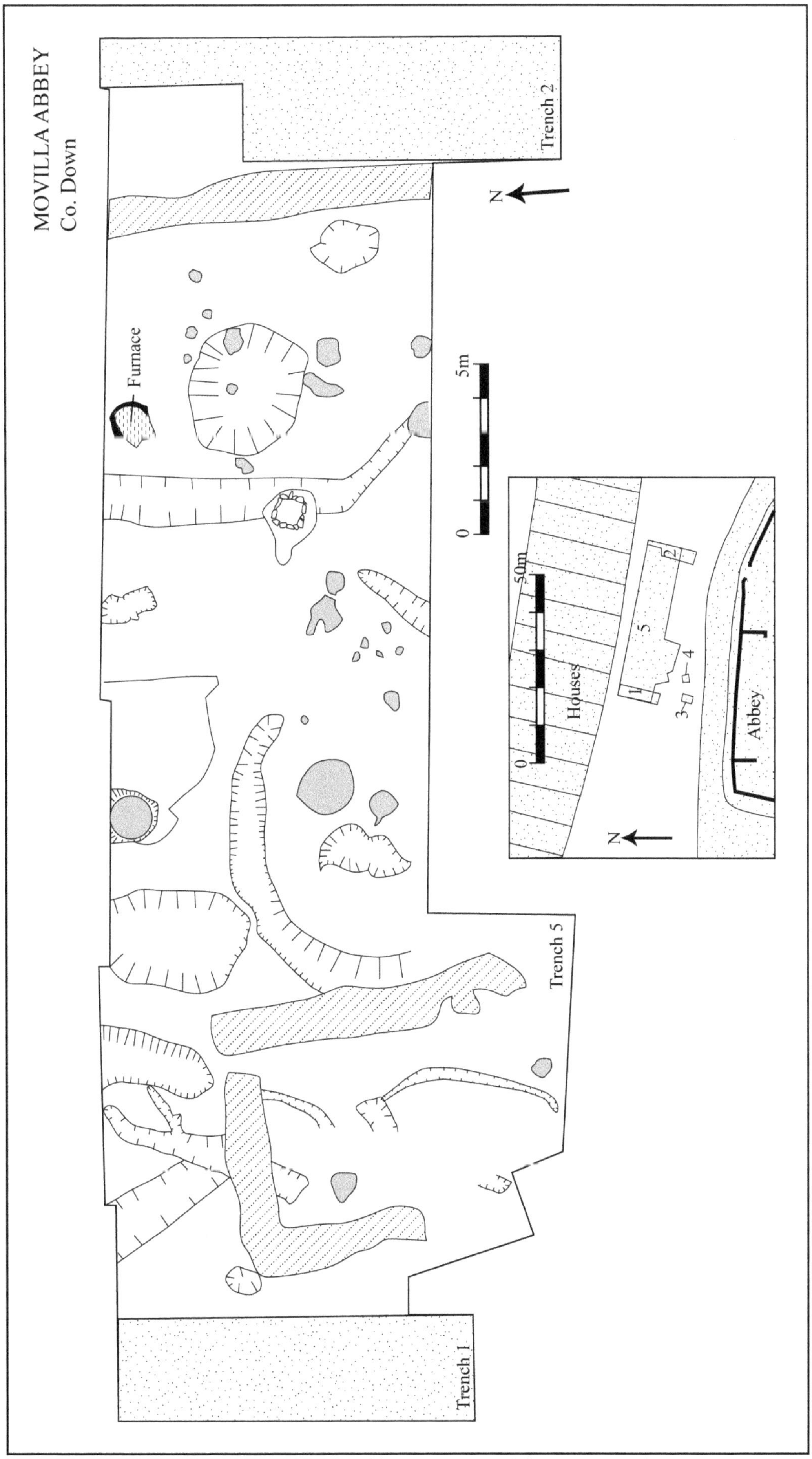

Fig. 87 Plan of Movilla Abbey, Co. Down (after Ivens 1984b).

8. MOVILLA ABBEY, Co. DOWN.
Townland Movilla *O.S. 6"* 006
Figs. 87–88 *Nat. grid* J 0504/0744

Site summary
This monastic site was founded by St. Finnian, who died in 579AD (Annals of Ulster). The early monastic settlement continued into the ninth/tenth century AD, while in the late twelfth/thirteenth century an Augustinian abbey was built on the site. Initial investigations took place prior to road-works and were followed by more comprehensive excavations in the area immediately north of the existing abbey ruins. Work revealed part of an early ecclesiastical site, either an industrial zone or an adjacent area, judging by the relevant artefacts found and the remains of a possible furnace. Due to recent housing developments very little of the original monastic site was available for excavation. Although the bones of cattle, sheep, goat, and pig were found there was a general lack of domestic habitation evidence.

Metalworking features
A possible furnace base was identified in the area excavated. It consisted of a shallow scoop cut into the natural subsoil. The scoop itself showed no signs of burning but contained much burnt material. A burnt layer partly sealed the top of the furnace. The relevant artefacts were found in association with this feature.

Relevant artefacts
185 crucible fragments, 1 slate motif-piece, 1 lignite motif-piece, a length of very fine beaded gold wire.

Comment
Ivens (1981) suggests that the shallow scoop did not function as a furnace due to the lack of evidence for burning, rather that it acted as a base upon which a hearth or furnace was constructed hence the burnt layer. It is possible, however, that the scoop was used as a furnace but it had been deliberately cleaned out for re-use. The lack of a burnt clay lining could be explained in this way. The furnace at Moynagh Lough (Bradley 1991) was also devoid of signs of burning and contained a clean fill.

Reference
Yates, M.J. 1983 'Preliminary excavations at Movilla Abbey, Co. Down, 1980.' *Ulster Journal of Archaeology* 46, 53–66.

Ivens, R.J. 1984b 'Movilla Abbey, Newtownards, Co. Down: excavations 1981.' *Ulster Journal of Archaeology* 47, 71–108.

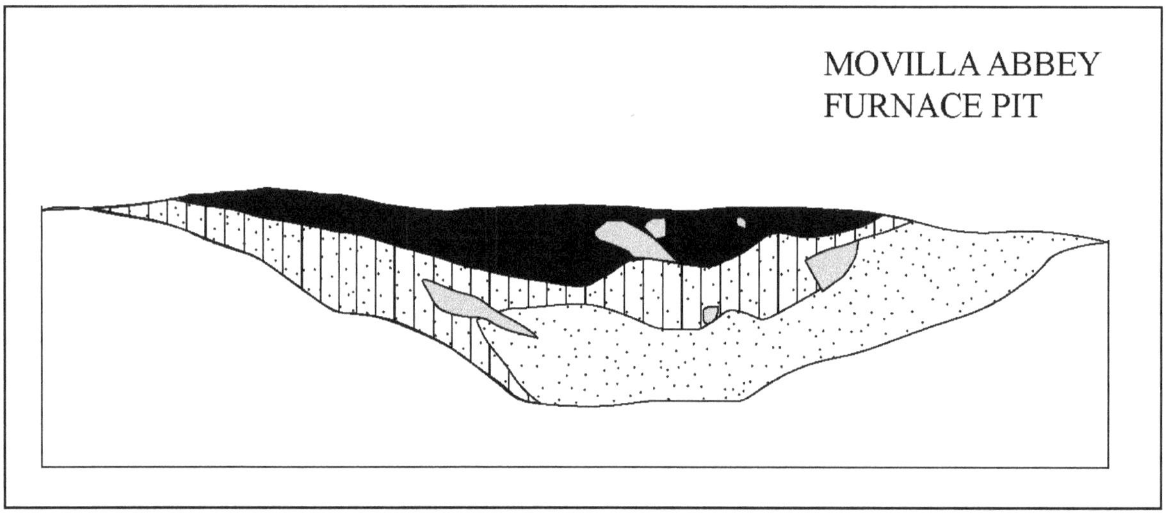

Fig. 88 Section drawing of furnace pit (after Ivens 1984b).

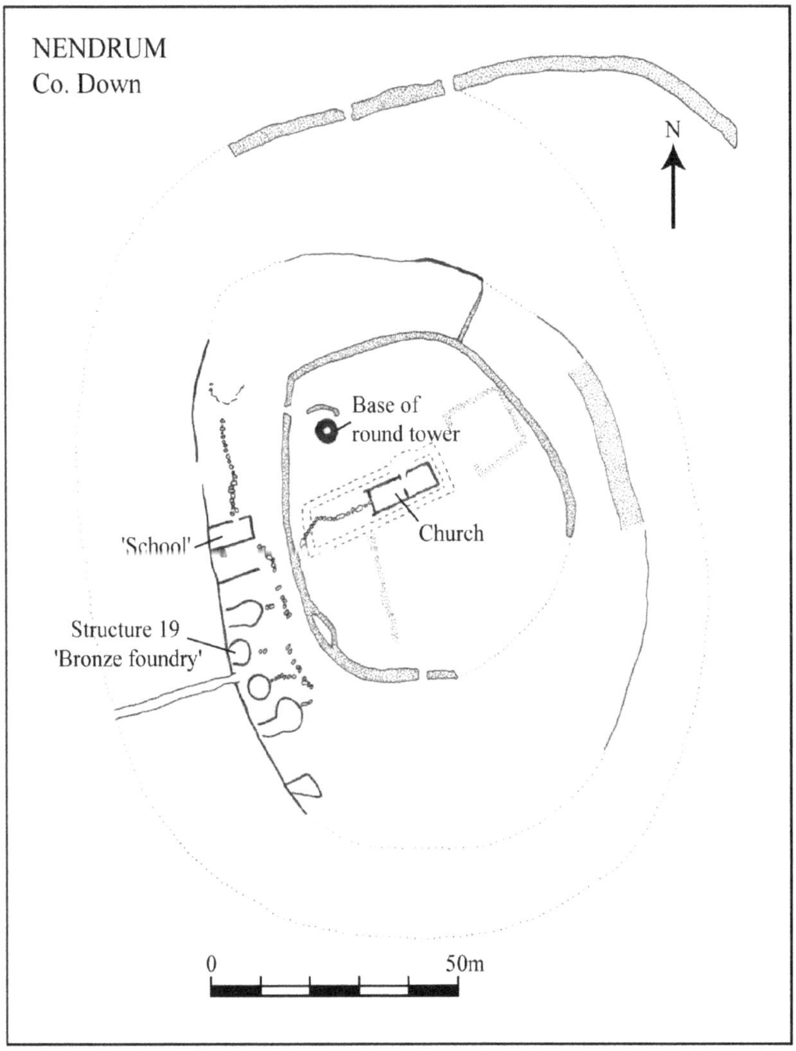

Fig. 89 Plan of Nendrum (after Lawlor1925).

9. NENDRUM, CO. DOWN.
Townland Mahee Island *O.S. 6"* 017
Fig. 89 *Nat. grid* J 5244/6363

Site summary
The monastic site of Nendrum was probably founded by St. Mochaoí before the end of the fifth century and survived into the tenth. The enclosure consists of three concentric stone ramparts, the central one enclosing an area of approximately one acre. Within the latter were found the remains of numerous burials, a round tower, a church and several cross-inscribed gravestones. The remains of six or seven stone structures were located inside the middle enclosure. These included, what the excavator termed, the 'school' which produced a large number of motif-pieces, and a non-ferrous workshop. Evidence of ironworking was found outside the middle rampart on the north of the site.

Metalworking features
Lawlor identified hut 19 (see Fig. 89), in the middle enclosure, as a bronze 'foundry'. It consists of a sub-circular stone structure abutting the middle rampart and it produced the relevant finds. It is located relatively near to the 'school' and may have had some connection with it. To the north of the 'school' was found a rectangular pit, dug into the ground and stone-lined. It measured 0.76 x 0.46m and the stones showed signs of burning. The purpose of the feature is uncertain, however it may have been used in metalworking.

Relevant artefacts
6 complete clay crucibles, numerous clay crucible sherds, 3 stone flat-bottomed vessels, stone mould for cresset lamps, stone mould for pin-heads, a number of polishing stones, approximately 30 motif-pieces, circular bronze ingot, iron pliers, arm of an iron compass, 4 iron styli.

Comment
Nendrum was clearly an important and well organised monastic site in Early Historic Ireland. Unfortunately, it was excavated in the early 1900s and Lawlor's publication in 1925 lacks detail and clarity.

Reference
Lawlor, H.C. 1925 *The Monastery of Saint Mochaoí of Nendrum*. Belfast; Belfast Natural History and Philosophical Society.

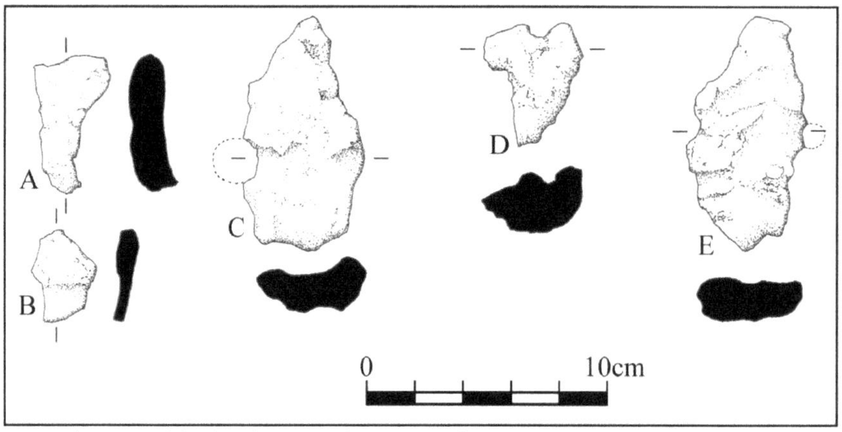

Fig. 90 A–B clay crucible sherds, C–E clay tuyère fragments from Reask (after Fanning 1981).

10. REASK, Co. KERRY.
Townland Reask — *O.S. 6"* 042
Figs. 36 and 90–91 — *Nat. grid* 03668/10435

Site summary
Excavations at Reask were undertaken to establish the nature and extent of the archaeological remains prior to conservation and preservation. The monastic site is enclosed by a strong cashel-like wall, the interior divided into a domestic and a sacred area (southern part of enclosure) by a stone wall. The enclosure is roughly oval in shape and measures 43–45m in diameter. The stone remains of six beehive huts, one rectangular structure, and a small rectangular oratory were identified as well as a lintel grave cemetery. The site is dated from the fifth to the seventh century AD and produced evidence of both iron and non-ferrous working.

Metalworking features
Evidence of ironworking was found in both hut G and outside huts C/D which are on opposite sides of the site. Hut G also produced the finds listed below which would suggest non-ferrous metalworking was also undertaken here. Two features within hut G may have been used in bronze-working.

A shallow pit and associated drain feature appear to have initially been used as a furnace for ironworking and then re-used as a forging hearth and/or a heating hearth. The latter is represented by two layers of charcoal and ash. A later and deeper pit with associated gully feature just inside the southern wall of the hut was used in a similar fashion.

Relevant artefacts
8 clay crucible fragments (2 from a flat-bottomed vessel, 1 from a pyramidal vessel), 5 clay tuyère fragments, possible stone ring-mould fragment.

Comment
The comparatively small size of the site probably explains the occurrence of both ferrous and non-ferrous metalworking evidence in the same area. Extra-mural activity at Reask is represented by a small corn-drying kiln built against the outer wall face of the main enclosure. The site could have equally had extra-mural metallurgy area.

Reference
Fanning, T. 1981 'Excavation of an Early Christian cemetery and settlement at Reask, Co. Kerry.' *Proceedings of the Royal Irish Academy* 81C, 67–172.

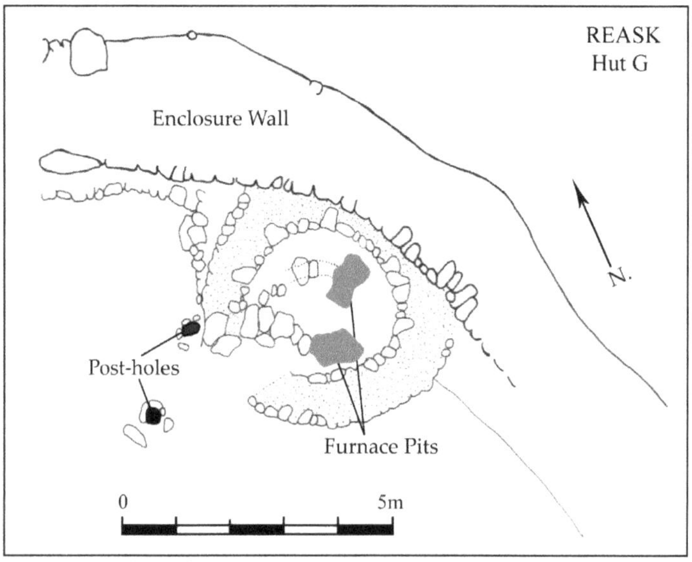

Fig. 91 Plan of structure G at Reask (after Fanning 1981).

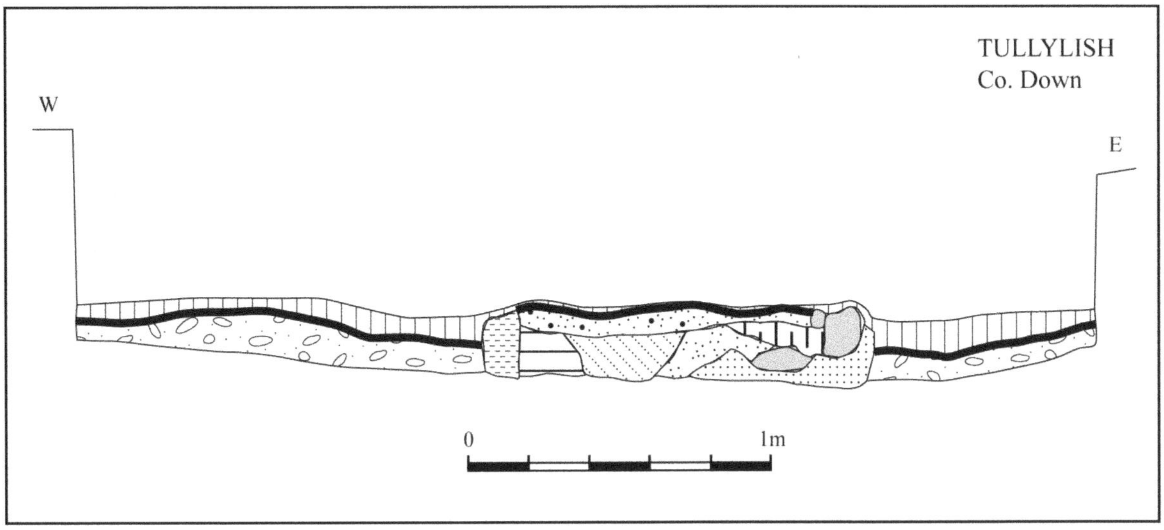

Fig. 92 Section through hearth, south face (after Ivens 1987).

11. TULLYLISH, Co. DOWN.
Townland Tullylish *O.S. 6"* 026
Figs. 92–93 *Nat. grid* J 0830/4858

Site summary
Rescue excavations at Tullylish were undertaken prior to construction work and revealed a small section of a large monastic enclosure, at least 120m in diameter, which may have started out as a large ringfort. The site is enclosed by two large ditches with the remains of a bank between, and was possibly founded in the sixth/seventh century AD. The area excavated lies to the north of the existing graveyard and church remains. It consists of a section of the bank and two ditches on the north-west of the enclosure. Despite the absence of clear dating, four major phases were identified, beginning with the digging of the inner ditch (late Iron age/Early Historic period). This was followed by undated occupation associated with the largely infilled ditch. Next, the outer ditch was constructed and the industrial area in the inner ditch used (Early Historic, no definite date). Finally, the outer ditch was infilled (fourteenth century).

Metalworking features
Industrial debris on the site is associated with trenches 2 and 3 and is roughly contemporary with the construction of the outer ditch. The remains of a timber workshop were uncovered in trench 2 and consisted of stone-packed postholes and stone settings. This structure was built either in the ditch or over it. A layer of charcoal-rich earth accumulated within the structure while it was in use. Most of trench 2 was then infilled, while industrial activity continued on its eastern margin and in trench 3. Trench 3 was covered with thick spreads of charcoal-rich earth containing considerable amounts of slag of an unspecified nature. These spreads centred around a large rectangular hearth which consisted of a sunken fire-pit edged with kerb stones. The hearth had been cleared out many times resulting in several lenses of ash in its immediate vicinity. Adjacent to this hearth were located several 'hotplates'. These consisted of alternate layers of charcoal and pads of sandy clay. The upper pad of one hotplate was baked to a yellow-red colour and formed a shallow, flat-based, vertical-sided dish.

Relevant artefacts
9 clay crucible fragments, 25 small clay mould fragments.

Comment
The location of the industrial area right on the edge of the enclosure reflects the nature of the activities undertaken there. The large quantities of slag found around the hearth in trench 3 might suggest that it was more than a simple hearth. The hotplates could have been used in the indirect heating of objects or vessels.

Reference
Ivens, R.J. 1987 'The Early Christian monastic enclosure at Tullylish, Co. Down.' *Ulster Journal of Archaeology* 50, 55–121.

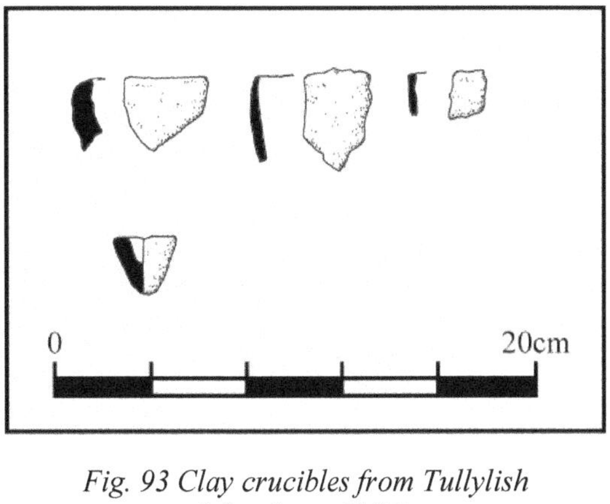

Fig. 93 Clay crucibles from Tullylish (after Ivens 1987).

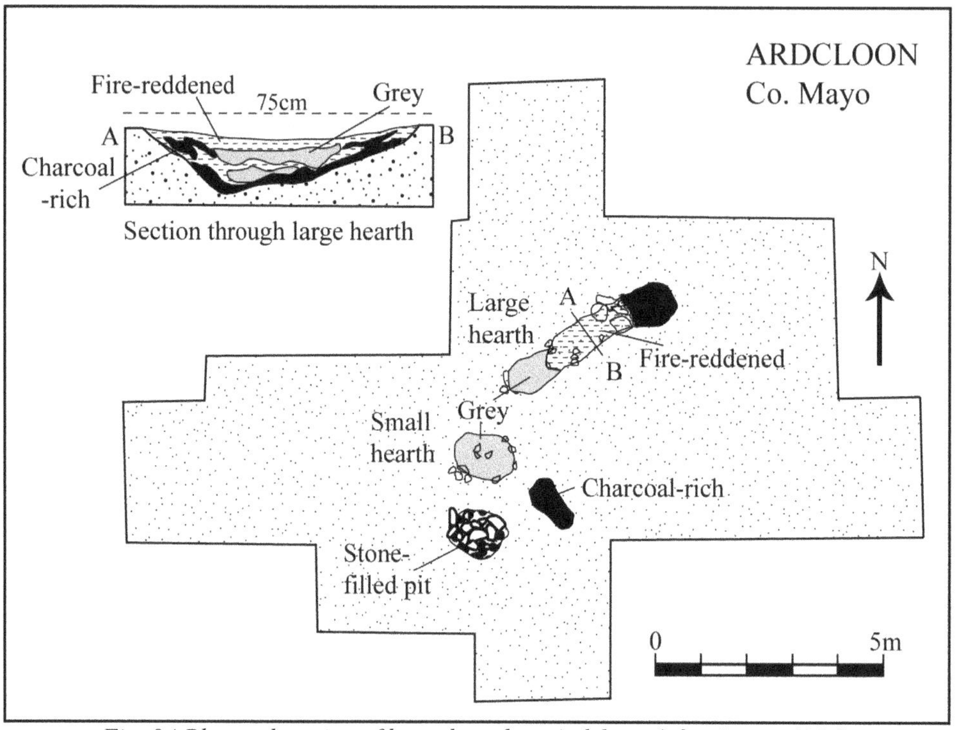

Fig. 94 Plan and section of large hearth at Ardcloon (after Rynne 1956).

12. ARDCLOON, Co. MAYO.
Townland Ardcloon *O.S. 6"* 070
Figs. 43 and 94 *Nat. grid* 12687/29739

Site summary
Rescue excavations uncovered a ringfort, 58m in diameter, with no definite hut or house structures distinguishable in the interior. Two hearths and a stone-lined pit were identified in the centre of the enclosure. The site was constructed and inhabited during the late seventh/early eighth century AD (dated by the discovery of a bronze pin of that date).

Metalworking features
A large hearth or furnace was found in the interior, in a pit measuring 4.5m in length by 0.7m in width. This pit was dug into the ground, increasing in depth to 0.18m at the north-east end. This deep area was divided from the rest by large flat stones and its fill contained charcoal.

Relevant artefacts
3 pieces of lead ore.

Comment
Two long, thin trenches were excavated along the north-south and east-west diameters of the ringfort. As metalworking areas tend to occur along the edges of a site such evidence may easily have been missed. Alternatively, it is possible that this site dealt with only one of the non-ferrous metals, lead. Perhaps the central furnace/hearth was used for lead smelting, the three pieces of ore found adjacent to it.

Reference
Rynne, E. 1956 'Excavation of a ring-fort at Ardcloon, Co. Mayo.' *Journal of the Royal Society of Antiquaries of Ireland* 86, 203–214.

13. BALLYCATTEEN, Co. CORK.
Townland Ballycatteen *O.S. 6"* 124
Fig. 42 *Nat. grid* 15820/04592

Site summary
A larger than average ringfort producing evidence of successive houses in the interior. The excavator interpreted the small amount of habitation refuse as indicative of intermittent activity on the site. Within the area excavated twelve hearths were investigated, two of which were located on the margins of the site. The site is dated *c.*600 AD.

Metalworking features
The discovery of iron slag and formless scraps of iron in some of these is indicative of ironworking. Unfortunately there is no comparable non-ferrous evidence.

Relevant artefacts
9 clay crucible fragments.

Comment
The excavation of Ballycatteen was undertaken because of its comparison with Garranes (Ó Ríordáin 1942). It is, in fact, larger than Garranes although both have trivallate defences. Garranes measures 110m in external diameter, Ballycatteen 119m. The material evidence from Garranes, however, reflects a higher status than that from Ballycatteen.

Reference
Ó Ríordáin, S.P. and Hartnett, P.J. 1943 'The excavation of Ballycatteen Fort, Co. Cork.' *Proceedings of the Royal Irish Academy* 49C, 1–44.

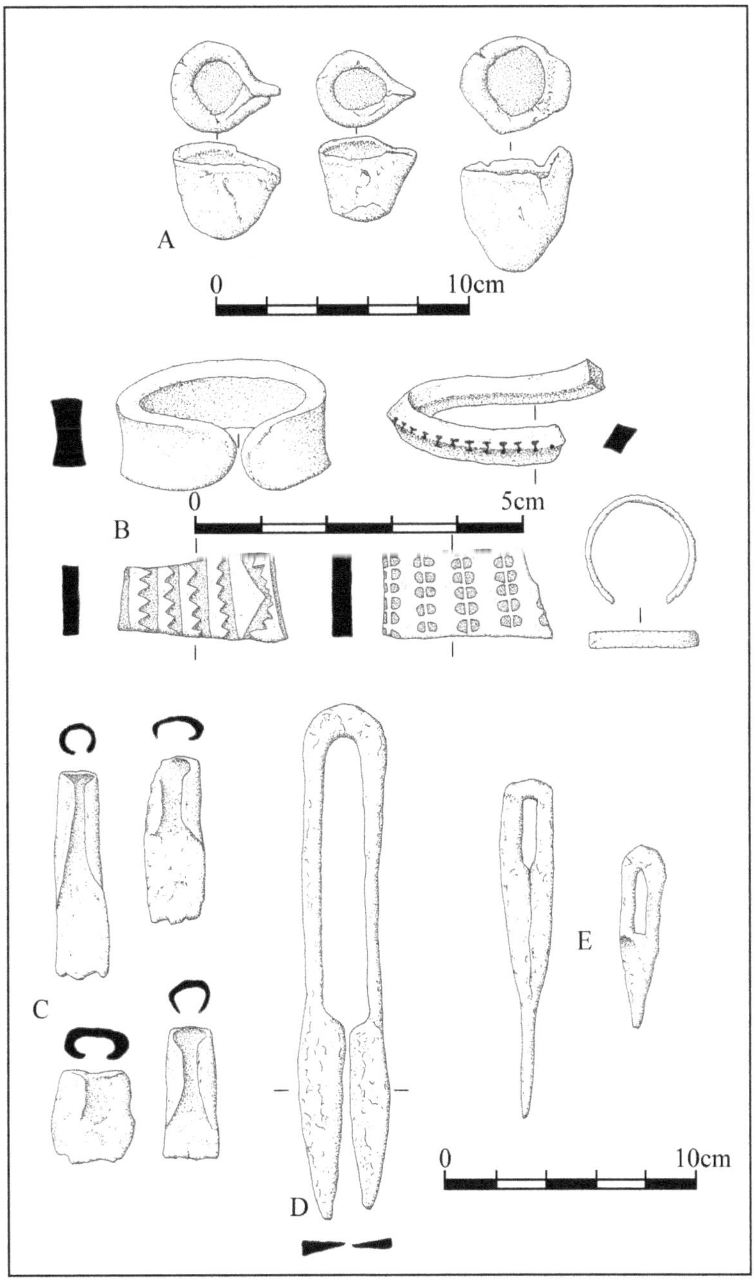

Fig. 95 Finds from Carraig Aille; A clay crucibles, B silver hoard, C iron socketed and pronged tools, D iron shears, E iron pointed and slotted tools (after Ó Ríordáin 1949).

14. CARRAIG AILLE II, Co. LIMERICK.
Townland Lough Gur *O.S. 6"* 032
Figs. 44 and 95–96 *Nat. grid* 16534/14063

Site summary
A large cashel located on a ridge, 48m in diameter with ramparts 4m thick and up to 2m high originally. The excavation revealed a series of houses, hearths and stone pavements in the interior. The site is dated from the eighth to the late tenth/early eleventh century AD. Evidence of iron-working was found on the site as well as indications of non-ferrous metalworking.

Metalworking features
Nine hearths were discovered around the internal edges on the fort, some of which may have been used for metalworking. There was no distinct concentration of finds to indicate a metalworking area.

Relevant artefacts
3 complete handled clay crucibles, 50 clay crucible fragments, 2 hammer stones, 3 strands of twisted bronze wire, a length of bronze wire, 3 fragments of bronze cake, waste bronze, 3 pointed and slotted iron tools, 4 iron chisels, 2 iron awls, 1 iron punch, pronged iron tool, iron shears, Viking silver hoard (ring, 3 bracelet fragments, ingot, ingot fragment), silver ingot, silver ring.

Reference
Ó Ríordáin, S.P. 1949 'Lough Gur excavations; Carraig Aille and the 'Spectacles'.' *Proceedings of the Royal Irish Academy* 52C, 6–111.

Fig. 96 Silver hoard from Carraig Aille (photo; Dept. of Archaeology, NUI, Galway).

15. CASTLESKREEN, Co. DOWN.
Townland Castleskreen *O.S. 6"* 037
Nat. grid J 0473/0403

Site summary
Castleskreen is a univallate circular rath, with a diameter of approximately 35m. The rath may have been occupied as early as the seventh century while in the late twelfth century the site was re-built for use as a motte. Limited excavation identified three distinct phases. The first phase is marked by the digging of a ditch approximately 2m deep. In phase 2 an earthen bank with an interior palisade was added. Both phases 1 and 2 are dated to the Early Historic period. Phase 3 sees the construction of the motte and is dated to the end of the twelfth century. Occupation appears to have been continuous though evidence from the lower levels is scarce due to damage caused in the building of the motte. The crucible fragments came from phase 2.

Metalworking features
None identified.

Relevant artefacts
Clay crucible fragments of the pyramidal type vessel.

Reference
Dickinson, C.W. and Waterman, D.M. 1959 'Excavation of a rath with motte at Castleskreen, Co. Down.' *Ulster Journal of Archaeology* 22, 67–82.

Dickinson, C.W. and Waterman, D.M. 1960 'Excavations at Castleskreen, Co. Down.' *Ulster Journal of Archaeology* 23, 63–77.

16. CATHAIR FIONNÚRACH, Co. KERRY.
Townland Ballnavenooragh *O.S. 6"* 034
Nat. grid 04293/11070

Site summary
A circular cashel, 25.5m in internal diameter, enclosing a drystone house comprising two conjoined circular rooms. The enclosure wall survives to a height of 2.5m and width of 3.5m, an entrance on the west. A shallow ditch runs along the interior of the enclosure wall, with a wall niche on the north-west, and a collapsed wall chamber on the south-east. The site was first investigated by William Darcy in 1893, with excavations in 1994–1997 concentrating on the entrance area and the larger room of the house (6.5m in internal diameter). This room was at least partly paved and had a drain, two hearths, two pits and a possible anvil stone. The smaller room, 4.5m in internal diameter, provided access to an L-shaped souterrain, and also had a hearth. The site was used possibly as early as the sixth century AD and continued to be used, in some form, into the thirteenth century.

Metalworking features
Possible anvil stone in the centre of the larger room, with furnace pit immediately south of it, and charcoal spreads in the vicinity.

Relevant artefacts
Fragments of clay crucible.

Comment
While the clay crucible fragment points to non-ferrous metalworking at the site, the lack of published information prevents the definite association of the anvil and furnace with this activity.

Reference
Gibbons, E. 1995 'Cathair Fionnúrach, Ballnavenooragh.' In *Excavations* 1994, 42–3.

17. FELTRIM HILL, Co. DUBLIN.
Townland Feltrim *O.S. 6"* 012
Nat. grid 32006/24456

Site summary
An oval cashel measuring 25m by 35m, situated on the crest of a hill, the west end of which was quarried 1946–7. This led to a full-scale rescue excavation in 1947. The interior of the cashel was divided into two levels, with the main habitation located on the lower level. Evidence of this included occupation debris and several hearths, though no traceable house plans. The site is dated from the eighth to the tenth century AD.

Metalworking features
None identified.

Relevant artefacts
Iron chisel, working end of a similar chisel, possible compass leg, silver ingot.

Reference
Hartnett, P.J. and Eogan, G. 1964 'Feltrim Hill, Co. Dublin: A Neolithic and Early Christian site.' *Journal of the Royal Society of Antiquaries of Ireland* 94, 1–38.

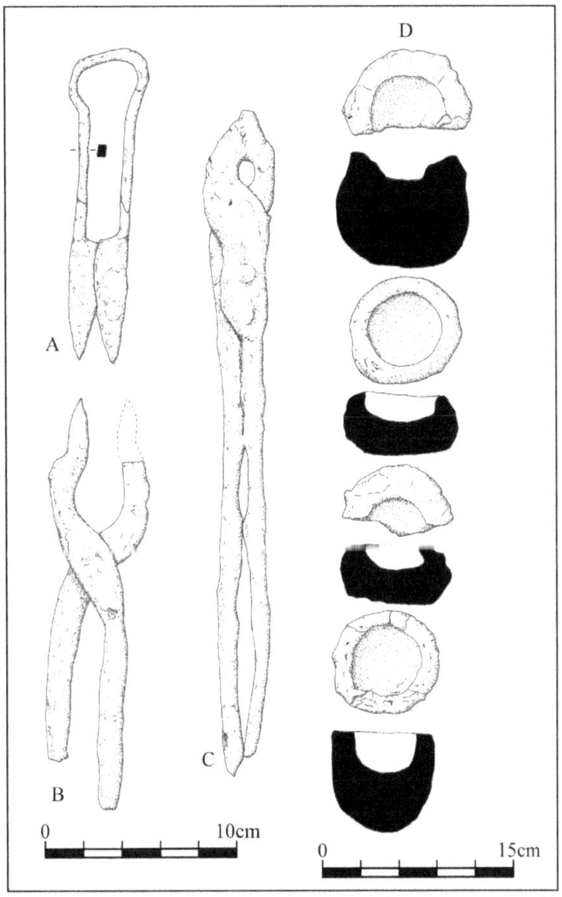

Fig. 97 Finds from Garranes; A iron shears, B–C iron pincers/tongs, D crucibles (after Ó Ríordáin 1942).

18. GARRANES, Co. CORK.

Townland Garranes *O.S. 6"* 084
Figs. 39 and 97 *Nat. grid* 14733/06400

Site summary

Garranes is an impressive trivallate ringfort, tentatively identified as *Rath Raithleann* the seat of a branch of the ruling Eóganacht dynasty. It measures 110m in external diameter. The excavation revealed a well defended entrance consisting of a complex series of gates. The site is dated to the late fifth/early sixth centuries AD though, unfortunately, only a portion of the interior was excavated. Numerous postholes were found in the interior though no definite house plans could be identified from these. One of the largest areas excavated, Site D, proved to be a metalworking area leading the excavator to suggest that the site had been exclusively occupied by a group of craftsmen.

Metalworking features

Site D is located on the south side of the fort, along the edge of the innermost bank. It is marked by a dense black layer containing charcoal and the relevant finds. This layer is 0.15m–0.43m thick and measures approximately 34m by 7m. Several postholes were identified though there were no discernible structures. The possible remains of a hut, however, were identified, a setting of stones forming an irregular arc of a circle. Its location in a clearly industrial area might suggest its use as a workshop.

Relevant artefacts

39 complete clay crucibles, 2500 clay crucible fragments (pyramidal and flat-bottomed vessels),1 clay tuyère, vitrified clay fragments possibly from a furnace, 30 clay moulds or fragments of (for rings and pins), 6 (at least) stone ingot moulds, 1 possible stone crucible, bronze pin fresh from mould, unfinished bronze pin-head, unfinished rectangular bronze object, fragment of a bronze casting, bronze casting jet, a length of bronze wire, bronze ingot (from stone mould no.1), much scrap and waste bronze, 2 iron pincers, 1 iron shears, 3 iron awls, 1 lead ring, 2 lumps of tin.

Comment

Both the size and trivallate nature of Garranes, in association with the literary references, indicate the high status of the site. However, they also argue against the site having been the exclusive home of a community of craftsmen. Metalworkers were held in high regard in Early Historic Ireland but they were not of such high standing as to allow them control over such an important site. It must also be remembered that a large portion of the interior remains unexcavated thus preventing a balanced view of the site and its economy.

Reference

Ó Ríordáin, S.P. 1942 'The excavation of a large earthen ringfort at Garranes, Co. Cork.' *Proceedings of the Royal Irish Academy* 47C, 77–150.

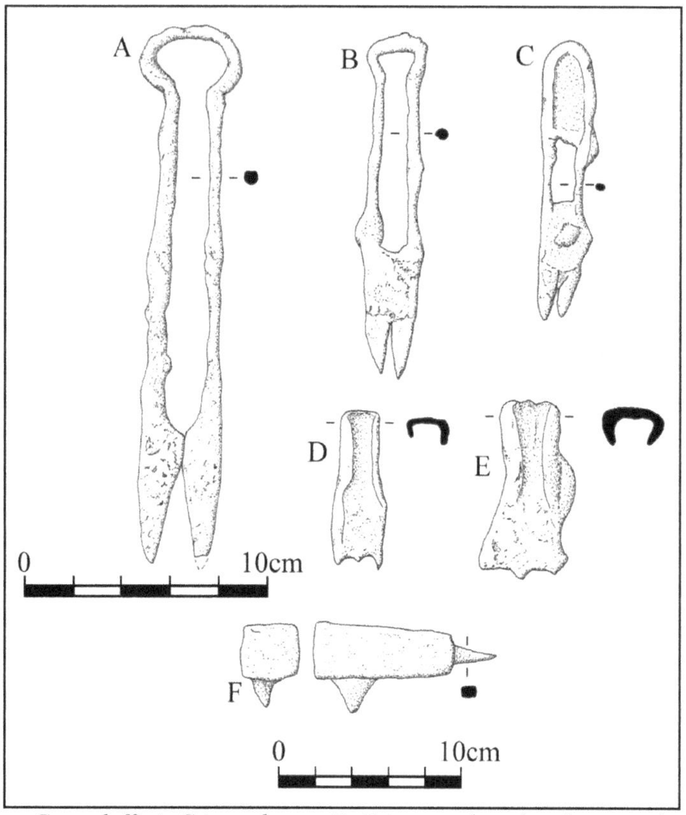

Fig. 98 Finds from Garryduff; A–C iron shears, D–E iron socketed and pronged tools, F iron anvil (after O'Kelly 1963).

19. GARRYDUFF, Co. CORK.

Townland Garryduff — *O.S. 6" map* 055

Figs. 40 and 98–99 — *Nat. grid* 19249/08633

Site summary

Garryduff is a cashel consisting of a large stone rampart and external rock-cut fosse measuring 20m in internal diameter. The excavation of the interior uncovered the remains of two houses, several hearths and an iron-working furnace. Two phases of occupation were indicated by two black layers which were in places clearly separated from each other. The two phases are separated by a collapse of the bank, the majority of relevant evidence coming from the earlier phase. A seventh/early eighth century date is suggested for the site.

Metalworking features

Two paved areas in the south of the fort were associated with two hearths and several post-holes, a possible metalworking area. Evidence of ironworking on the site consisted of a smelting furnace, iron slag, and stocks of iron ore. These were found on the northern edge of the site, directly opposite the possible non-ferrous working area.

Relevant artefacts

3 complete crucibles, 24 crucible fragments, fragment of a clay 'blow-pipe', 10 stone motif-pieces, bronze tweezers, a length of bronze wire, 2 fragmentary iron shears, 4 complete iron shears, 9 iron awls, 2 iron chisels, 3 iron dividers/callipers,10 iron socketed and pronged tools, 1 iron anvil, 2 iron ladles.

Comment

Although Garryduff is not very large in diameter, its high status is evident in the careful and impressive construction of its rampart and rock-quarried ditch.

Reference

O'Kelly, M.J. 1963 'The excavations of two earthen ringforts at Garryduff, Co. Cork.' *Proceedings of the Royal Irish Academy* 63C, 17–150.

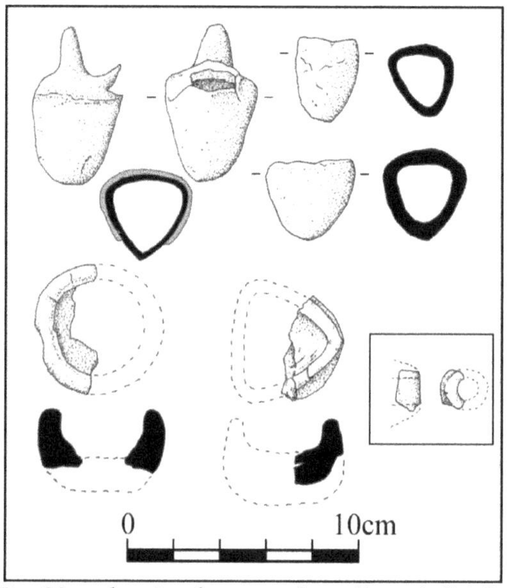

Fig. 99 and tuyère fragment from Garryduff (after O'Kelly 1963).

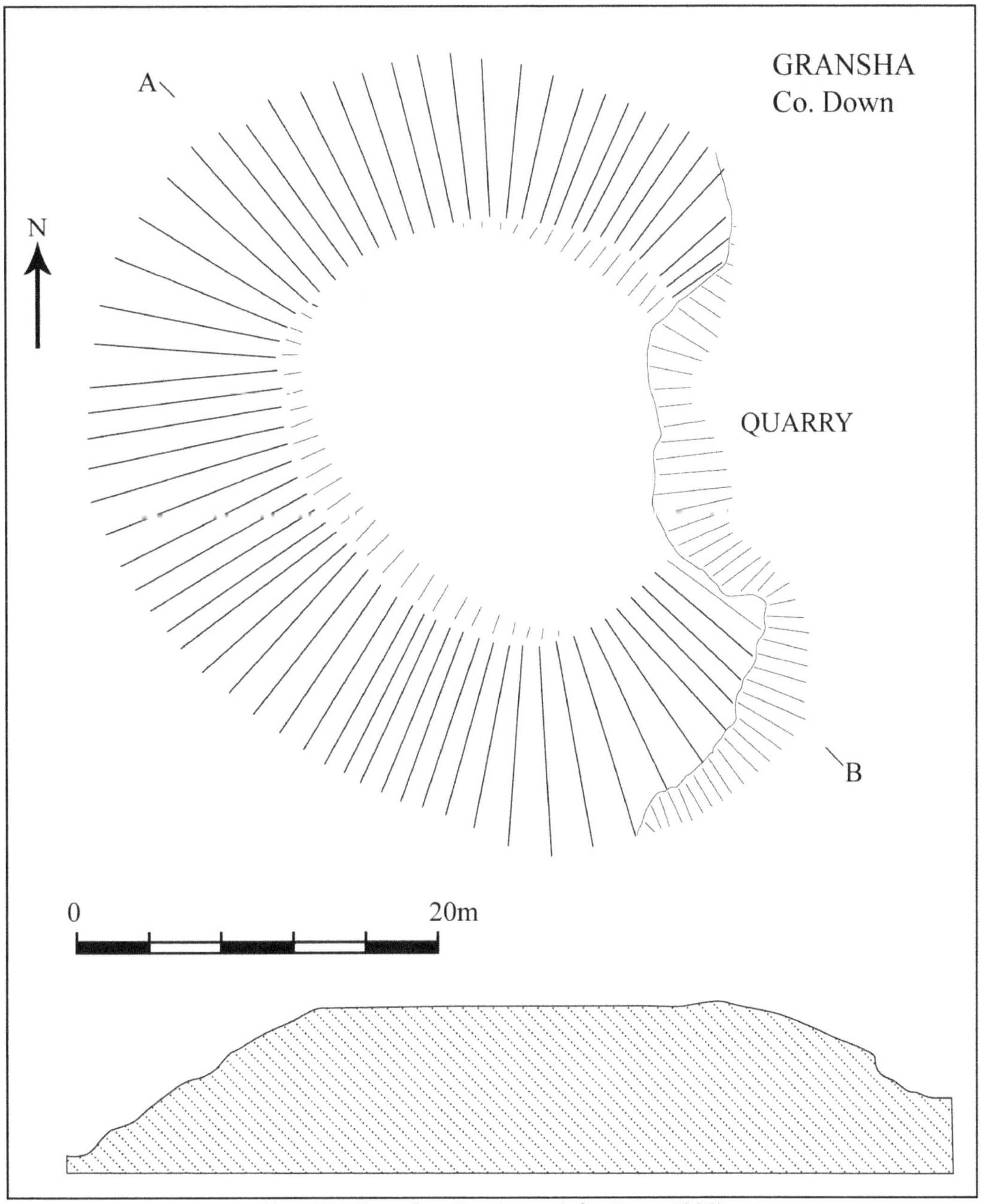

Fig. 100 Gransha, Co. Down (after Lynn 1985).

20. GRANSHA, Co. DOWN.
Townland Gransha
Fig. 100 *Nat. grid* J 531/769

Site summary
A raised ringfort built on a natural gravel hillock, artificially raised to 6.7m in height. The summit measures 22m in diameter. A substantial portion of the site was quarried away, with limited rescue excavations in 1972 and again in 1982. Three phases of activity were identified. Phase 1 saw the utilisation of the natural mound, originally *c.*35m in diameter. It was defined by the trace of low bank with a burned timber palisade and ditch outside it. Little occupation material was recovered, the phase dated to the seventh century. Phase 2 was also enclosed by the trace of a bank, with possible structural remains within, dated to the early ninth century. Phase 3 was marked by a dramatic re-modelling of the ringfort, into a large, steep-sided and flat-topped mound in the tenth century. Remains included the stone footings of a rectangular building, two hearths and approximately thirty pits.

Metalworking features
None identified

Relevant artefacts
Bronze scraps, clay moulds, tweezers, *c.*40 shale motif-pieces, stylus, slotted and pointed tool, hones.

Reference
Lynn, C.J. 1985a 'Excavations on a mound at Gransha, Co. Down, 1972 and 1982: an interim report.' *Ulster Journal of Archaeology* 48, 81–90.

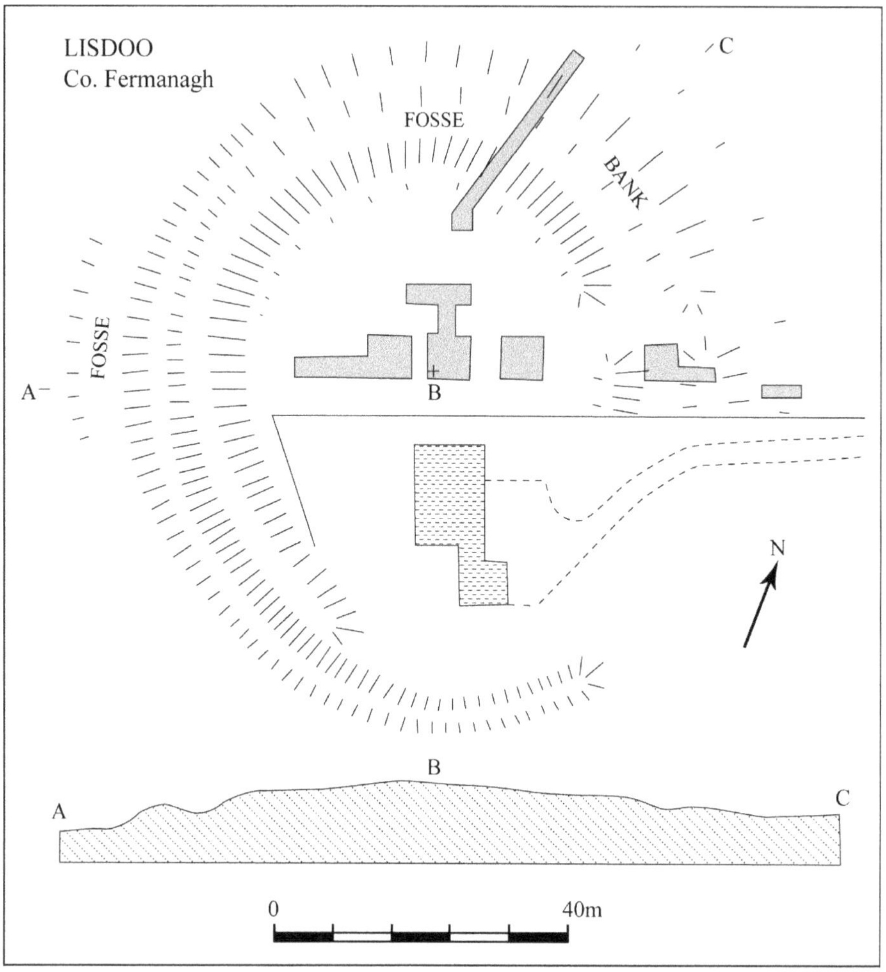

Fig. 101 Excavated areas at Lisdoo (after Brannon 1981–1982).

21. LISDOO, Co. FERMANAGH.
Townland Lisnaskea *Irish grid map* 246
Fig. 101 *Nat. grid* H 363 332

Site summary
Bivallate ringfort located on the summit of a small hill, southern half destroyed by building in 1977. Rescue excavation undertaken prior to further construction. 104m in external diameter, 5m-wide entrance on north-east. Inner ditch nearly 2m deep with a maximum width of 8m. The ditch produced a radiocarbon date of 1655±45 BP.

Metalworking features
None identified.

Relevant artefacts
6 crucible sherds, iron awl.

Reference
Brannon, N.F. 1981-2 'A rescue excavation at Lisdoo fort, County Fermanagh.' *Ulster Journal of Archaeology* 44–45, 53–59.

22. LISDUGGAN, Co. CORK.
Townland Lisduggan North *O.S. 6"* 023
Nat. grid 14274/10327

Site summary
Prior to quarry development, rescue excavations were undertaken at three ringforts on the south-facing slope of Knocknanuss Hill. Ringforts 1 and 2 were excavated in 1972, ringfort 3 in 1973. Ringforts 1 and 2 were adjacent, their ditches possibly conjoined, both *c.*50m in diameter. The remains of three houses and three hearths were uncovered in the south-west quadrant of Ringfort 1, while Ringfort 2 produced no structural evidence. It did, however, produce the relevant non-ferrous metalworking evidence. Univallate Ringfort 3 was situated 500m south-west of Ringforts 1 and 2, measured 45m in diameter and enclosed a central rectangular structure. All three are roughly dated to the seventh century AD.

Metalworking features
None identified.

Relevant artefacts
3 whetstones, clay crucible fragment, small unworked piece of amber.

Reference
Twohig, D.C. 1990 'Excavation of three ringforts at Lisduggan North, Co. Cork.' *Proceedings of the Royal Irish Academy* 90C, 1–32.

23. MARLINSTOWN, Co. WESTMEATH.
Townland Marlinstown *O.S. 6"* 019
Nat. grid 04634/05275

Site summary
A four-week rescue excavation was undertaken prior to road construction. Oval enclosure atop a natural glacial knoll. Measures 52m by 46m enclosed by an earthen bank and external ditch, with some evidence of an internal ditch. The bank measured 1m in maximum surviving height, the outer ditch 1.8m deep. Six burials were excavated in the south-west quadrant, all extended east-west, with no grave goods.

Metalworking features
None identified.

Relevant artefacts
Clay crucible fragments.

Reference
Keeley, V.J. 1990 'Marlinstown.' In I. Bennett (ed.), 55.

24. RAHEENNAMADRA, Co. LIMERICK.
Townland Raheennamadra *O.S. 6"* 041
Figs. 102–103 *Nat. grid* 17338/12976

Site summary
A large earthen ringfort 38–40m in external diameter enclosed by a low bank and shallow fosse. The excavation in the interior revealed the remains (post-trenches) of a round structure 6-7m in diameter. A large hearth pit was found in the interior of the structure, measuring 1m x 0.7m and 0.2m deep. The site was dated to approximately 650 AD. Also associated with this hut structure was a souterrain.

Metalworking features
None identified.

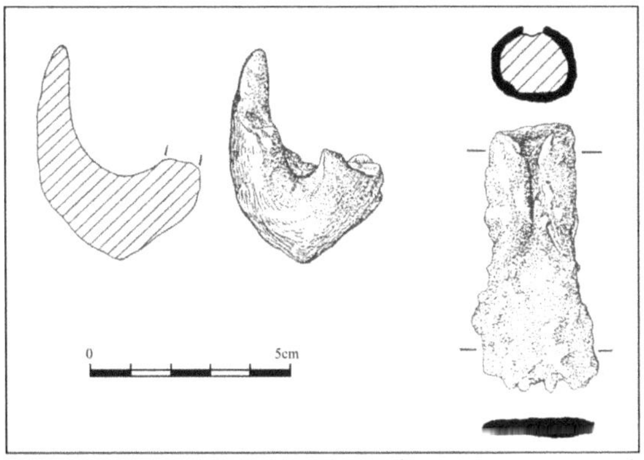

Fig. 102 Crucible fragment and iron pronged tool from Raheennamadra (after Stenberger 1967).

Relevant artefacts
Clay crucible fragment of the pyramidal type, an iron socketed and pronged implement.

Comment
Non-ferrous working at this site does not appear to have been very substantial though it is possible that the evidence may not have survived.

Reference
Stenberger, M. 1967 'A Ringfort at Raheennamadra, Knocklong, Co. Limerick.' *Proceedings of the Royal Irish Academy* 65C, 37–54.

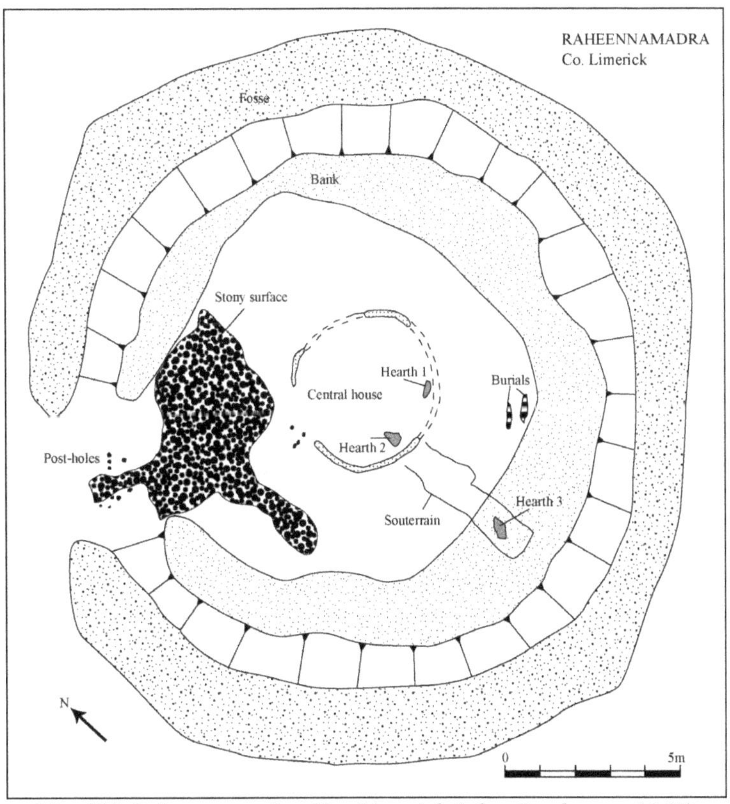

Fig. 103 Raheennamadra, Co. Limerick (after Stenberger 1967).

Fig. 104 Aerial view of Rathgurreen, Co. Galway.

25. RATHGURREEN, Co. GALWAY.

Townland Cottage *O.S. 6" map* 095

Figs. 104–105 *Nat. grid* 13787/22002

Site summary

Excavations by M.V. Duignan in 1948 and 1949 identified a two-phase bivallate ringfort of the Early Historic Period at Rathgurreen, Co. Galway. Approximately one quarter of the site was dug, comprising the south-western quadrant (interior) and the entrance area. Well-preserved dumps of animal bone an sea-shells were discovered throughout the excavated area, with a number of different crafts (such as the preparation of purple dye) represented in the artefactual assemblage. The latter also included a fragment of an oil lamp of Roman origin. Two main phases of occupation were identified, though evidence of Phase I was minimal. Radiocarbon analyses date the occupation of the site from the fourth or fifth century AD to at least the ninth century.

Metalworking features

The remains of a smelting furnace were identified in the innermost ditch enclosing the site on the south. Finds of iron-smelting slag within the fill suggests that the furnace was used for iron smelting, though this does not preclude the possibility of non-ferrous metalworking in this area also.

Relevant artefacts

19 clay crucible sherds, 6 vitrified ceramic fragments, possible tuyère fragment.

Comment

The surviving non-ferrous metalworking evidence dates to Phase 1 of the occupation. The crucible sherds are quite small and uninformative, probably derived from bag-shaped or pyramidal vessels and not flat-bottomed examples. No moulds were discovered, although this may be due to the limited survival of Phase 1 evidence. Bronze-working was, most likely, also undertaken in Phase 2, though perhaps located on the opposite (unexcavated) side of the site to the Phase 2 ironworking. Raw material for this work had to be imported to the site, either in the form of ingots or scrap bronze, or individual units of tin and copper (ore, ingots, cakes, scrap).

Reference

Comber, M. 2002 'M.V. Duignan's excavations at the ringfort of Rathgurreen, Co. Galway, 1948–9.' *Proceedings of the Royal Irish Academy* 102C, 137–197.

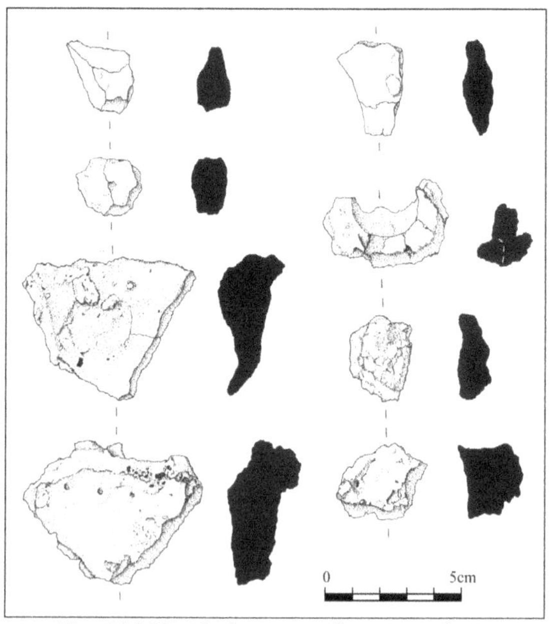

Fig. 105 Vitrified ceramic fragments from Rathgurreen.

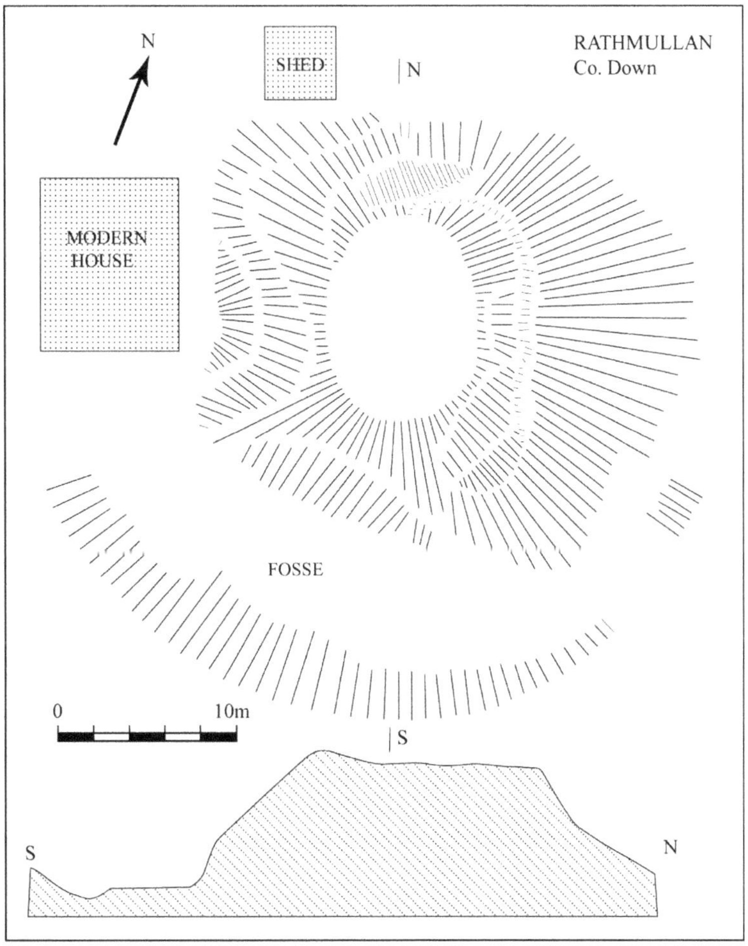

Fig. 106 Rathmullan, Co. Down (after Lynn 1982-1982).

26. RATHMULLAN, Co. DOWN.
Townland Rathmullan Lower *O.S. 6"* 044
Fig. 106 *Nat. grid* J 478/373

Site summary
A large raised rath with a small Anglo-Norman motte built on top, 8m in height and approximately 28m in diameter at base. Excavations in 1978 investigated the motte levels and one-quarter of the rath. Seven phases of activity were identified. Phases 1 and 2 were marked by the construction and use of circular houses within a ringfort, phases 3 and 4 by the use of a souterrain and rectangular structures with drystone footings on a raised ringfort. Phases 5 to 7 date to motte activity. Radiocarbon analyses dated Phase 1 mid-fifth to early seventh century AD, and Phase 3 to the ninth/tenth century AD.

Metalworking features
None identified.

Relevant artefacts
Clay crucible, bronze scraps, whetstones and hones, possible hammer-stone.

Reference
Lynn, C.J. 1981–2 'The excavation of Rathmullan, a raised rath and motte in County Down.' *Ulster Journal of Archaeology* 44–45, 56–171.

Lynn, C.J. 1985 'The excavation of Rathmullan, Co. Down: Addenda.' *Ulster Journal of Archaeology* 48, 130–131.

27. SLUGGARY, Co. LIMERICK.
Townland Sluggary *O.S. 6"* 13
Nat. grid 15601/16630

Site summary
A bivallate ringfort, polygonal in plan. Measures 40m in internal diameter (north-south), and 70m in external diameter. A 2m-wide cutting was excavated across the centre of the ringfort, prior to development.

Metalworking features
None identified.

Relevant artefacts
Small bronze strip, clay mould fragments.

Reference
Shee, E. 1974–1975 'Sluggary'. In I. Bennett (ed.), 22–23.

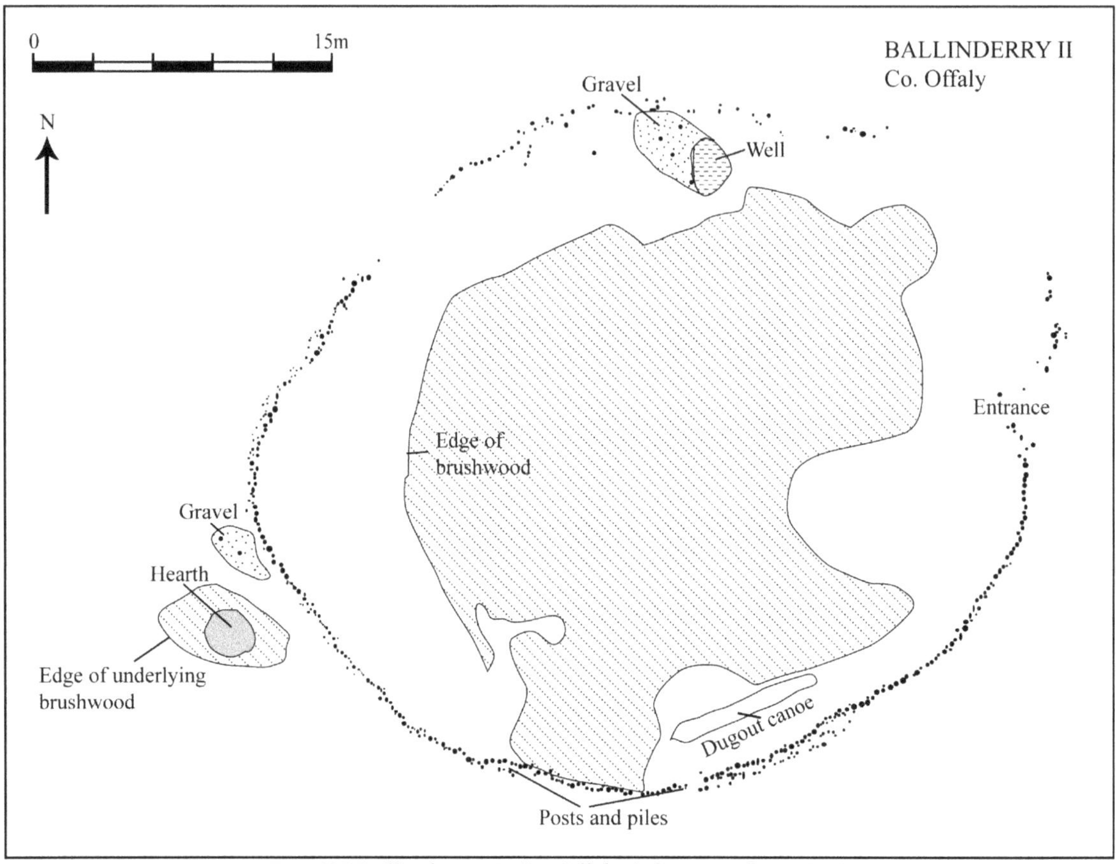

Fig. 107 Ballinderry II, Co. Offaly (after Hencken 1942).

28. BALLINDERRY II, Co. OFFALY.
Townland Ballynahinch *O.S. 6"* 01
Figs. 107–108 *Nat. grid* 22158/23898

Site summary
This crannóg, excavated by the Harvard expedition in 1933, is approximately 35m in diameter and dated somewhere between the seventh and early ninth century AD. The site began as a Late Bronze Age lakeside habitation and was abandoned when the level of the lake rose. This habitation is dated from the fourth to the first century BC. The site was reoccupied in Early Historic times when the lake levels fell again. Unlike many other crannógs, Ballinderry II is only partly artificial and is an extension of a natural island. No definite structures were identified in the interior of the Early Historic crannóg though some areas of timber flooring were uncovered.

Metalworking features
In addition to a domestic hearth found in the interior, a further hearth was located between the palisade and the water's edge. This would have been an ideal place for metalworking, keeping it separate from the domestic settlement. There was no concentration of relevant finds to indicate a possible industrial area.

Relevant artefacts
43 clay crucible fragments (majority of the pyramidal type), 8 clay mould fragments, 2 sandstone ingot moulds, 1 shale motif-piece, 2 iron ladles.

Comment
It would seem likely, from the relatively small quantity of relevant finds and the lack of an identifiable metalworking area, that non-ferrous working at Ballinderry was not carried out on a very large scale.

Reference
Hencken, H.O'N. 1942 'Ballinderry Crannóg no. 2.' *Proceedings of the Royal Irish Academy* 47C, 1–76.

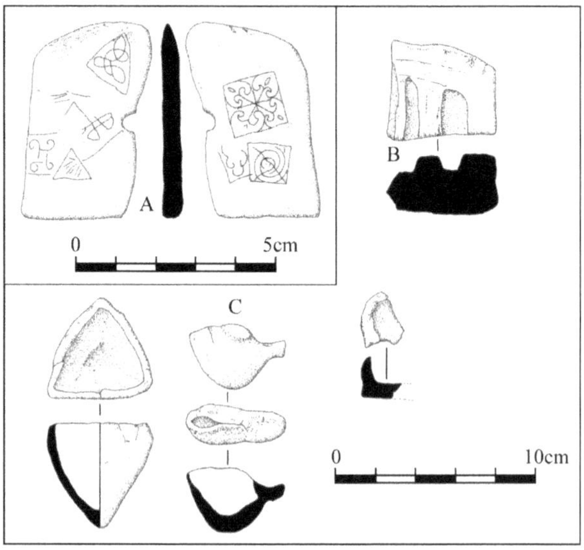

Fig. 108 Finds from Ballinderry II; A stone motif-piece, B stone mould, C crucibles (after Hencken 1942).

29. BALLYDOOLOUGH, Co. FERMANAGH.
Townland Ballydoolagh *O.S. 6"* 022
Nat. grid H 2849/4817

Site summary
An unexcavated crannóg, investigated in 1870 when it appeared above water after a period of drought. The maximum diameter of the site is 24m and the remains of a rectangular timber structure were found in the interior. This structure measured 5.2 x 5.3m. The site also produced finds of iron slag.

Metalworking features
None identified.

Relevant artefacts
1 complete clay crucible, fragments of sheet bronze.

Reference
Wood-Martin, W.G. 1886 *The Lake Dwellings of Ireland*, 181–183.

30. BALLYJAMESDUFF, Co. CAVAN.
Townland Exact location unknown *O.S. 6"* 038

Site summary
There is no available information on this site. It is not mentioned by Wood-Martin in his 1886 work. A crucible from the site was, however, examined by Moss in 1927.

Metalworking features
None recorded.

Relevant artefacts
A complete pyramidal clay crucible.

Reference
Moss, R.J. 1927 'A chemical examination of the crucibles in the collection of Irish antiquities of the Royal Irish Academy.' *Proceedings of the Royal Irish Academy* 37C, 175–193.

31. BALLYMENA, Co. ANTRIM.
Townland No modern record of the site

Site summary
An unexcavated crannóg revealed by turf-cutters. Located approximately 4 miles from the town of Ballymena, this crannóg is not exceptionally large.

Metalworking features
None recorded.

Relevant artefacts
Clay crucible fragments, pointed and socketed iron tool, possible stone ingot mould.

Reference
Wood-Martin, W.G. 1886 *The Lake Dwellings of Ireland*, 169.

32. CLEA LAKES, Co. DOWN.
Townland Clea Lakes *O.S. 6"* 024
Nat. grid J 509/549

Site summary
Limited excavations only were undertaken at Clea Lakes in 1956. Located approximately 183m from the shore, the crannóg substructure consists of a mound of stone, artificially deposited peat, sand and clay. The stone footing of a building was uncovered in the interior.

Metalworking features
None identified.

Relevant artefacts
2 clay crucible fragments, 1 socketed iron gouge, whetstone, bronze fragment.

Reference
Collins, A.E.P. and Proudfoot, B. 1959 'A trial excavation in Clea Lakes Crannóg, Co. Down.' *Ulster Journal of Archaeology* 22, 92–102.

33. CLONES, Co. MONAGHAN.
Townland No modern record of site

Site summary
Two crannógs were 'excavated' near Clones around 1900 and are both located near the edge of Drumacrittin lake. No structural details are recorded in the antiquarian report.

Metalworking features
None recorded.

Relevant artefacts
Crannóg no.1; 1 fragment of a clay crucible, 1 sandstone ingot mould, 1 slate motif-piece, 2 iron chisels, 1 iron hammer-head, 1 wooden mallet. Crannóg no. 2; 2 possible iron files, 1 small fragment of tin.

Comment
D'Arcy's work at Clones is typical of such antiquarian investigations. He recovered and described the artefacts from the two crannógs, however he recorded no structural details.

Reference
D'Arcy, S.A. 1900 'An account of the excavation of two lake-dwellings in the neighbourhood of Clones.' *Journal of the Royal Society of Antiquaries of Ireland* 30, 204–236.

34. CLONEYGONNELL, Co. CAVAN.

Townland Clonagonnell — *O.S. 6"* 025
Nat. grid 23979/30291

Site summary
The crannóg was investigated in 1862 just after a section of it had been cut by the railway. The crannóg was found to have a maximum diameter of 37m while the interior had a timber surface.

Metalworking features
None recorded.

Relevant artefacts
1 stone ingot mould, 1 possible slate mould with 3 circular cavities, 1 hammer-head, several pieces of slag (of unknown nature), 4 clay crucibles.

Reference
Wood-Martin, W.G. 1886 *The Lake Dwellings of Ireland*, 197–199.

35. CORRANEARY, Co. CAVAN.

Townland Corraneary — *O.S. 6"* 023
Fig. 109 — *Nat. grid* 26492/30534

Site summary
The crannóg was excavated by T.J. Barron in 1937 and 1938. The structure of the crannóg consists of a large mound of stone covered by a layer of timber beams and brushwood. The surface area measures approximately 9m x 11m. In the interior were located the remains of a stockade and a hearth.

Metalworking features
None identified.

Relevant artefacts
5 clay crucibles, 2 clay mould fragments.

Reference
Davies, O. 1942 'Contributions to the study of crannógs in south Ulster.' *Ulster Journal of Archaeology* 5, 14–30.

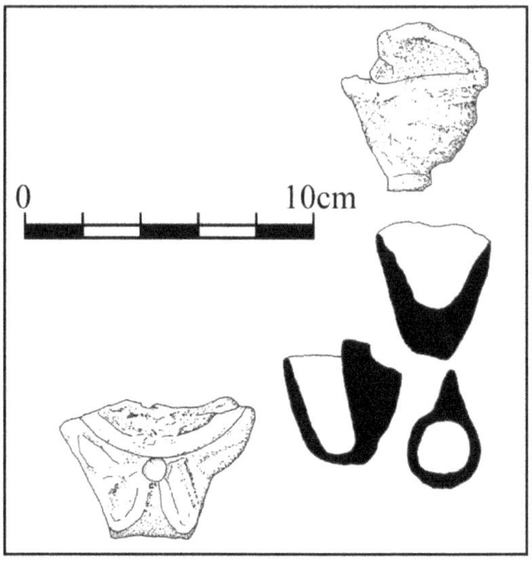

Fig. 109 Crucible and mould fragment from Corraneary (after Davies 1942).

36. CRAIGYWARREN, Co. ANTRIM.

Townland Craigywarren — *O.S. 6"* 032
Fig. 110 — *Nat. grid* D 0118/0090

Site summary
The crannóg is roughly circular in shape, approximately 19m by 16m and is located 73m out from the original shore. The body of the crannóg consists of layers of heather, small branches, tree trunks, and heavier branches. In the interior were found the remains of a hut (8m square) and a hearth. A bronze brooch from the site is dated to around 900 AD.

Metalworking features
None identified.

Relevant artefacts
1 hammer stone, 2 clay crucibles, a sandstone motif-piece, an iron chisel.

Reference
Knowles, W.J. and Coffey, G. 1907 'Craigywarren Crannóg.' *Proceedings of the Royal Irish Academy* 26C, 109–118.

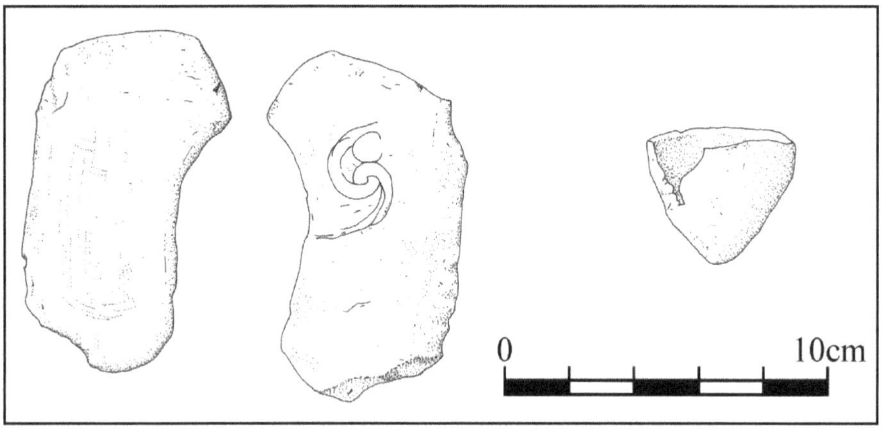

Fig. 110 Stone motif-piece and clay crucible from Craigywarren (after Knowles and Coffey 1907).

37. DRUMKEERY LOUGH, Co. CAVAN.
Townland Exact location unknown *O.S. 6"* 028

Site summary
Crannóg built on an existing small island which was artificially raised. In the interior were identified a rough stone pavement, a kitchen-midden, and on the northern edge the remains of a landing stage.

Metalworking features
None identified.

Relevant artefacts
1 clay crucible, a possible clay mould.

Reference
Wood-Martin, W.G. 1886 *The Lake Dwellings of Ireland*, 200–202.

38. LAGORE, Co. MEATH.
SEE CHAPTER 6 and APPENDIX 3.

39. LOUGH FAUGHAN, Co. DOWN.
Townland Lough Faughan *O.S. 6"* 037
Figs. 111–112 *Nat. grid* J 446/411

Site summary
Circular crannóg, 34m in diameter, 1m thick. The body of the crannóg consists of layers of brushwood, hazel, alder, birch and peat. The northern half of the site was excavated. No structural remains identified. A series of hearths was uncovered, some possibly industrial. The site also produced evidence of iron-working.

Metalworking features
None identified.

Relevant artefacts
1 complete clay crucible, clay crucible fragments, clay mould for a pin with 'thistle-head', iron shears, slotted and pointed iron tool, slag (unspecified nature).

Reference
Collins, A.E.P. 1955 'Excavations in Lough Faughan crannóg, Co. Down.' *Ulster Journal of Archaeology* 18, 45–80.

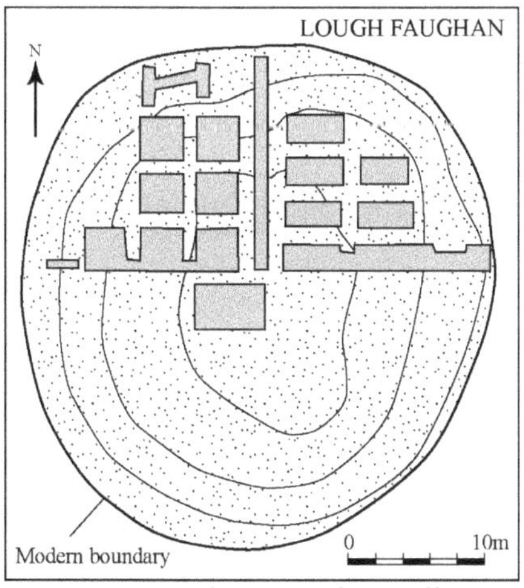

Fig. 111 Excavated areas (after Collins 1955).

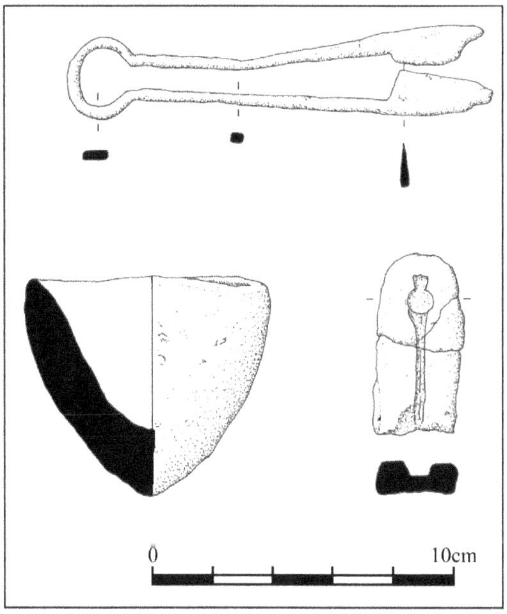

Fig. 112 Shears, crucible and mould from Lough Faughan (after Collins 1955).

40. LOUGH MOURNE, Co. ANTRIM.
Townland No modern record of site

Site summary
The body of the crannóg consists of layers of stone, moss, timbers, heather and brushwood. The remains of a hut with a stone floor were identified in the interior.

Metalworking features
None recorded.

Relevant artefacts
2 stone crucibles.

Reference
Wood-Martin, W.G. 1886 *The Lake Dwellings of Ireland*, 171–173.

41. LOUGHREA, Co. GALWAY.
Townland Loughrea *O.S. 6"* 105
Nat. grid 16070/21541

Site summary
The crannóg is known as Shore Island and appears to have been connected to the mainland by a causeway. In the interior were identified areas of basket-flooring and partitions. The site, however, is unexcavated.

Metalworking features
None recorded.

Relevant artefacts
Clay crucible fragment, iron shears, 'A vessel of hammered iron, that had been used for smelting purposes' (Wood-Martin 227).

Reference
Wood-Martin, W.G. 1886 *The Lake Dwellings of Ireland*, 226–227.

42. LOUGHTARMIN, Co. ANTRIM.
Townland Lisnahunshin *O.S. 6"* 31
Nat. grid D 0070/0809

Site summary
Circular in shape, approximately 64m in diameter. It was uncovered in 1820 when the lake was drained. The remains of a stone house were identified.

Metalworking features
None recorded.

Relevant artefacts
A small anvil, several small hammers.

Comment
The relevant artefacts can only be tenuously identified with non-ferrous metalworking.

Reference
Wood-Martin, W.G. 1886 *The Lake Dwellings of Ireland*, 165–166.

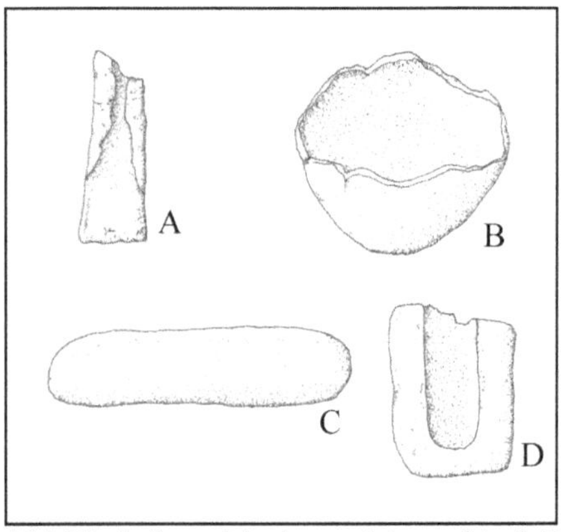

Fig. 113 Finds from Moylarg; A iron chisel, B clay crucible, C bronze ingot, D stone mould (after Buick 1894, no scale).

43. MOYLARG, Co. ANTRIM.
Townland Moylarg *O.S. 6"* 032
Fig. 113 *Nat. grid* D 0058/0083

Site summary
This crannóg was originally an island approximately 27m in diameter. The site is surrounded by wooden piles, while a hearth was found in the interior. The site produced evidence of iron-working in the form of two pieces of the iron ore haematite.

Metalworking features
None identified.

Relevant artefacts
Clay crucible fragments, clay ingot mould, 1 bone chisel, 1 horn hammer-head, bronze ingot (from clay mould), possible socketed iron chisel, fragmentary iron shears, possible iron file, socketed iron awl or 'borer', lead ring model.

Reference
Buick, R. 1893 'The Crannóg of Moylarg.' *Journal of the Royal Society of Antiquaries of Ireland* 23, 27–43.

Buick, R. 1894 'The Crannóg of Moylarg. 2nd Paper.' *Journal of the Royal Society of Antiquaries of Ireland* 24, 315–332.

44. MOYNAGH LOUGH, Co. MEATH.
Townland Moynagh *O.S. 6"* 05
Figs. 114–115 *Nat. grid* 28195/28555

Site summary
The crannóg has been dendro-dated to the seventh and eighth centuries AD. Excavation revealed a palisade, two round houses, a non-ferrous furnace and two metalworking areas. Other activities include farming, carpentry, iron- leather- and bone-working. Four sequential phases of occupation were identified in the interior. The bulk of the non-ferrous metalworking evidence is dated to phase X, *c.*715– *c.*748.

Metalworking features
Two metalworking areas and an associated furnace were identified. The furnace consisted of a clay-lined, bowl-shaped hollow with a circular mouth. It measured 0.14m deep with a maximum diameter of 0.35m, and contained a deliberately clean fill of sand and tiny pebbles. Its location in a metalworking area that produced finds including 67 crucible sherds (pyramidal and oval-mouthed) and over 600 mould fragments suggests its use in non-ferrous metalworking. The excavator felt that the relatively small amount of slag found (1kg) indicated the use of the furnace in melting metal rather than smelting ore.

Metalworking area 1 is located on the western side of the site and was defined by a spread of charcoal measuring 4.2m x 4m, 0.06m thick. Associated finds included crucible sherds, mould fragments, an iron stake for beating sheet metal, and three motif-pieces. Metalworking area 2 is located on the eastern side of the site and measures 5.3m x 5m. The furnace (described above) was situated in this area, with a pebbled surface and a pink clay spread. The pebbled area may have been a stacking place for cooling moulds, considering the proximity of the large mould dump. The pink clay spread and associated roughly cobbled area is the most likely location for post-casting work. Fragments of heating trays were found here, one still bearing the annular impression of an object worked on it. The pink clay may have been used in the manufacture of the moulds and crucibles used here.

Relevant artefacts
70 clay crucibles and fragments, over 600 clay mould fragments including those for 2 penannular brooches, mounts, studs and other miscellaneous items, clay heating tray fragments, clay tuyères, stone ingot mould, 3 bone motif-pieces, 1 stone motif-piece, 1 copper-alloy ingot, 1 iron tongs, iron stake, gold wire.

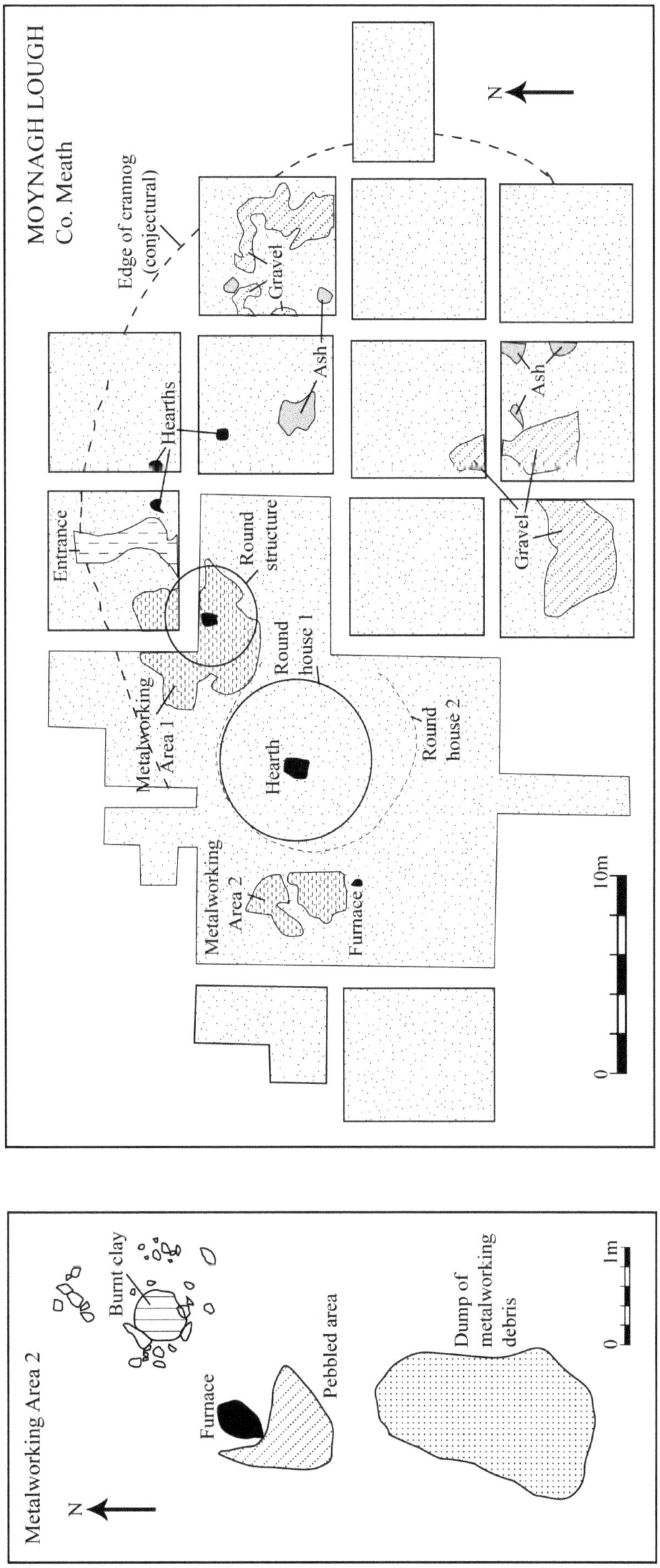

Fig. 114 Major features in Metalworking Area 2 and general site plan of Moynagh Lough (after Bradley 1991).

Comment
Non-ferrous metalworking appears to have been an important aspect of the economy at Moynagh Lough though it must be remembered that other activities were also undertaken; ironworking, agriculture, carpentry, leather and bone-working. All of these point to the high status of the inhabitants.

Reference
Bradley, J. 1986 'Excavations at Moynagh Lough 1984.' *Ríocht na Midhe* 7, 4, 79–92.

Bradley, J. 1991 'Excavations at Moynagh Lough, Co. Meath.' *Journal of the Royal Society of Antiquaries of Ireland* 121, 5–26.

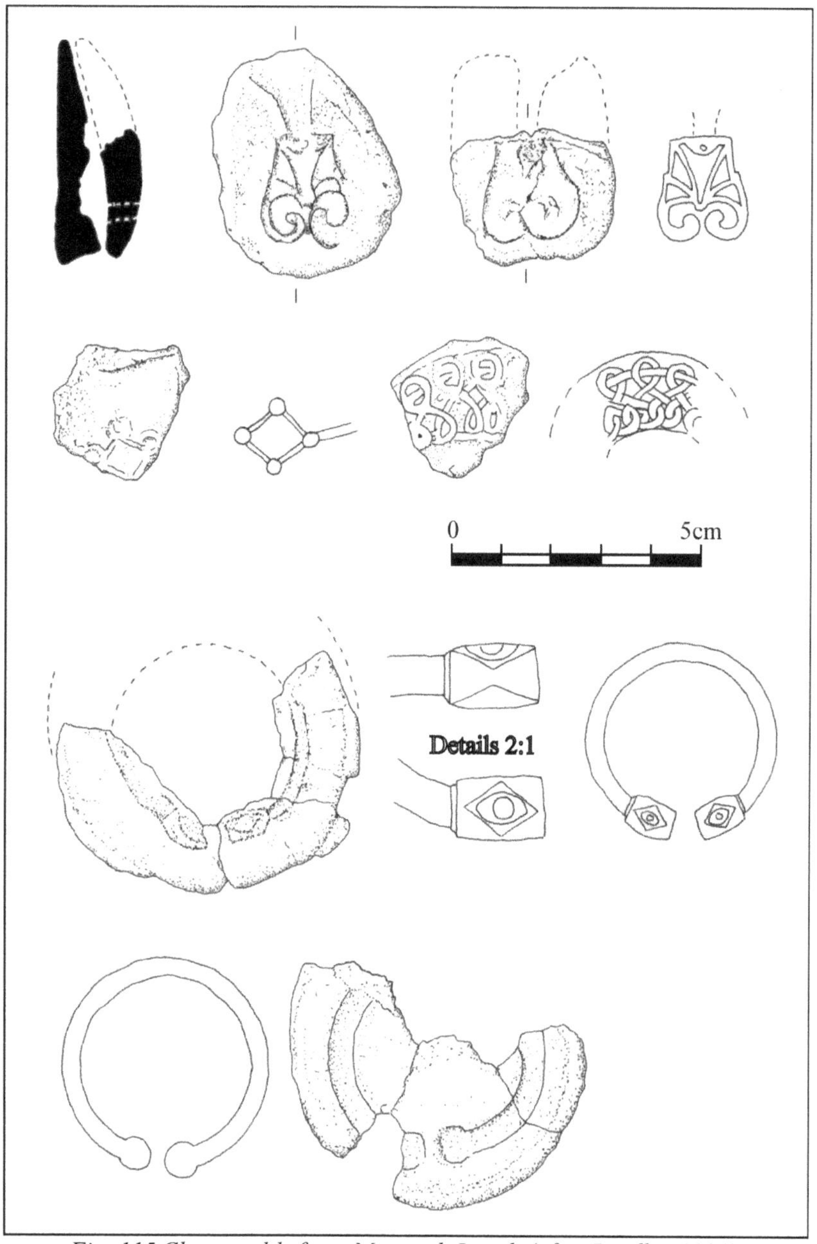

Fig. 115 Clay moulds from Moynagh Lough (after Bradley 1991).

45. RANDALSTOWN, Co. ANTRIM.

Td. Kilknock/Ballybollen/Gillistown *O.S. 6"* 043
Nat. grid J (location uncertain)

Site summary
A large crannóg consisting of heavy beams laid horizontally; no further details.

Metalworking features
None recorded.

Relevant artefacts
Numerous clay crucibles, iron tongs, possible iron anvil ('a rough lump of iron somewhat smoothed on one side, and weighing 50 or 60 pounds.' Wood-Martin 168).

Reference
Wood-Martin, W.G. 1886 *The Lake Dwellings of Ireland*, 167–168.

46. RATHTINAUN, Co. SLIGO.

Townland Lough Gara *O.S. 6"* 047
Fig. 116 *Nat. grid* 17209/29953

Site summary
Rathtinaun, crannóg no. 61, is one of approximately 185 crannógs found at Lough Gara, Co. Sligo. It consists of an oval-shaped mound of stone and timber 2.5m thick. The site is multi-period, dating from the Late Bronze Age through to the Early Historic period. Period 3 dates to the seventh century AD, Period 4 and 5 possibly eighth/ninth century AD. Non-ferrous metalworking is evident from all of these periods.

Metalworking features
None identified.

Relevant artefacts
Pyramidal clay crucibles, bag-shaped clay crucibles, bivalve clay mould fragments, bronze *pressblech* die, wooden spatula (possibly used in the lost wax process).

Comment
From initial reports the crannóg at Rathtinaun appears, in many ways, to be similar to both Ballinderry II (Hencken 1942) and Lagore (Hencken 1950). The possible *pressblech* die is the first of its kind to be found in Ireland.

Reference
Raftery, Unpublished.

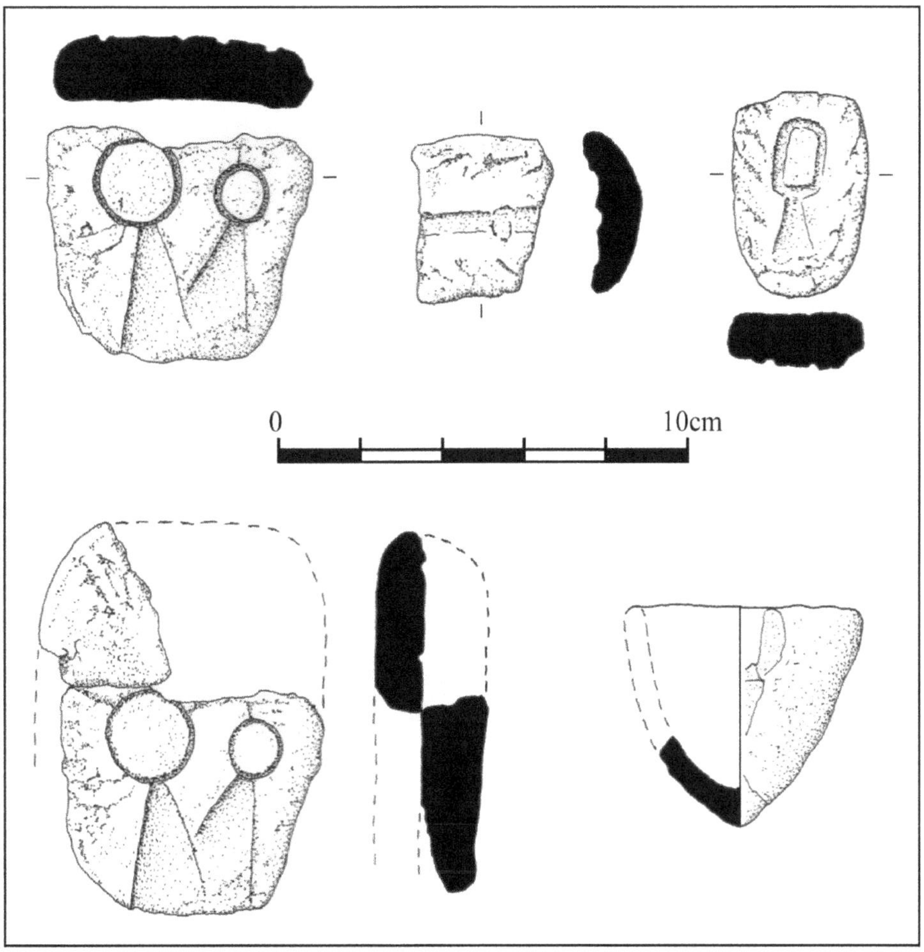

Fig. 116 Clay moulds and clay crucible from Rathtinaun (after Raftery pers. comm.).

47. THE MIRACLES, Co. FERMANAGH.

Townland Tullycreevy *O.S. 6"* 021
Nat. grid H 0144/0487

Site summary
The remains of this crannóg were uncovered in 1875 during turf cutting. A wooden hut stood in the centre of the crannóg before being destroyed by workmen. Iron slag was found on the site.

Metalworking features
None recorded.

Relevant artefacts
Complete clay crucible, stone chisel.

Reference
Wood-Martin, W.G. 1886 *The Lake Dwellings of Ireland*, 192–193.

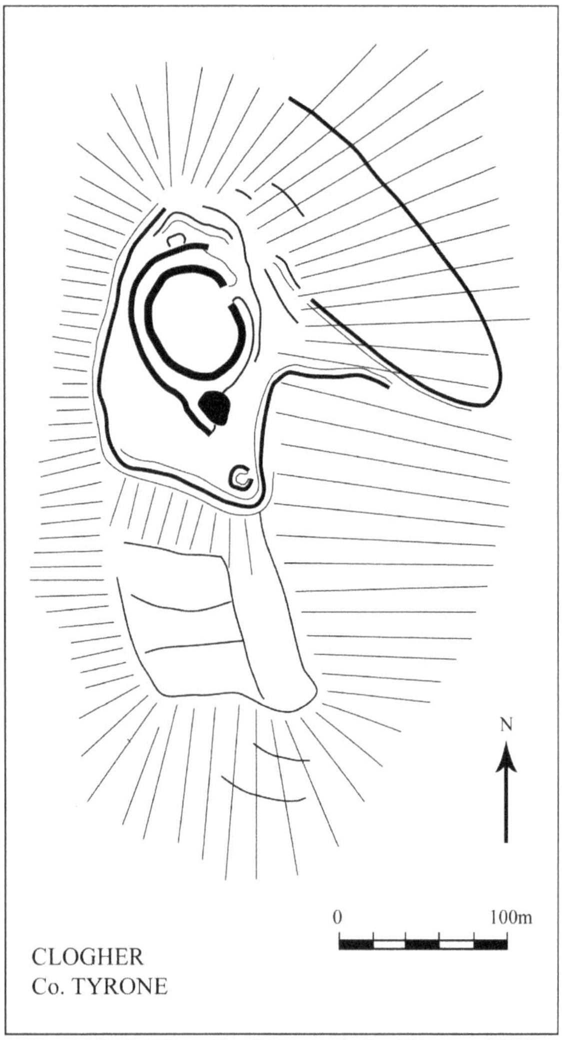

Fig. 117 Clogher, Co. Tyrone (after Warner 1973).

48. CLOGHER, Co. TYRONE.
Townland Clogher/Demesne *O.S. 6"* 058
Fig. 117 *Nat. grid* H 5387/5133

Site summary
The relevant Early Historic settlement is located within the confines of Clogher hillfort. From the fifth to the eighth century AD the site remained largely unenclosed and undefended. In the eighth century, however, a deep 'V'-sectioned ditch was dug with the material from it forming an internal bank. No available information regarding internal features.

Metalworking features
No available information.

Relevant artefacts
Complete clay crucible, clay crucible stand, gold-working crucible, gold-rubbing stone, scrap bronze bracelet, bronze brooch terminal waster, lead brooch pin model.

Comment
Very little information is, as yet, available regarding the excavations at Clogher. It would appear that the site was an important royal settlement, comparable with Garranes in terms of dating and the extent of non-ferrous metalworking. The latter can be inferred from the artefacts mentioned above and the brooch 'factory' information provided by Kilbride-Jones in his *Zoomorphic Penannular Brooches*.

Reference
Warner, R.B. 1973 'The excavations at Clogher and their context.' *Clogher Record* 8, 1, 5–12.

49. DALKEY ISLAND, Co. DUBLIN.
Townland Dalkey Island *O.S. 6"* 023
Fig. 118 *Nat. grid* 32767/22656

Site summary
Archaeological evidence on Dalkey Island dates from the Mesolithic through to the Medieval period. There is a substantial gap in the artefact evidence between the Late Bronze Age and the Early Historic Period, though the same pattern was not evident in the site stratigraphy. Remains from these two periods were found in a layer of dark black soil, distinguishable only by the concentration of prehistoric or Early Historic artefact types. Excavations were limited and took place between 1956 and 59, mainly focusing on the interior of a promontory fort which is located on the north-west corner of the island. The fort encloses approximately 0.25 hectares and is defined by a bank with external ditch. The interior is divided into two roughly flat platforms, one higher than the other. The majority of Early Historic evidence came from the upper level of the fort and consisted of a range of artefacts and a small number of features, both pre- and post-bank in date.

Metalworking features
No features were specifically associated with non-ferrous metalworking, though two hearths, an oven or kiln and an iron smelting pit were dated to the Early Historic Period. The smelting furnace was located in the lower part of the site and was roughly circular in shape with a maximum depth of 0.25m. It contained charcoal, two halves of a baked clay tuyère and several separate lumps of iron slag.

Relevant artefacts
11 clay crucible fragments, a length of bronze wire, a circular bronze ingot, slate motif-piece, a clay tuyère.

Comment
The difficulty in distinguishing the Bronze Age levels from the Early Historic may have resulted in the incorrect dating of some finds, for example some of the clay mould fragments. Also, no scientific analyses were undertaken on the slag to confirm its identification as iron smelting slag.

Reference
Liversage, G.D. 1968 'Excavations at Dalkey Island, Co. Dublin. 1956–1959.' *Proceedings of the Royal Irish Academy* 66C, 53–233.

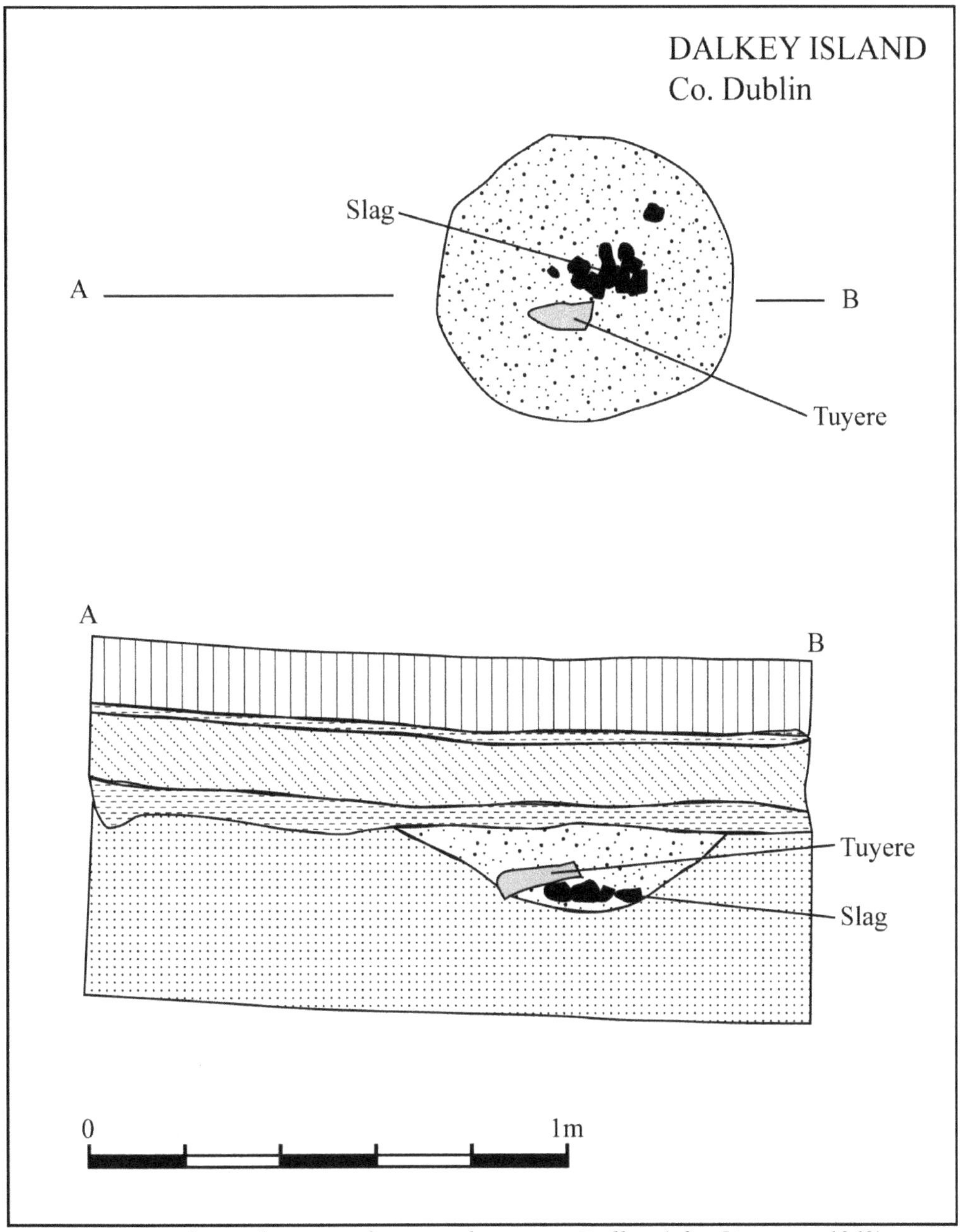

Fig. 118 Plan and section of iron-smelting pit at Dalkey (after Liversage 1968).

50. DOOEY, Co. DONEGAL.

Townland Dooey *O.S. 6"* 065
Fig. 119 *Nat. grid* 17601/40170

Site summary
A four phase settlement was identified in the sandhills at Dooey, Co. Donegal. The first phase consisted of an unprotected area, approximately 56m x 64m, with a series of internal fire-pits. The second phase saw the surrounding of the central area with a fosse. This fosse filled up and the site spread out over the top of it in phase three. The final phase consists of a cemetery in the central area.

Metalworking features
None identified.

Relevant artefacts
Clay crucible fragments, clay mould fragments, antler motif-piece, bronze awl, iron awls, iron chisels, iron hammer-head, iron pincers, iron pronged tools, lead alloy brooch pin.

Comment
The relevant finds listed above suggest that non-ferrous metalworking was an important part of the Dooey economy.

Reference
Ó Ríordáin, A.B. and Rynne, E. 1961 'A settlement in the sandhills at Dooey, Co. Donegal.' *Journal of the Royal Society of Antiquaries of Ireland* 91, 58–64.

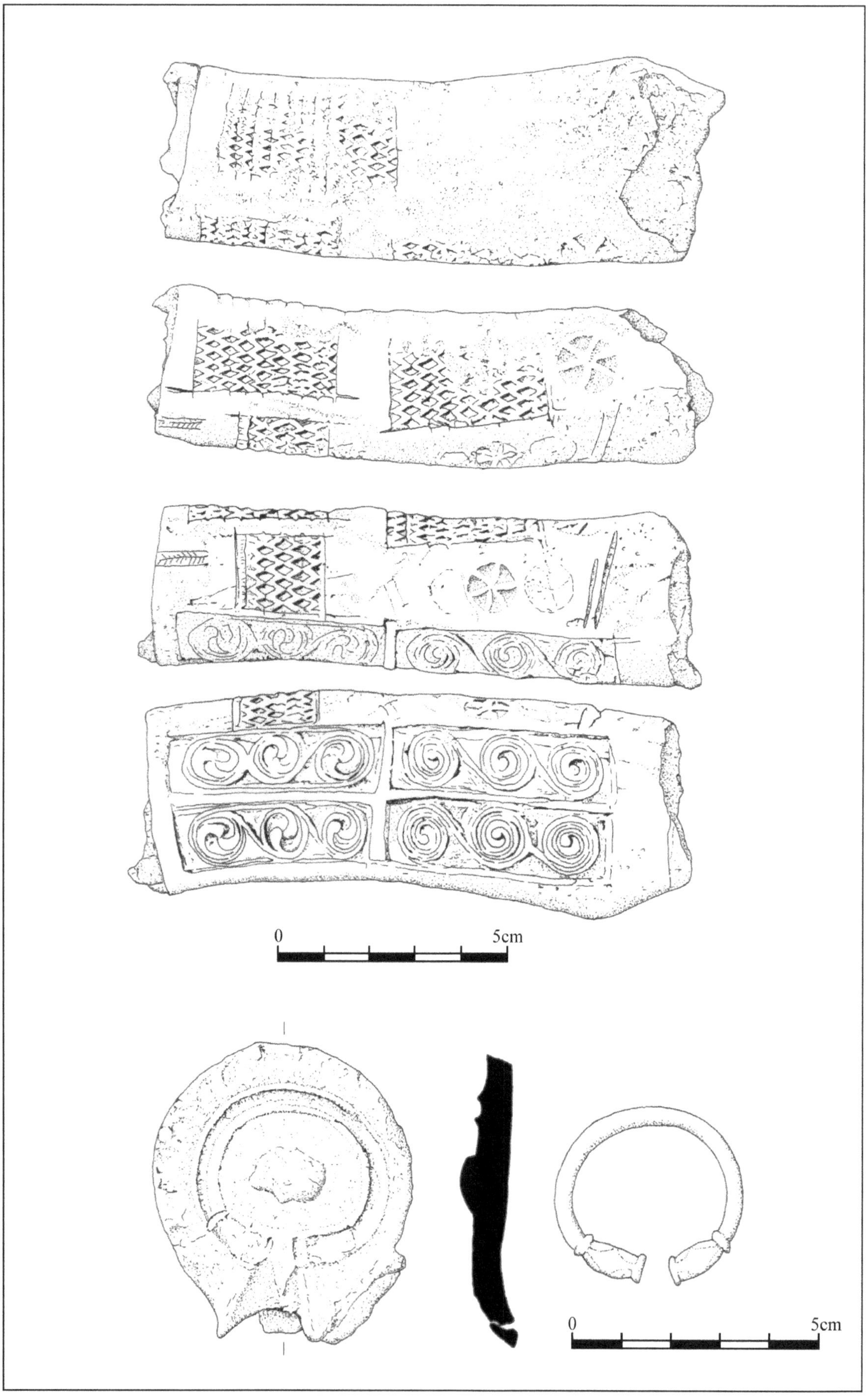

Fig. 119 Finds from Dooey; antler motif-piece and clay mould with reconstruction of cast object (after Ó Ríordáin and Rynne 1961).

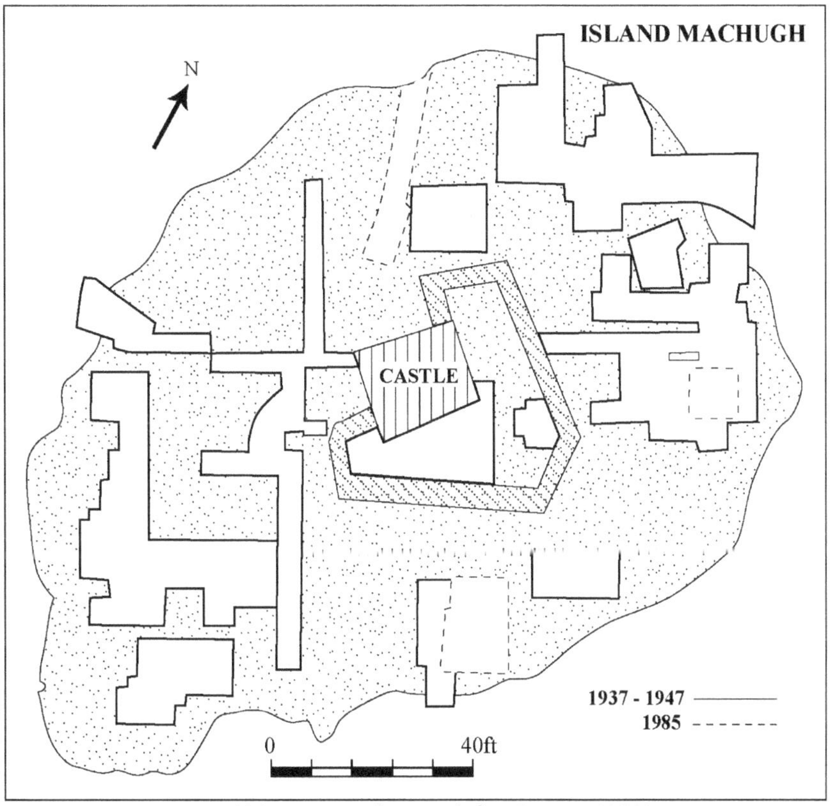

Fig. 120 Excavated areas (after Davies 1950).

51. ISLAND MACHUGH, Co. TYRONE.

Townland Barons Court — *O.S. 6"* 017

Fig. 120 — *Nat. grid* H

Site summary

A poorly stratified island site, ranging in date from the Late Bronze Age to the Medieval period. No structures identified as Early Historic in date.

Metalworking features

None identified.

Relevant artefacts

Clay crucible fragments.

Reference

Davies, O. 1950 *Excavations at Island MacHugh*. Supplement to the Proceedings and Reports of the Belfast Natural History and Philosophical Society.

52. KNOWTH, Co. MEATH.

Townland Knowth — *O.S. 6"* 019

Fig. 38 — *Nat. grid* 29968/27343

Site summary

A high-status settlement atop the passage-tomb mound at Knowth. Identified as the seat of the kings of northern Brega, Lagore's counterpart. Earlier Iron Age settlement was delimited by two large fosses, surrounding the base and top of the mound. There is no evidence of any Early Historic banks or palisades and the occupation is marked by a series of houses, souterrains and hearths.

Metalworking features

The excavation produced evidence of iron-working (on the northern side of the site) with evidence of non-ferrous metalworking found opposite it. The non-ferrous evidence includes a metalworking hearth. This consisted of a large rectangular stone (30 x 12cm), set on edge, around which was a concentration of charcoal. Associated finds include two complete heating trays and a fragment of a third, and clay crucible fragments.

Relevant artefacts

Clay crucible fragments, gold-working clay crucible, 2 complete clay heating pans, 1 clay heating pan fragment.

Comment

The quantity of relevant artefacts recovered from Knowth was not as great as from Lagore, its royal counterpart, nevertheless, the importance of its non-ferrous working is indicated by the discovery of a gold-working crucible. Lagore produced no such crucible though the gold filigree mount from the site (Hencken 1950, 86–87) may have been manufactured there.

Reference

Eogan, G. 1974 'Report on the excavations of some passage graves, unprotected inhumation burials and a settlement at Knowth, Co. Meath.' *Proceedings of the Royal Irish Academy* 74C, 11–112.

Eogan, G. 1977 'The Iron Age – Early Christian settlement at Knowth, Co. Meath.' in Markotic (ed.) *Ancient Europe and Mediterranean Studies*, 69–76.

APPENDIX III

LAGORE CRANNÓG
- A CATALOGUE OF THE NON-FERROUS METALWORKING EVIDENCE

CRUCIBLES

Unless otherwise stated, all crucibles and crucible fragments are of a very hard, coarse grey clay. They were manufactured from wet clay, shaped by hand, dried and baked. The size of these vessels ranges from 29mm in height, 32mm x 30mm at the mouth (no. 24) to 105mm high and 90mm x 85mm at the mouth (no. 3). These are housed in the National Museum of Ireland artefact store (habitat C17:5). For an explanation of crucible residues and a discussion of their significance see Chapters 3 and 4.

The Museum holds 279 crucibles and crucible fragments. This equals approximately 97% of the total number excavated. The remaining 3% have been lost. Of the 279, 38 are of the pyramidal type, 10 are Bag-shaped, 23 are flat-bottomed, 27 are base-sherds, 85 are rim sherds, 77 are body sherds, 10 are vessel fragments, 2 are handled vessels, 5 are vessel projections, with one polypod vessel and one crucible lid. Approximately 10% are complete, with the remainder in fragmentary condition.

The following catalogue entries all begin by stating the artefact type, followed by the National Museum of Ireland registration number and the excavation number (where recorded). Each artefact received a registration number with the prefix E14: from the Museum, these numbers differing from those used by Hencken in his excavation report. Several artefacts were recovered from the site in the nineteenth and early twentieth century prior to excavation. These are recorded as 'Old Finds' and their catalogue number given. The following abbreviations are used; W – Wilde catalogue, Wk. – Wakeman catalogue, P. – Petrie catalogue.

A description of each artefact and its approximate find spot is given, followed by the relevant measurements and illustration.
Pyramidal crucible measurements include maximum depth, maximum height and thickness of crucible at rim.
Bag-shaped measurements include internal diameter; maximum depth; external diameter; maximum height and thickness at rim.
Both *flat-bottomed* vessels and *basal sherds* record maximum depth; maximum height; diameter (internal and external for basal sherds, external for flat-bottomed crucibles); thickness of base and thickness at rim/top.
Rim and body sherd measurements include length; breadth and height, while the maximum height, breadth and thickness at rim/top are recorded for vessel fragments.

The following abbreviations are used: H/height, L/length, B/breadth, T/thickness, D/depth, Diam./diameter.

Pyramidal Crucibles

1. Pyramidal Crucible (E14:436 / 1312)
Fig. 121
Complete vessel with red vitreous material on the rim, both internally and externally. NW area. H 50mm. D 37mm. Rim T 9mm.

2. Pyramidal Crucible (E14:509)
Fig. 121
Incomplete vessel - missing a large rim sherd. Approx. 20mm above the base internally, a 'tide mark' is visible above which the walls of the crucible are blackened. D 52mm. H 70mm. Rim T 12mm.

3. Pyramidal Crucible (E14:372)
Fig. 122
Incomplete vessel reconstructed from five fragments. No visible residues internally. The lower half of the exterior is encrusted with red vitreous material. NE area. D 94mm. H 105mm. Rim T 8mm.

4. Pyramidal Crucible (E14:371)
Fig. 123
Virtually complete vessel - missing some rim sherds. Traces of green metallic residue visible on the interior. Externally, the base is encrusted with grey porous slag. A clay projection at the base of the crucible functions as a 'stand' or 'foot'. NE area. D 31mm. H 44mm. Rim T 8mm.

5. Pyramidal Crucible (E14:494 A, B, C, D)
Fig. 123
This incomplete vessel is reconstructed from three sherds, with a fourth unattached but obviously belonging to the same crucible. Externally the base is covered with grey porous slag. E14:494 D is a rim sherd, blackened internally with no residues visible externally. W area. Rim T 4mm.

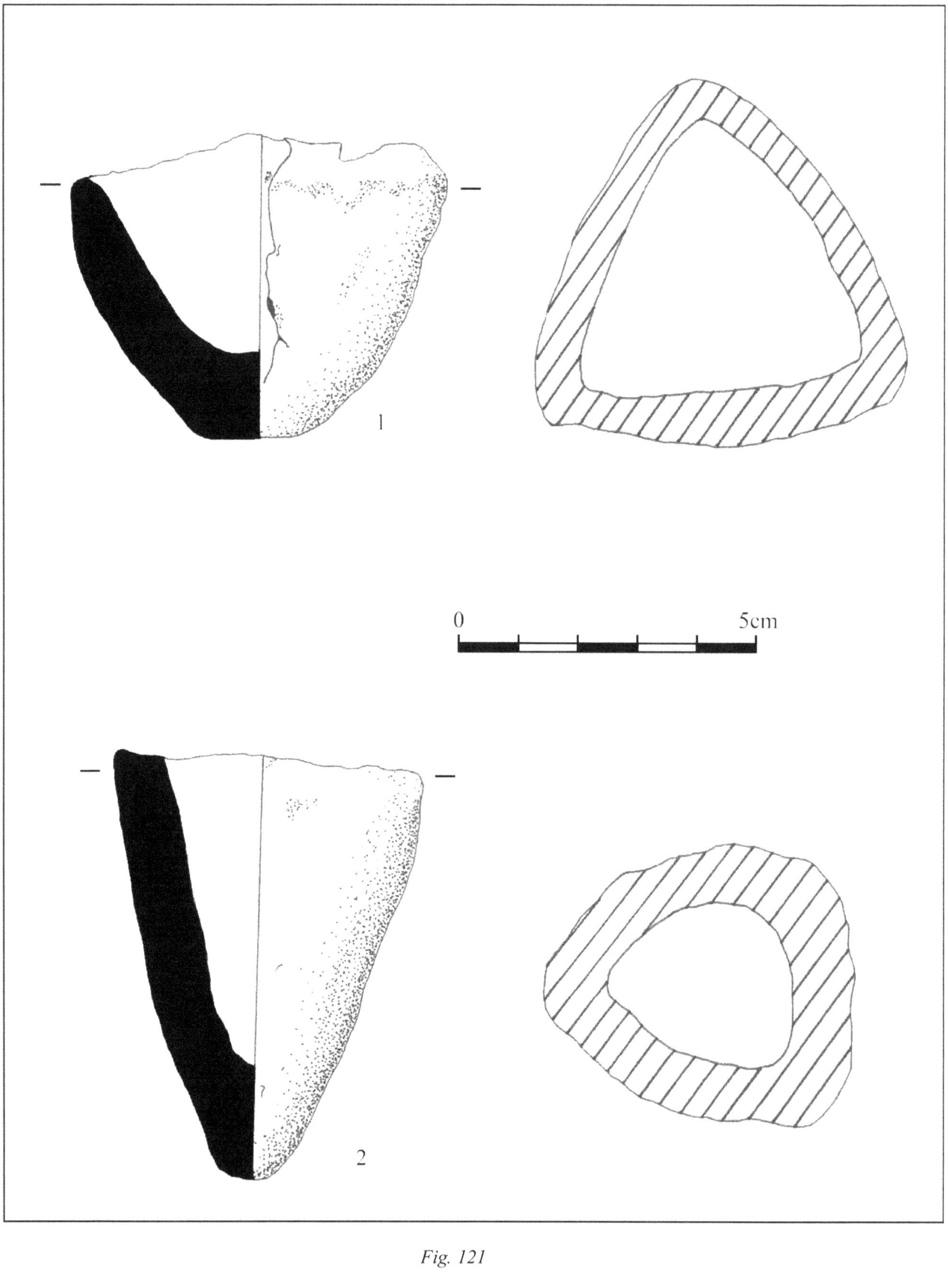

Fig. 121

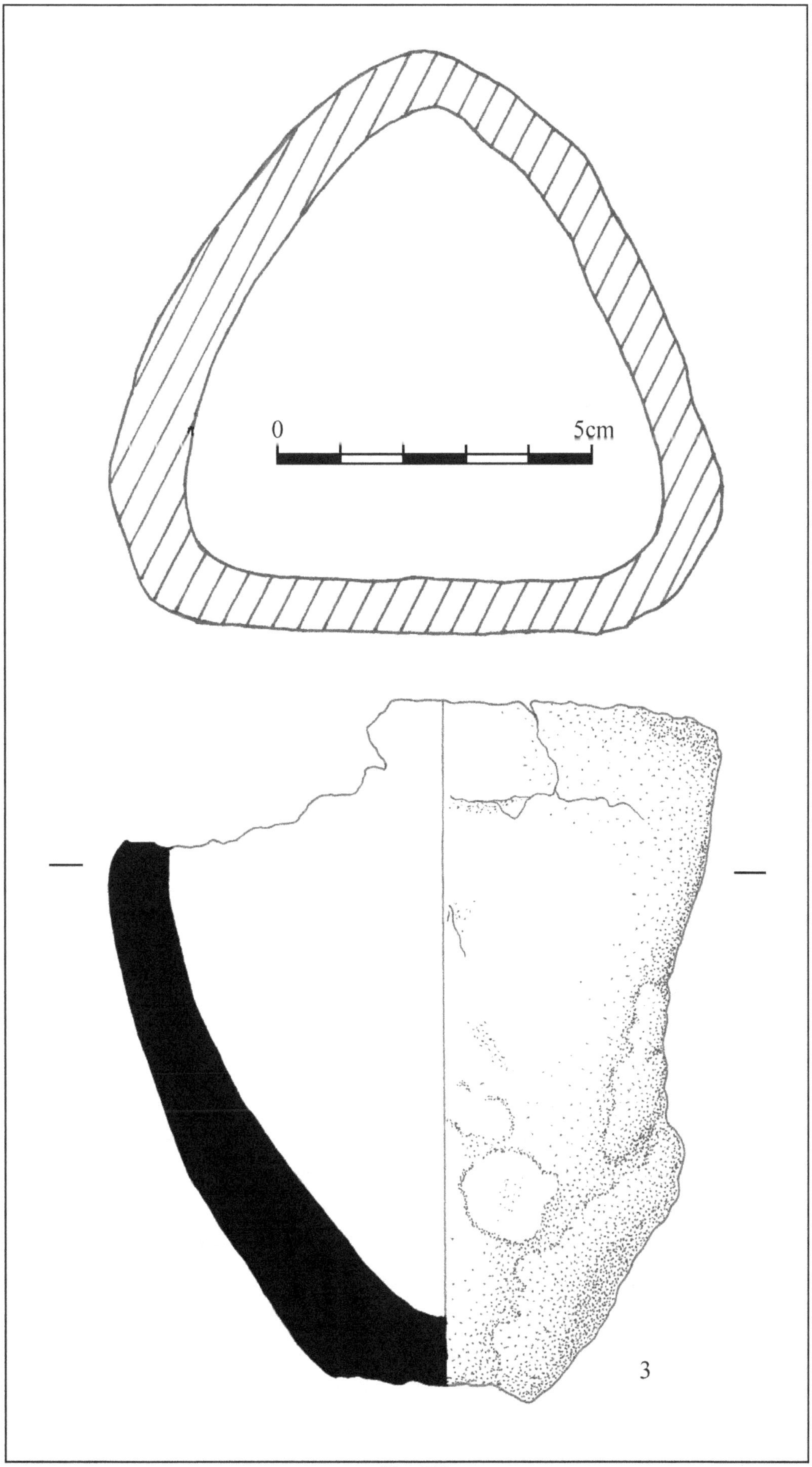

Fig. 122

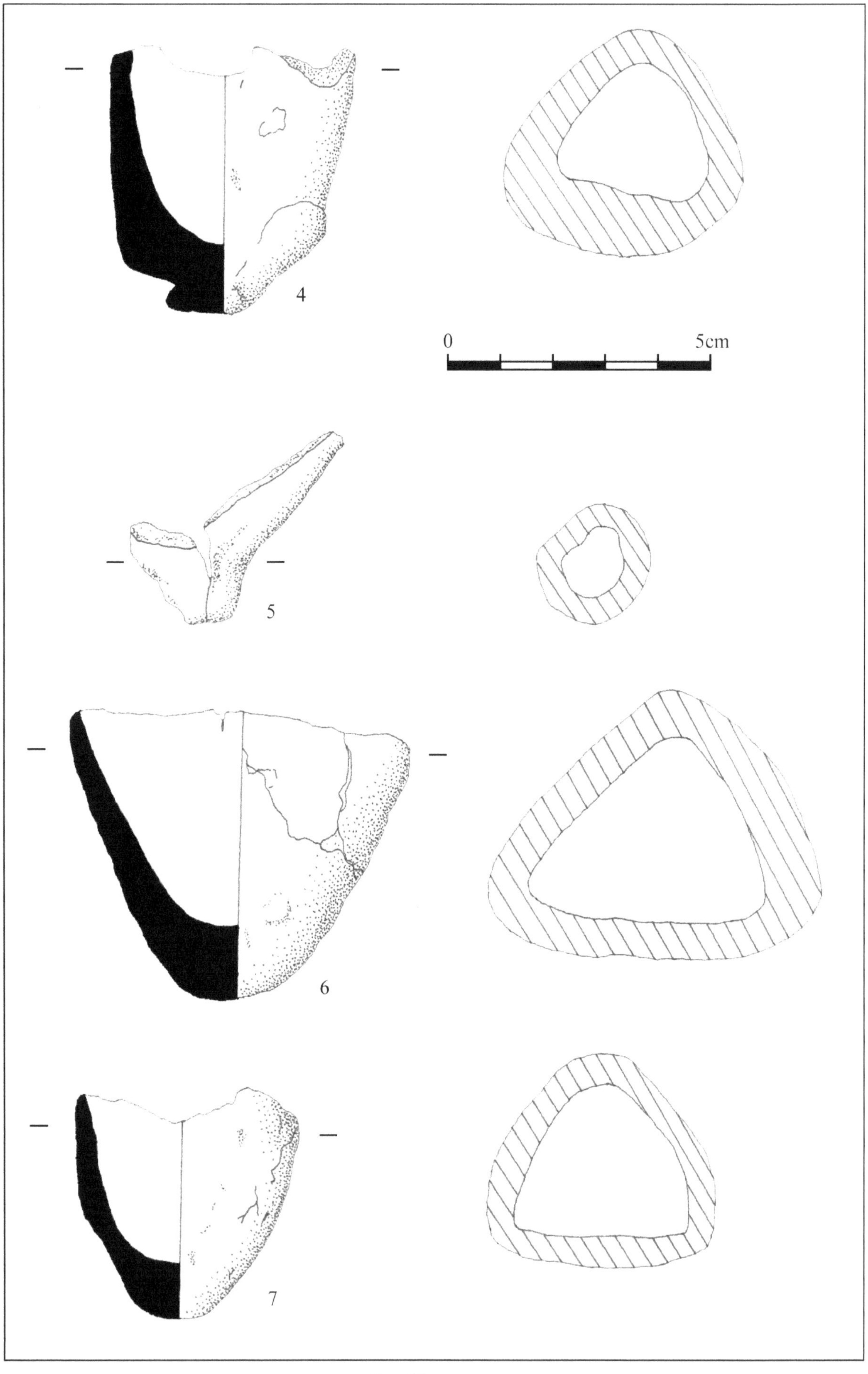

Fig. 123

6. Pyramidal Crucible (E14:242 / 1138)
Fig. 123
Complete vessel reconstructed from four sherds. The exterior and rim are covered with grey porous slag and traces of red vitreous material. E area, north of river. Surface find.
D 40mm. H 54mm. Rim T 8mm.

7. Pyramidal Crucible (E14:244)
Fig. 123
Virtually complete vessel - missing some rim sherds. Traces of green metallic residue visible on the interior. The exterior is covered with a thin layer of grey porous slag.
D 25mm. H 36mm. Rim T 5mm (including slag).

8. Pyramidal Crucible (E14:384)
Fig. 124
Fragment including base. No residues visible internally. Traces of grey porous slag visible on the exterior. NE area.
D of fragment 22mm. H 30mm. Rim T 4mm.

9. Pyramidal Crucible (E14:402)
Fig. 124
Complete vessel with light grey slag on the exterior. A clay projection at the base is possibly a deliberate feature to facilitate the handling of the vessel with tongs. NE area.
D 22mm. H 32mm. Rim T 5mm.

10. Pyramidal Crucible (E14:388)
Fig. 124
Complete vessel with no residues visible internally. Occasional areas of grey porous slag on the exterior. NE area.
D 32mm. H 42mm. Rim T 6mm.

11. Pyramidal Crucible (E14:373)
Fig. 124
Fragment. The interior surface is heat-reddened, and approx. 4% of the exterior base is covered with red vitreous material. The rim is encrusted with grey porous slag. NE area.
D 65mm. H 79mm. Rim T 9mm.

12. Pyramidal Crucible (E14:374)
Fig. 125
Fragment. The rim, internally and externally, is heavily slagged and covered with red vitreous material. NE area.
D 34mm. H 41mm. Rim T 4mm.

13. Pyramidal Crucible (E14:379)
Fig. 125
Fragment. Green metallic residues are visible on the blackened internal surface. The rim is covered with grey porous slag and red vitreous material. NE area.
D 29mm. H 43mm. Rim T 7mm.

14. Pyramidal Crucible (E14:470 A,C)
Fig. 125
Fragment consisting of two sherds. No visible residues internally, externally the base is encrusted with dark grey slag and red vitreous material. SE area. D 20mm. H 28mm. Rim T 6mm.

15. Pyramidal Crucible (E14:507 / 1057)
Fig. 125
Fragment. The interior is blackened and a thin layer of grey porous slag covers the exterior. WNW area. Disturbed area outside palisade.
D 40mm. H 48mm. Rim T 5mm.

16. Pyramidal Crucible (E14:322B and E14:477)
Fig. 125
Fragment including rim, consisting of two sherds. The exterior is covered with a thin layer of grey porous slag, the rim interior with red vitreous material. E area, north of river.
H 33mm. Rim T 3mm.

17. Pyramidal Crucible (E14:323 A, B and E14:327)
Fig. 125
Fragment consisting of three sherds. No visible residues internally. A small area of the exterior base is covered with dark grey slag. E area, north of river.
D 32mm. H 42mm. Rim T 3mm.

18. Pyramidal Crucible (E14:325)
Fig. 126
Base. Sufficiently flat-bottomed to allow it to stand independent of any support. No visible residues internally. Grey porous slag and red vitreous material covers the exterior. E area. north of river.
D 13mm. H 25mm. Rim T 9 mm.

19. Pyramidal Crucible (E14:33)
Fig. 126
Base. Relatively flat-bottomed; stands independent of any support. No visible residues internally. The exterior is covered with red vitreous material. E area, north of river.
D 3mm. H 14mm. Rim T 6mm.

20. Pyramidal Crucible (E14:341 / 1457)
Fig. 126
Rim sherd. No visible residues internally, glazed externally. E area, north of river. Found between palisades in brown peat layer 3. H 32mm. Rim T 3mm.

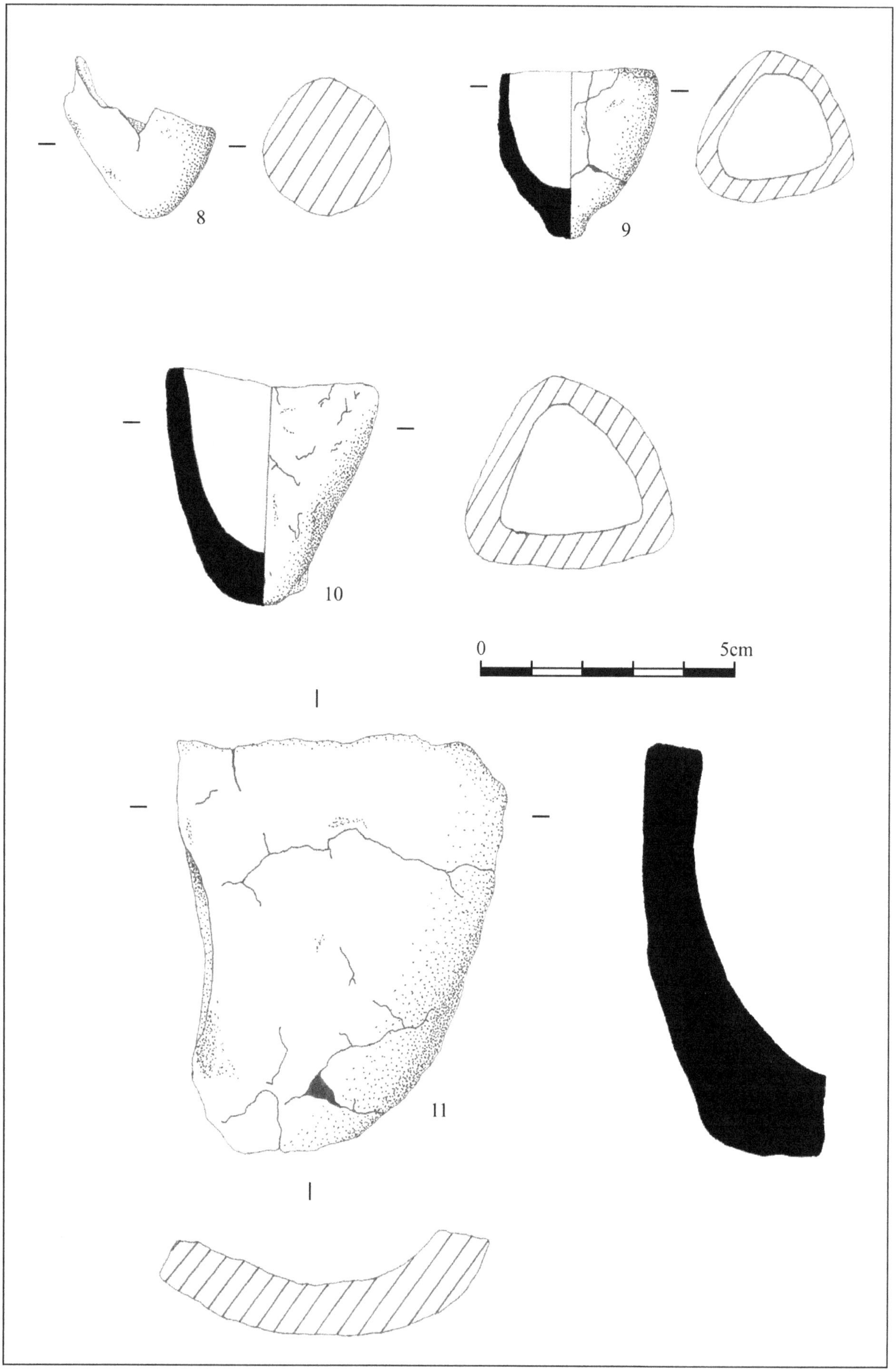

Fig. 124

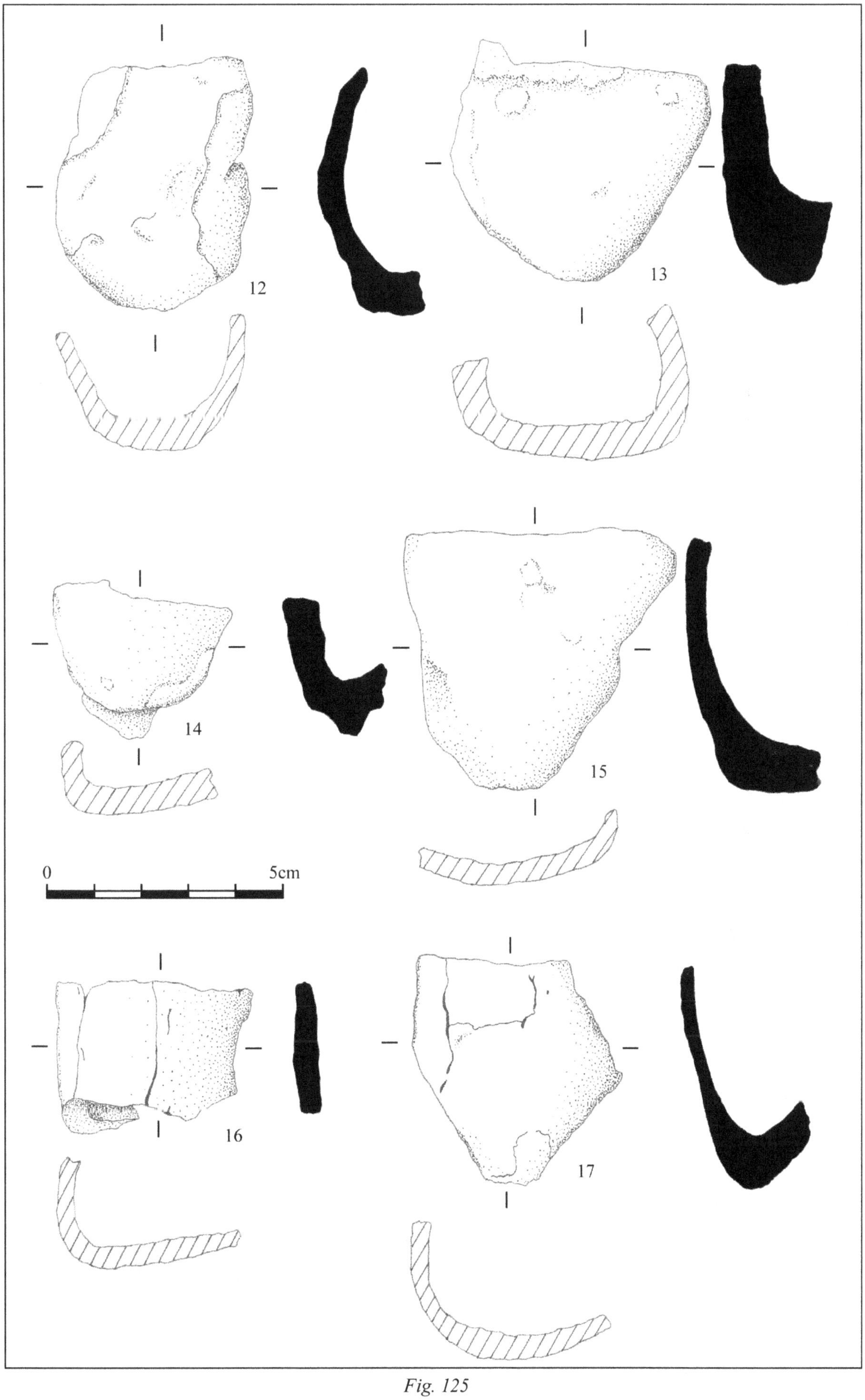

Fig. 125

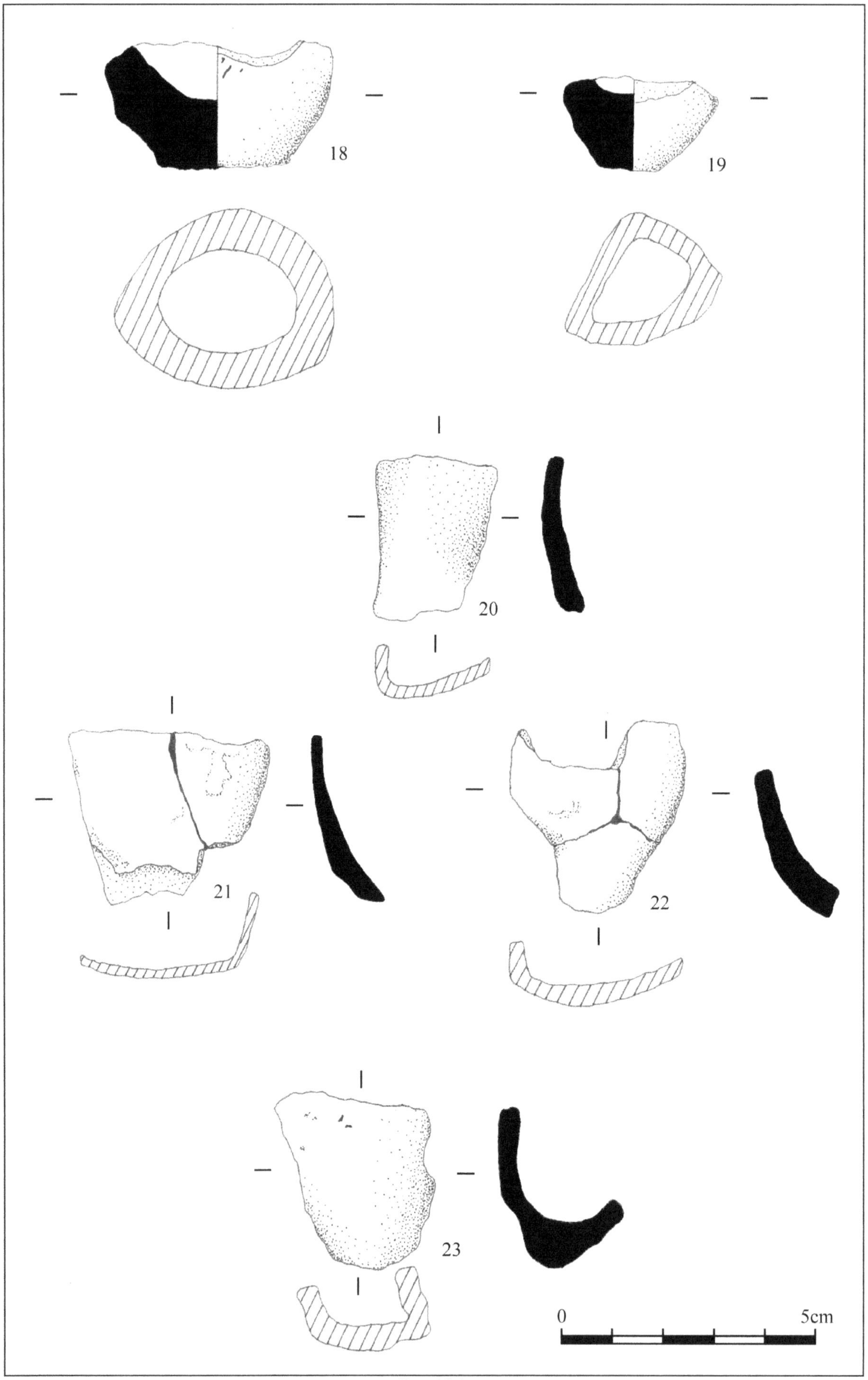

Fig. 126

21. Pyramidal Crucible (E14:302)
Fig. 126
Fragment including rim. The interior is slightly heat-reddened, the exterior rim encrusted with red vitreous material. E area, north of river. Surface find.
H 44mm. Rim T 2mm.

22. Pyramidal Crucible (E14:303)
Fig. 126
Fragment including rim. The exterior is glazed and bears trace red vitreous material. E area, north of river. Surface find.
H 47mm. Rim T 3mm.

23. Pyramidal Crucible (E14:491)
Fig. 126
Fragment including rim. The rim and exterior are coated with grey porous slag. W area.
H 43mm. Rim T 5mm.

24. Pyramidal Crucible (E14:442)
Fig. 127
Complete vessel with substantial amounts of green metallic residue visible on the interior. The exterior of the vessel is covered with grey porous slag and occasional red vitreous material, the interior of the rim with red vitreous material. SE area.
D 22mm. H 30mm. Rim T 3mm.

25. Pyramidal Crucible (E14:443)
Fig. 127
Virtually complete; missing some rim sherds. The exterior is covered with grey porous slag and faint red vitreous material. SE area.
D 38mm. H 53mm. Rim T 3mm.

26. Pyramidal Crucible (E14:448)
Fig. 127
Base. Approx. 30% of the exterior is covered with grey porous slag. SE area.
D 23mm. H 47mm. Rim T 8mm (5mm–12mm).

27. Pyramidal Crucible (E14:456)
Fig. 127
Base. The exterior is covered with grey porous slag and faint traces of red vitreous material. SE area.
D 19mm. H 29mm. Rim T 6mm (5mm–7mm).

28. Pyramidal Crucible (E14:457)
Fig. 127
Base. The exterior is heavily encrusted with grey porous slag and traces of red vitreous material. SE area.
D 28mm. H 53mm. Rim T 7mm.

29. Pyramidal Crucible (E14:508)
Fig. 128
Base. Approx. 10% of the exterior is covered with grey porous slag.
D 17mm. H 31mm. Rim T 7mm.

30. Pyramidal Crucible (E14:420, 426 and E14:430)
Fig. 128
Fragment including rim and base, consisting of three sherds. The exterior and rim bear grey porous slag and red vitreous material. SE area.
D 57mm. H 85mm. Rim T 7mm (6mm–8mm).

31. Pyramidal Crucible (E14:427)
Fig. 128
Base. No visible residues. SE area.
D 7mm. H 23mm. Rim T 6mm.

32. Pyramidal Crucible (E14:441 A, C)
Fig. 129
Fragment including rim, consisting of two sherds. The interior is blackened, the exterior bears trace red vitreous material. NW area.
D 51mm. H 66mm. Rim T 5mm.

33. Pyramidal Crucible (E14:486)
Fig. 129
Complete vessel. The rim, both internally and externally is covered with red vitreous material. SW area.
D 34mm. H 47mm. Rim T 8mm.

34. Pyramidal Crucible (E14:487)
Fig. 129
Virtually complete vessel; missing a large rim sherd. The base exterior is covered with grey porous slag. SW area.
D 24mm. H 39mm. Rim T 3mm.

35. Pyramidal Crucible (E14:245 / 1172)
Fig. 129
Virtually complete vessel - missing a rim sherd. The rim is covered with red vitreous material. The exterior is coated with grey porous slag and red vitreous material at the base. E area, outside palisade.
D 34mm. H 45mm. Rim T 5mm.

36. Pyramidal Crucible (E14:243 /1134)
Fig. 130
Complete vessel. Approx. 20% of the rim is covered with red vitreous material, while approx. 40% of the exterior is covered with grey porous slag. E area, north of river. Surface find.
D 41mm. H 55mm. Rim T 4mm.

37. Pyramidal Crucible (E14:370)
Fig. 130
Fragment including rim and base. The exterior is covered with grey porous slag and red vitreous material. NE area.
D 45mm. H 49mm. Rim T 4mm.

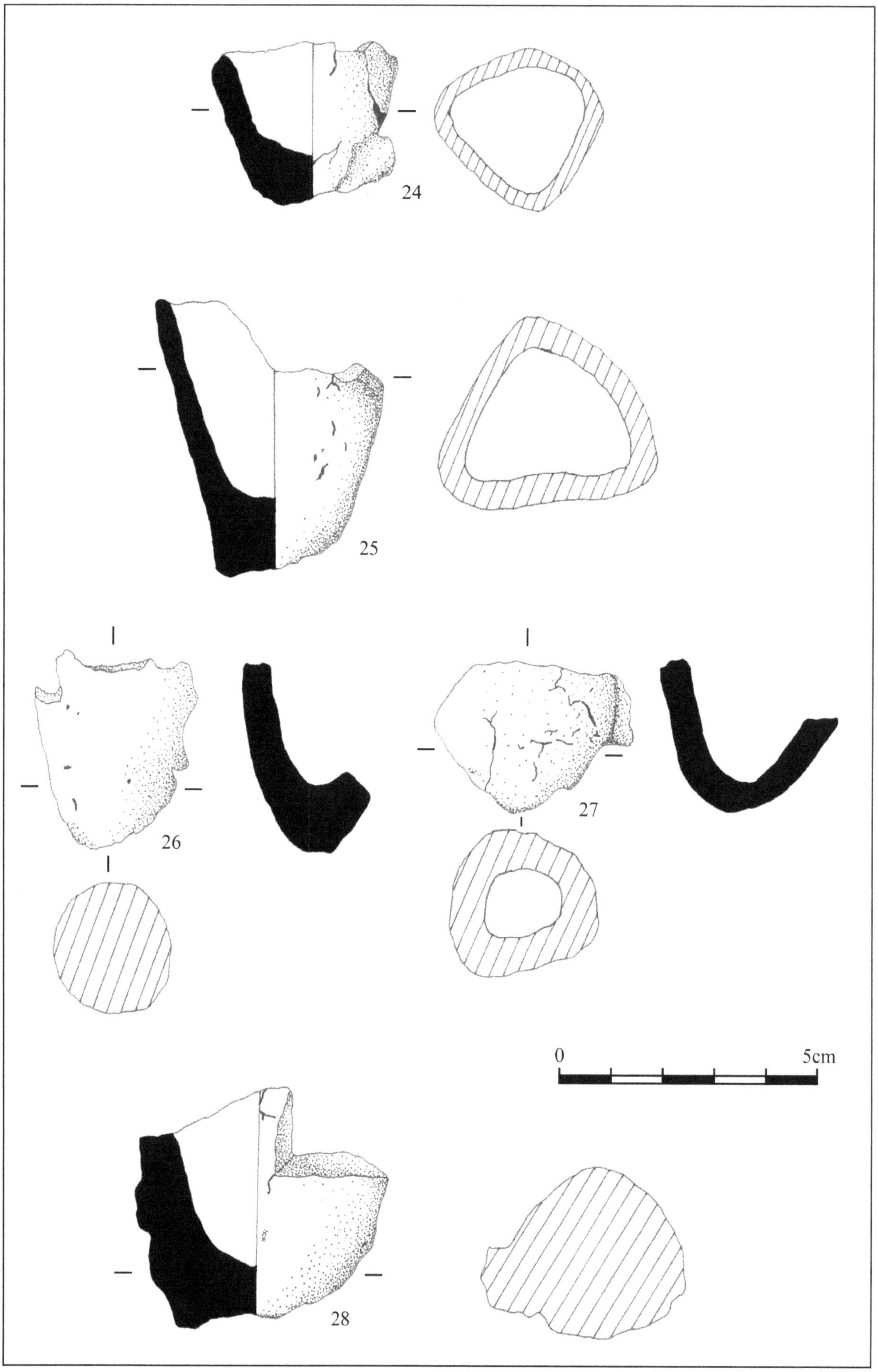

Fig. 127

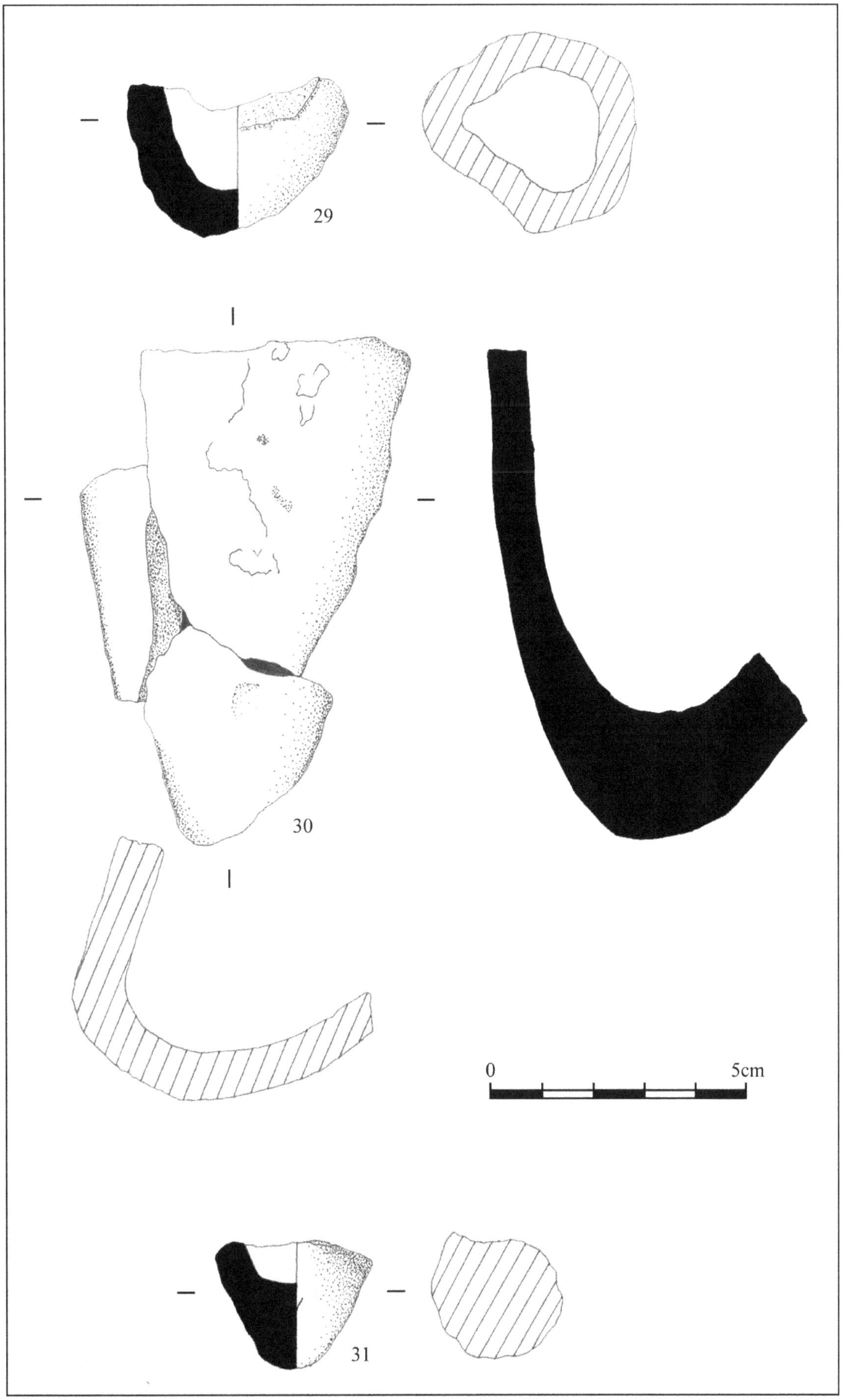

Fig. 128

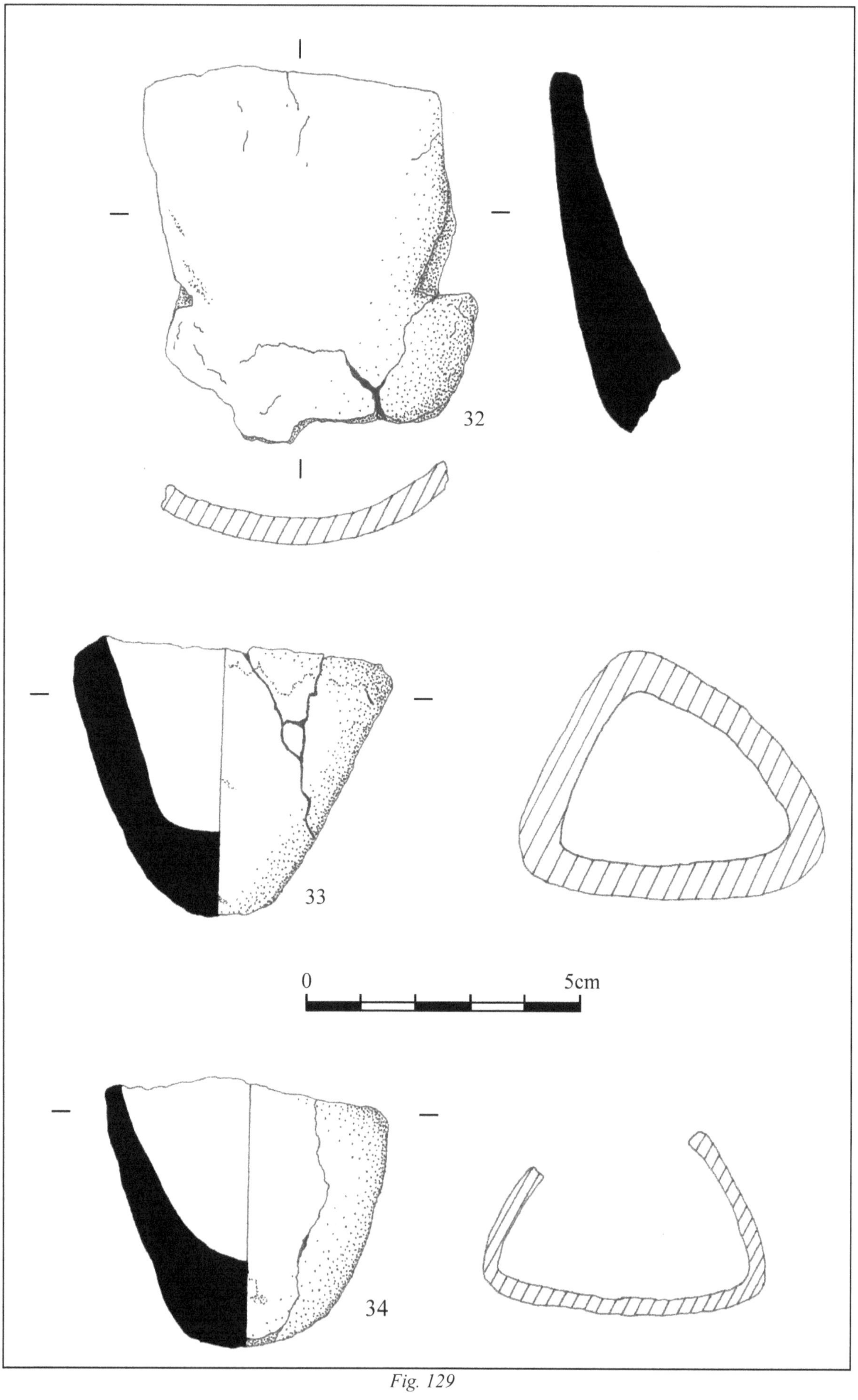

Fig. 129

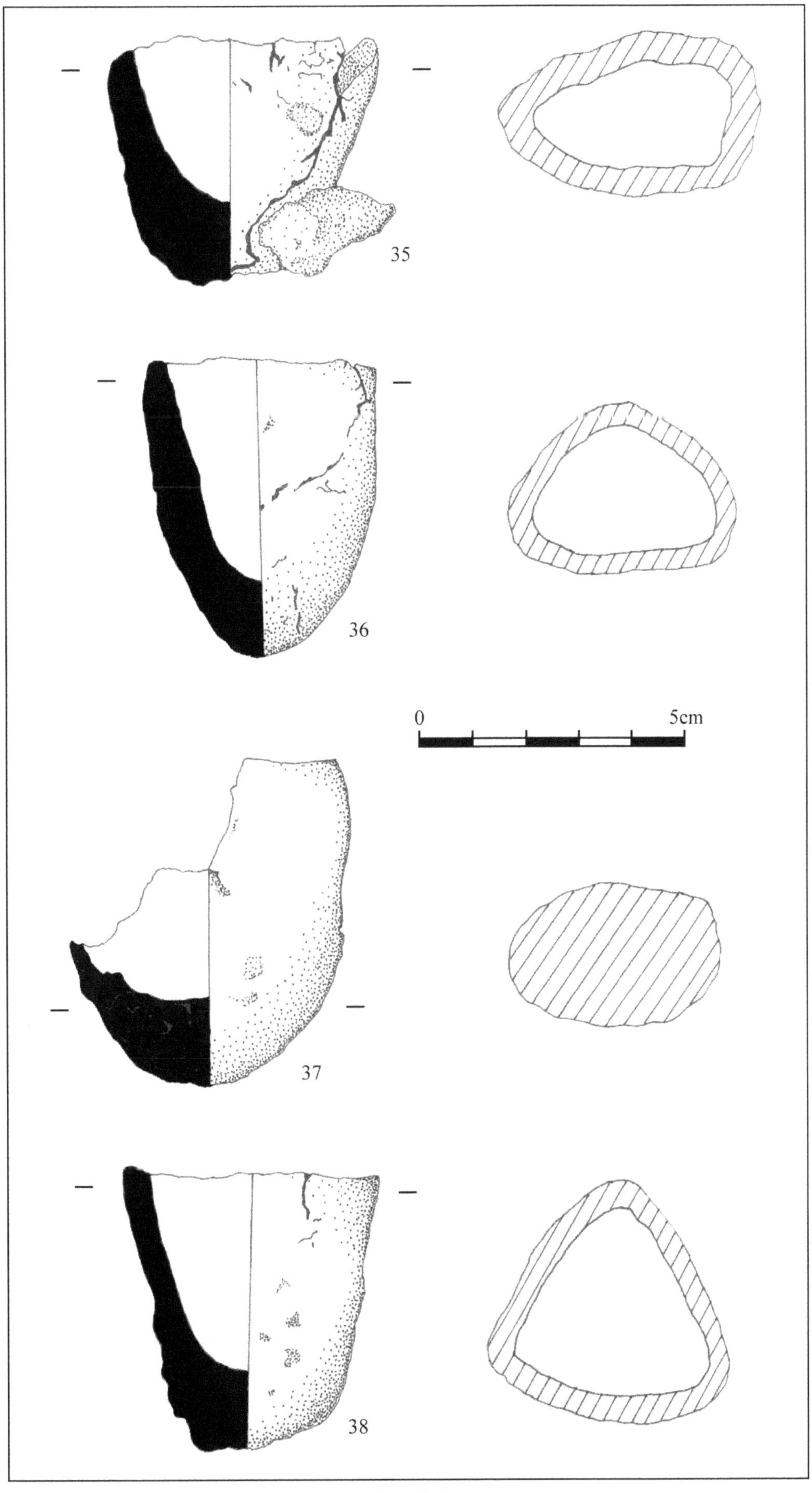

Fig. 130

38. Pyramidal Crucible (E14:498)
Fig. 130
A complete vessel with traces of green residue on the interior surface. The rim and exterior are encrusted with grey porous slag and red vitreous material.
D 36mm. H 51mm. Rim T 5mm.

Bag-shaped Crucibles

39. Bag-shaped Crucible (E14:440 / 496)
Fig. 131
Complete vessel reconstructed from nine sherds. Traces of green metallic residue visible internally. The exterior is covered with a layer of grey porous slag and occasional red vitreous material. The rim is heavily encrusted with slag and red vitreous material, so much so that it partially closes the crucible on two edges in lid-like fashion. NW area.
D 70mm. H 81mm. Rim T 12mm.

40. Bag-shaped Crucible (E14:369 and E14:470B)
Fig. 131
Virtually complete vessel - missing some rim sherds. Reconstructed from two fragments. Faint heat-reddening visible internally, while approx. 40% of the exterior is covered with red vitreous material. E area, south of river.
D 71mm. H 85mm. Rim T 10mm.

41. Bag-shaped Crucible (E14:428)
Fig. 132
Virtually complete vessel - missing some rim sherds. No visible residues internally. The exterior is heavily encrusted with grey porous slag and traces of red vitreous material. This encrustation partly overflows the rim on one side, indicating the almost complete submersion of this crucible in the furnace or hearth. SE area.
D 70mm. H 81mm. Rim T 6mm, 12mm including slag.

42. Bag-shaped Crucible (E14:42)
Fig. 132
Virtually complete vessel - missing some rim sherds. Green metallic residue visible internally, traces of red vitreous material externally. NW area.
D 69mm. H 80mm. Rim T 5mm,11mm including slag.

43. Bag-shaped Crucible (E14:431)
Fig. 133
Complete vessel. Green metallic residue visible internally. The exterior is heavily encrusted with slag and red vitreous material. The vessel is slightly bent out of shape, perhaps caused by handling with tongs. SE area.
D 59mm. H 74mm. Rim T 6mm, 11mm including slag.

44. Bag-shaped Crucible (E14:422)
Fig. 133
Virtually complete vessel; missing a rim sherd. No visible residues internally. The exterior is heavily encrusted with grey porous slag and red vitreous material. SE area.
D 79mm. H 90mm. Rim T 5mm, 11mm including slag.

45. Bag-shaped Crucible (E14:496 B)
Fig. 134
Sherd. No visible residues. W area.
H 28mm. Rim T 5mm.

46. Bag-shaped Crucible (E14:520)
Fig. 134
Fragment including rim. Approx. half of the exterior is covered with red vitreous material and dark grey slag. This spreads over the rim and onto the interior surface. A small drop of copper or bronze (approx. 6mm x 6mm) still adheres to the interior. Central, disturbed area.
D 48mm. H 60mm. Rim T 6mm.

47. Bag-shaped Crucible (E14:453)
Fig. 134
Complete vessel. Green residue visible internally. A layer of grey slag , approx. 6mm thick, covers the exterior of the vessel. The walls of the vessel are slightly indented, probably due to the use of a tongs to grip the crucible. SE area.
D 67mm. H 82mm. Rim T 10mm.

48. Bag-shaped Crucible (Viking Exhibit)
Fig. 135
A complete vessel with a small area of slag overhanging the rim. The interior bears trace green residue, the rim and exterior are covered with a layer of grey porous slag.
Rim T 7mm.

Flat-bottomed Crucibles

49. Flat-bottomed Crucible (E14:488 A,B)
Fig. 136
Fragment. The interior and rim are covered with dark grey slag, and traces of red and green vitreous material. W area.
D 15mm. H 24mm. Rim T 9mm.

50. Flat-bottomed Crucible (E14:518)
Fig. 136
Rim sherd. No visible residues externally. The interior is covered with black and red vitreous material. Disturbed central area.
H 24mm. Rim T 14mm.

51. Flat-bottomed Crucible (E14:259)
Fig. 136
Body sherd. No visible residues externally. The interior is covered with red vitreous material.
H. 20mm. Rim T 8mm.

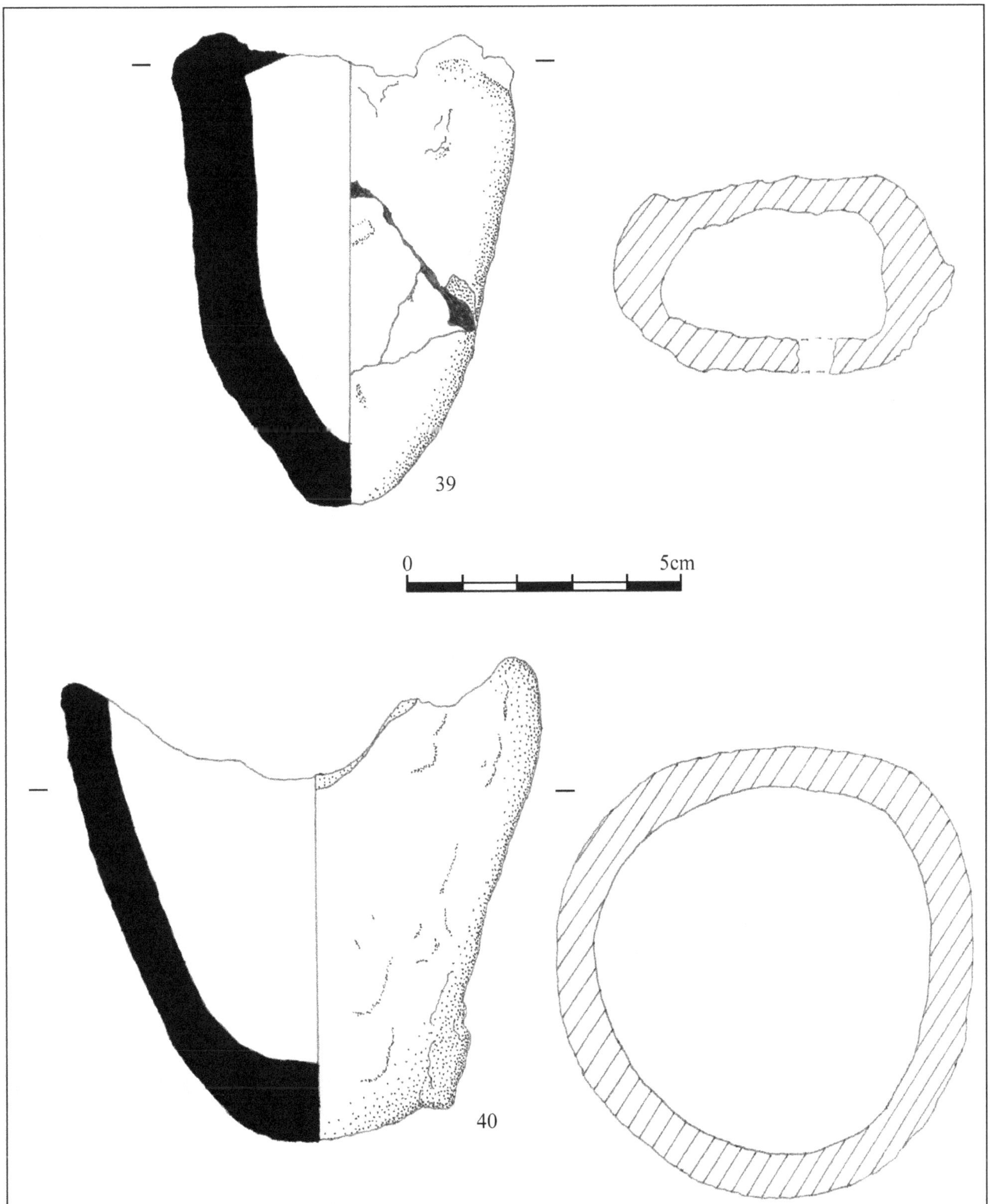

Fig. 131

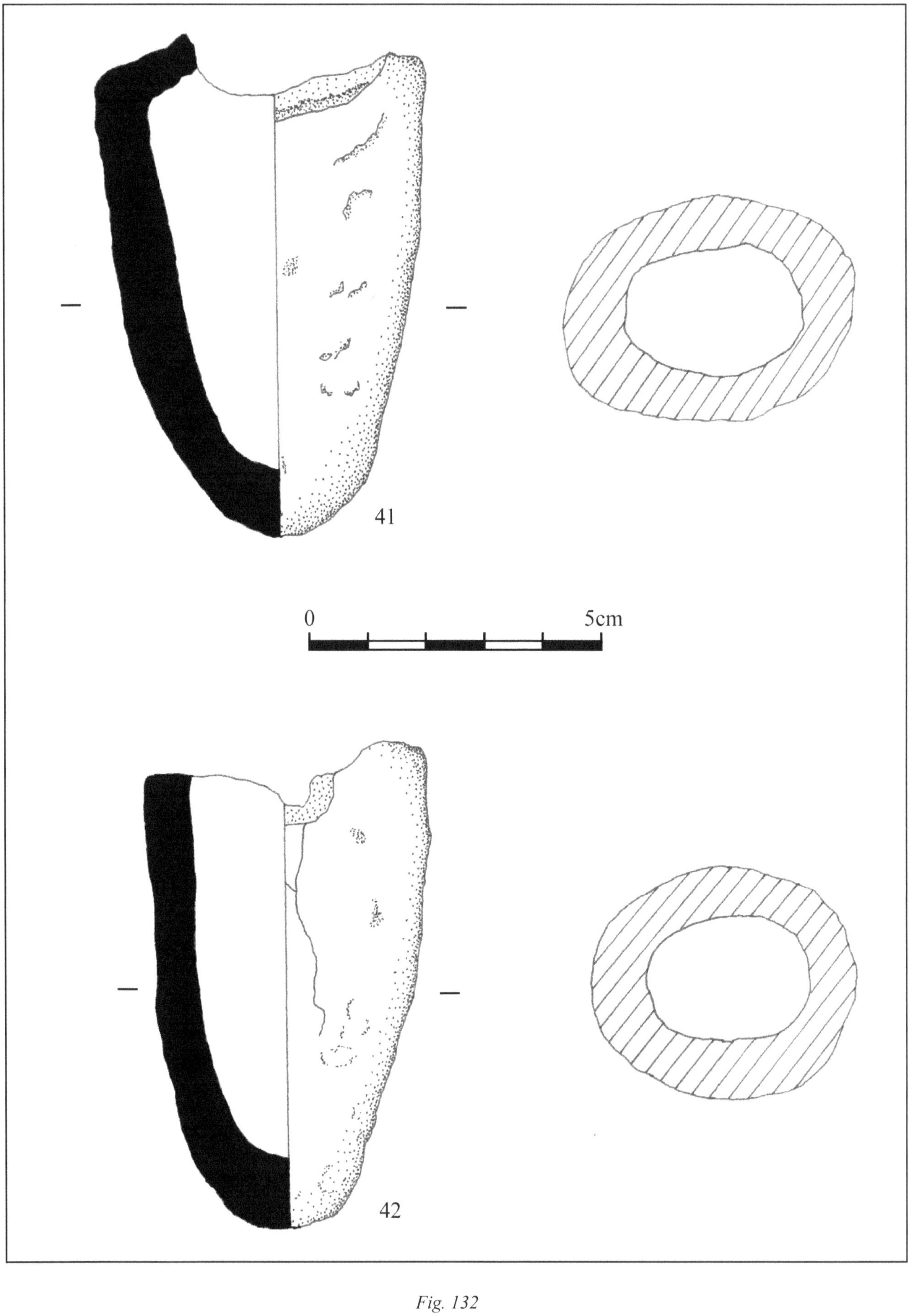

Fig. 132

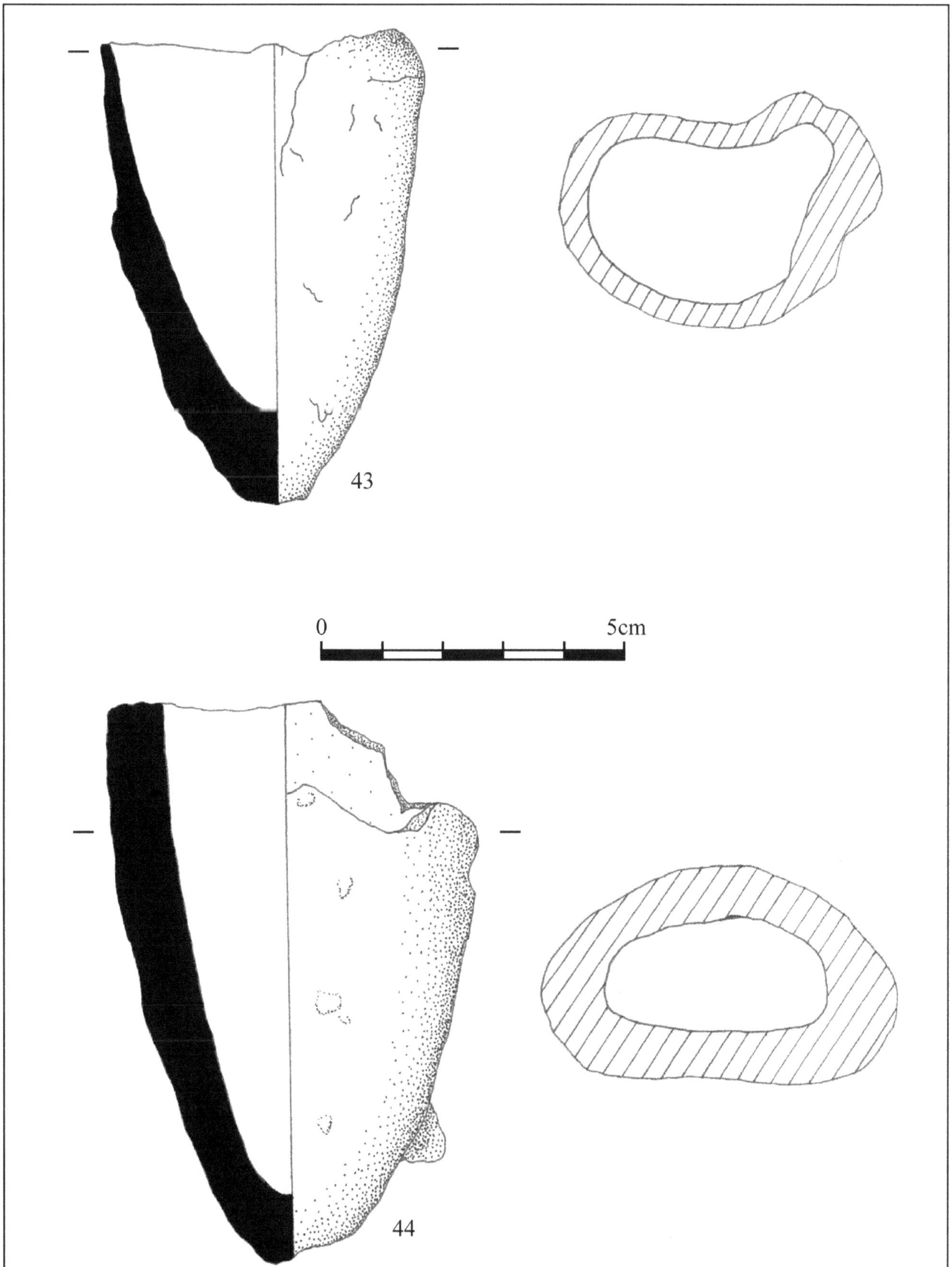

Fig. 133

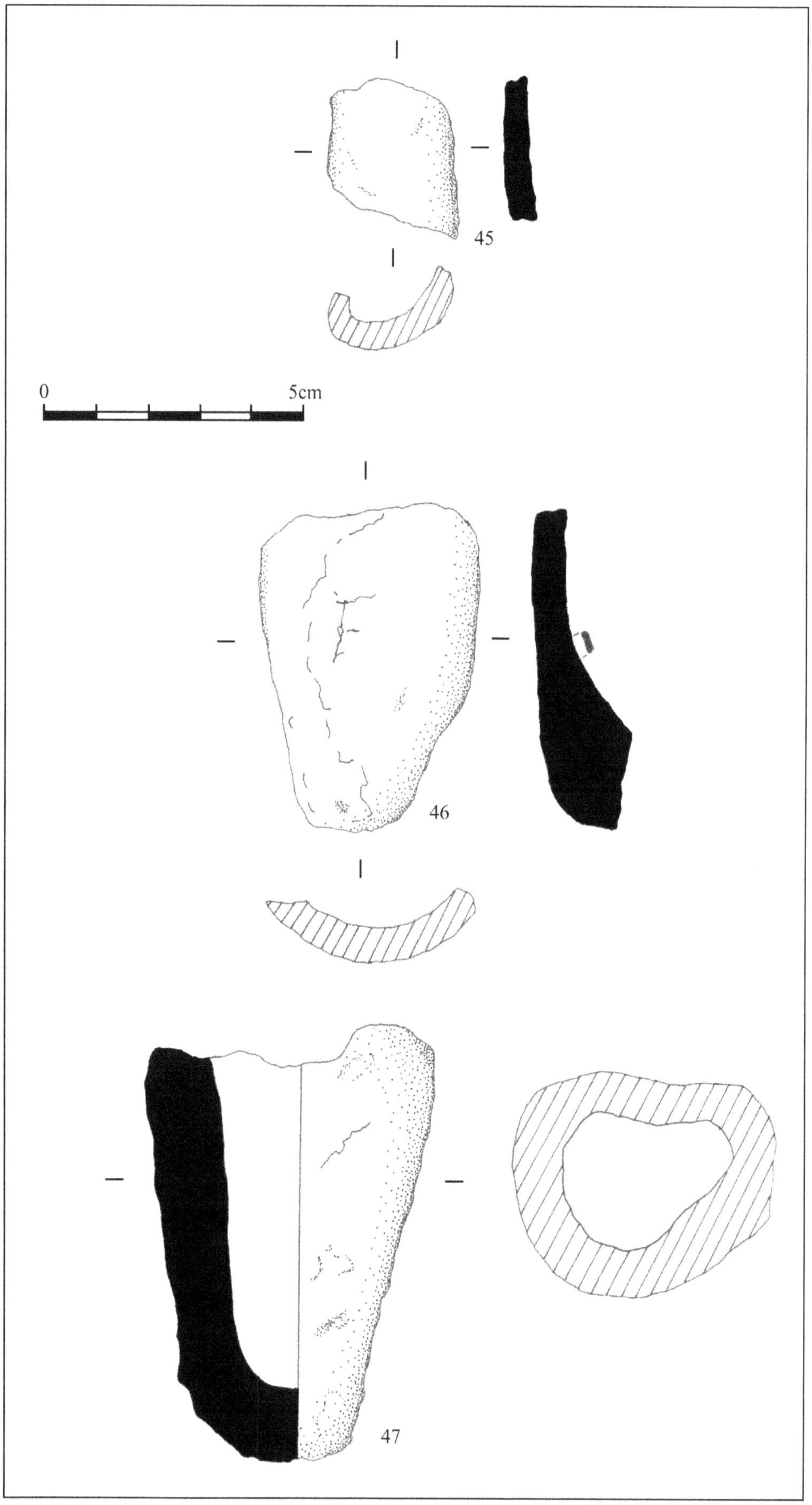

Fig. 134

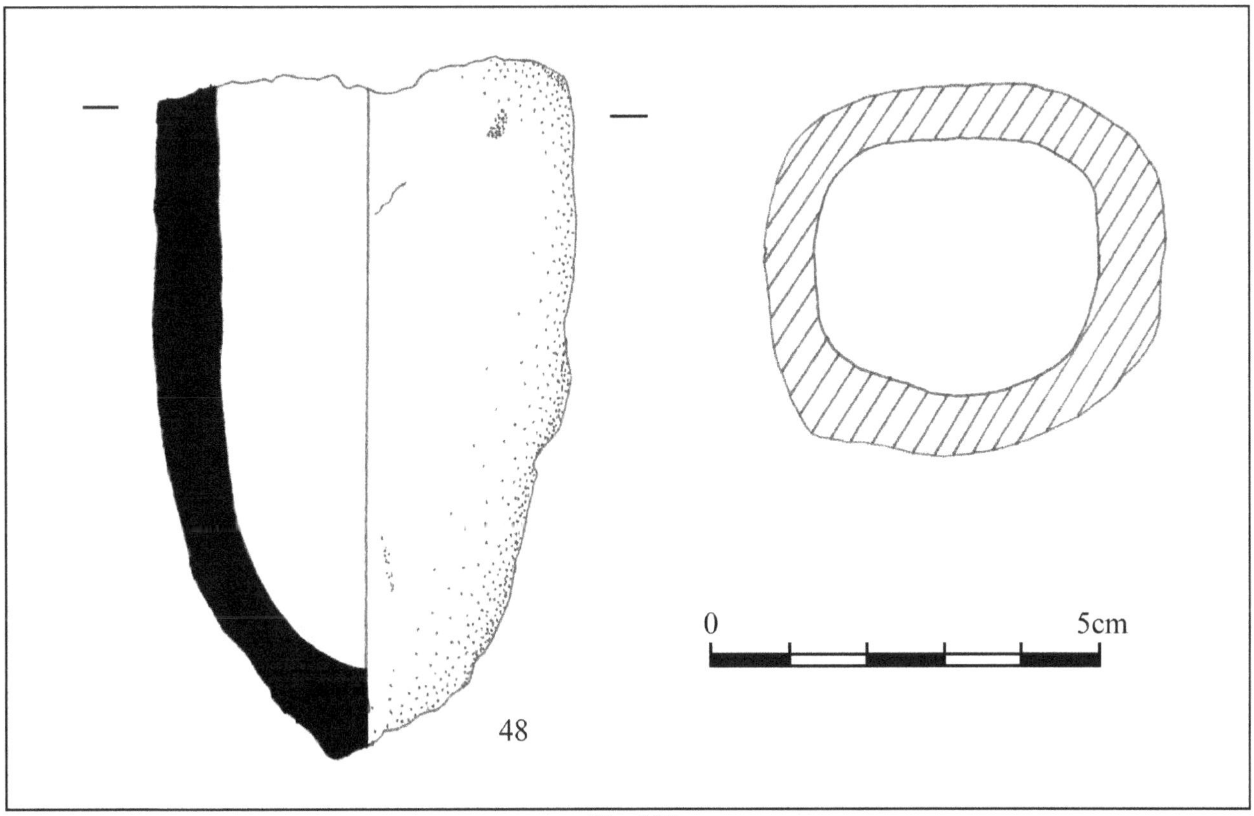

Fig. 135

52. Flat-bottomed Crucible (E14:401)
Fig. 136
Fragment. The rim and interior are covered with grey porous slag and red vitreous material. NE area.
D 9mm. H 14mm. Rim T 4mm.

53. Flat-bottomed Crucible (E14:435)
Fig. 136
Large fragment. The rim and interior are heavily coated with grey porous slag and red vitreous material. SE area.
D 11mm. H 20mm. Rim T 9mm.

54. Flat-bottomed Crucible (E14:280)
Fig. 136
Fragment. Red vitreous material covers the interior, the rim, spills down over one side and extends to the underside of the vessel. E area, north of river. Surface find.
D 8mm. H 13mm. Rim T 5mm.

55. Flat-bottomed Crucible (E14:257)
Fig. 137
Fragment. The interior is thickly coated with grey porous slag, the rim with red vitreous material which spills down one side.
D 11mm. H 16mm. Rim T 5mm.

56. Flat-bottomed Crucible (E14:270)
Fig. 137
Fragment. The interior of the rim is slightly blackened. No visible residues elsewhere. E area, north of river. Surface find.
D 10mm. H 17mm. Rim T 5mm.

57. Flat-bottomed Crucible (E14:288)
Fig. 137
Fragment. The interior and rim are covered with grey porous slag and red vitreous material. E area, north of river. Surface find.
D 5mm. H 10mm. Rim T 6mm.

58. Flat-bottomed Crucible (E14:292)
Fig. 137
Fragment. The interior and rim are covered with grey porous slag and red vitreous material. E area, north of river.
D 7mm. H 12mm. Rim T 6mm.

59. Flat-bottomed Crucible (E14:445)
Fig. 137
Fragment. A thin layer of grey porous slag covers the interior. The rim is coated with red vitreous material. SE area.
D 8mm. H 15mm. Rim T 4mm.

60. Flat-bottomed Crucible (E14:452)
Fig. 137
Fragment. The interior is coated with dark grey slag. SE area.
D 6mm. H 12mm. Diam. 38mm. Rim T 5mm.

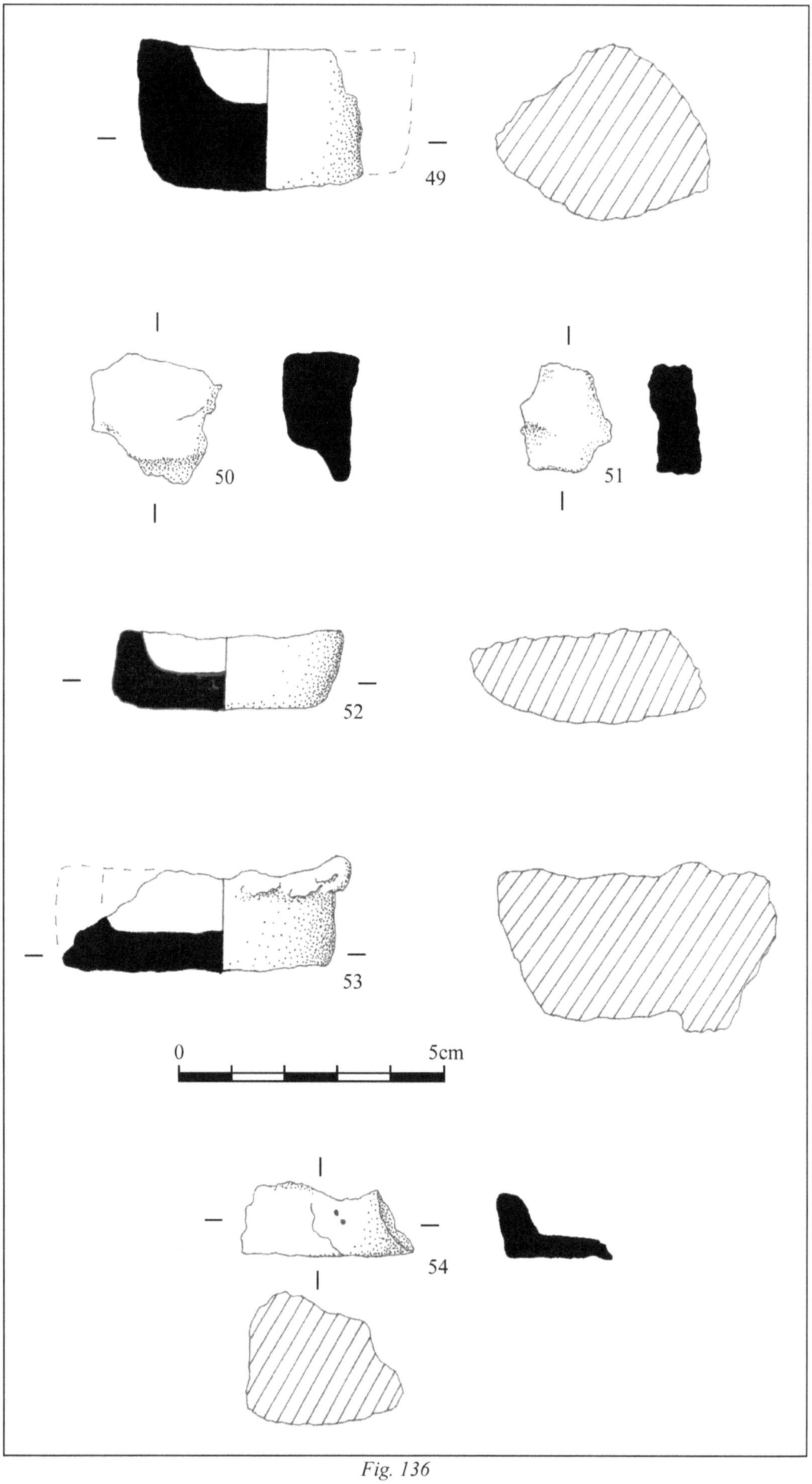

Fig. 136

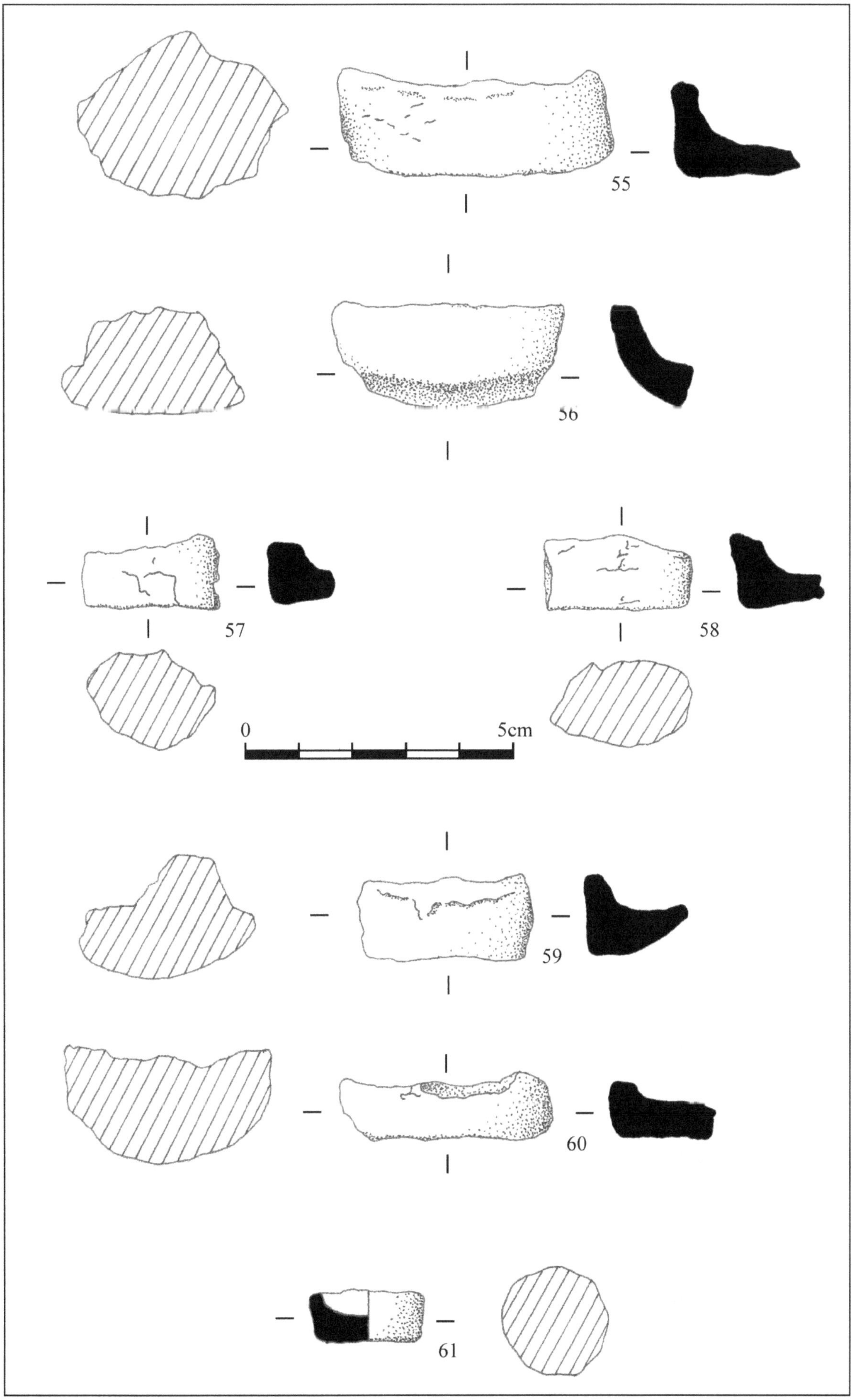

Fig. 137

61. Flat-bottomed Crucible (E14:453)
Fig. 137
Complete vessel. The interior is covered with a thin film of dark grey slag with red vitreous material near the rim. SE area.
D 5mm. H 10mm. Diam. 25mm. Rim T 2mm.

62. Flat-bottomed Crucible (E14:470 B)
Fig. 138
Fragment. A film of dark grey and red vitreous material covers the interior. SE area.
D 2mm. H 8mm. Rim T 4mm.

63. Flat-bottomed Crucible (E14:432)
Fig. 138
Rim sherd. Black vitreous material visible internally. SE area.
H 13mm. Rim T 10mm.

64. Flat-bottomed Crucible (E14:437)
Fig. 138
Sherd including rim. The interior is covered with red vitreous material. SE area.
D 10mm. H 16mm. Rim T 5mm.

65. Flat-bottomed Crucible (E14:439)
Fig. 138
Complete vessel. The interior is coated with red vitreous material, in which a circular impression is visible, possibly caused by the base of another crucible. NW area.
D 3mm. H 13mm. Diam. 40mm. Rim T 5mm.

66. Flat-bottomed Crucible (E14:366)
Fig. 138
Fragment including rim. Made of orange coloured clay as opposed to the hard grey of the other vessels. Grey and red vitreous material covers the interior, the rim, and spills over one side down onto the base. E area.
D 8mm. H 14mm. Rim T 5mm.

67. Flat-bottomed Crucible (E14:523)
Fig. 138
Fragment of a large vessel including rim. The interior is faintly heat-reddened. Disturbed central area.
D 32mm. H 42mm. Rim T 5mm.

68. Flat-bottomed Fragment (E14:416)
Fig. 139
Fragment including rim. The rim and interior are thickly slagged with two pieces of charcoal embedded in red vitreous material on the interior surface. NW area.
D 17mm. H 33mm. Rim T 13mm.

69. Flat-bottomed Crucible (E14:522)
Fig. 139
Complete vessel. The exterior is blackened. The rim and interior are heavily coated with red vitreous material, in which an impression is visible, possibly caused by the base of another crucible. There is also black vitreous material on the interior. Disturbed central area.
D 5mm. H 20mm. Diam. 42mm. Rim T 6mm.

70. Flat-bottomed Crucible (E14:414)
Fig. 139
Approx. half of a Flat-bottomed crucible. The rim bears grey porous slag while the interior is covered with black vitreous material. NE area.
D 8mm. H 16mm. Diam. 61mm. Rim T 6mm.

71. Flat-bottomed Crucible (E14:415)
Fig. 139
A complete vessel with black vitreous material and trace red vitreous material on the rim and interior. NE area.
D 13mm. H 21mm. Diam. 40mm. Rim T 5mm.

Crucible Bases

72. Crucible Base (E14:493)
Fig. 140
Clear glaze externally. W area.
D 6mm. H 20mm. Ext. Diam. 25mm. Base T 14mm. Rim T 7mm.

73. Crucible Base (E14:246)
Fig. 140
Light glaze externally with a trace of red vitreous material.
D 13mm. H 26mm. Ext. Diam. 37mm. Base T 13mm. Rim T 7mm.

74. Crucible Base (E14:377)
Fig. 140
Fragment. The exterior is covered with red vitreous material. NE area. NE area.
D 25mm. H 32mm. Base T 10mm. Rim T 7mm.

75. Crucible Base (E14:383)
Fig. 140
Fragment. The exterior is encrusted with grey porous slag and red vitreous material. NE area.
D 33mm. H 43mm. Base T 10mm. Rim T 6mm.

76. Crucible Base (E14:386)
Fig. 140
Fragment. The exterior is coated with red vitreous material. NE area.
D 19mm. H 38mm. Base T 19mm. Rim T 9mm.

77. Crucible Base (E14:392)
Fig. 140
Fragment. The interior is blackened. The exterior bears one small area of grey porous slag right at the bottom. NE area.
D 21mm. H 29mm. Base T 8mm. Rim T 4mm.

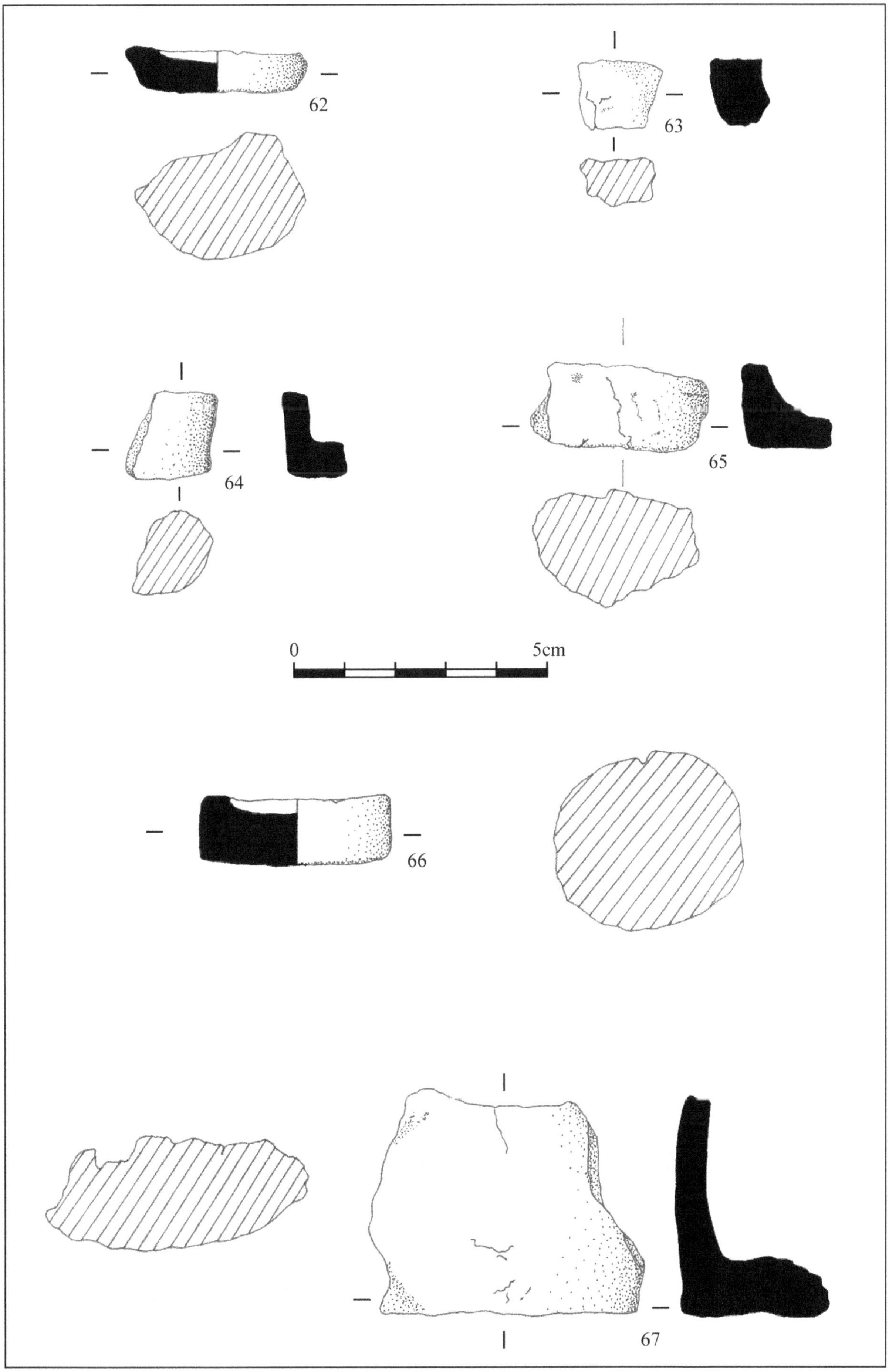

Fig. 138

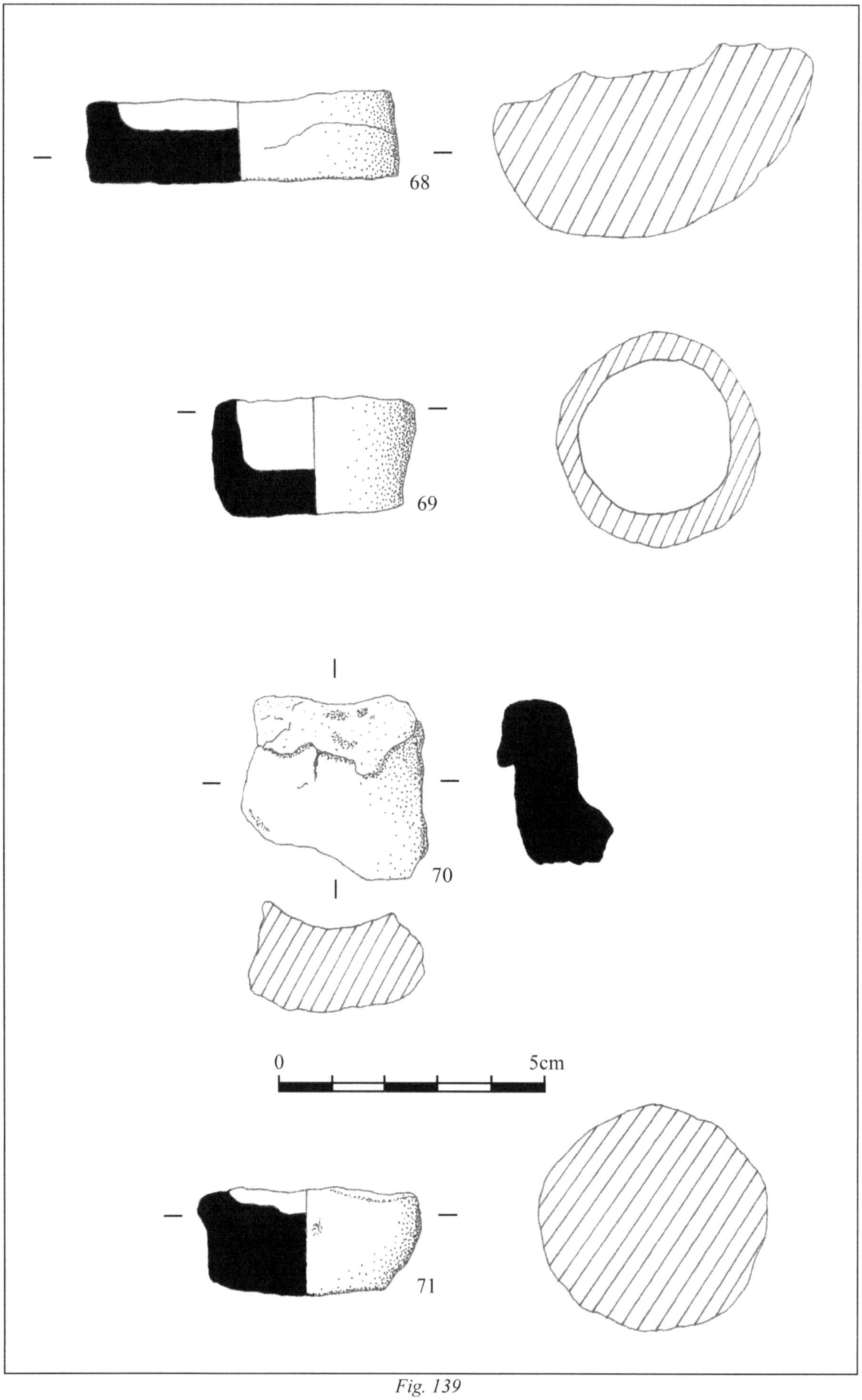

Fig. 139

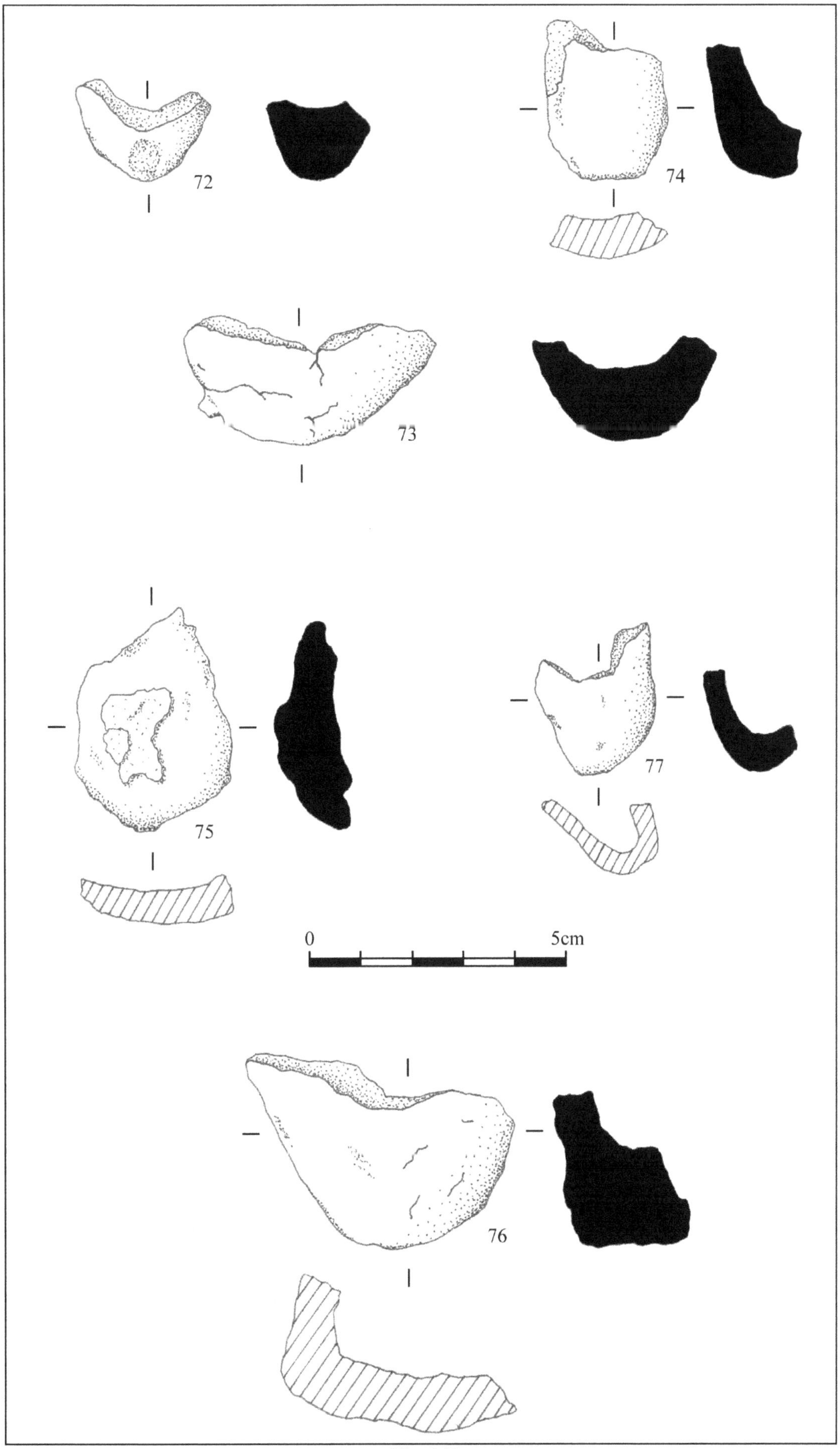

Fig. 140

78. Crucible Base (E14:517)
Fig. 141
The exterior is lightly glazed.
D 13mm. H 21mm. Ext. Diam. 33mm. Base T 8mm. Rim T 8mm.

79. Crucible Base (E14:470 E)
Fig. 141
Fragment. The exterior is completely covered with red vitreous material. SE area.
D 9mm. H 18mm. Base T 9mm. Rim T 2mm.

80. Crucible Base (E14:347)
Fig. 141
No visible residues.
D 8mm. H 15mm. Ext. Diam. 18mm. Base T 7mm. Rim T 5mm.

81. Crucible Base (E14:352 / 1307)
Fig. 141
Fragment. Glazed externally. E area, north of river. Found between palisades in brown peat layer 3.
D 10mm. H 15mm. Base T 5mm. Rim T 4mm.

82. Crucible Base (E14:279)
Fig. 141
The exterior is heavily coated with grey porous slag and red vitreous material. E area, north of river.
D 1mm. H 11mm. Ext. Diam. 30mm. Base T 10mm. Rim T 10mm.

83. Crucible Base (E14:358 A, B)
Fig. 141
Fragment. The exterior is encrusted with grey porous slag. E area, north of river.
D 18mm. H 32mm. Base T 14mm. Rim T 9mm.

84. Crucible Base (E14:363 A / 1511)
Fig. 141
The interior is slightly blackened. No visible residues externally. E area, north of river. Found in peat under brushwood layer 2.
D 11mm. H 21mm. Ext. Diam. 21mm. Base T 10mm. Rim T 5mm.

85. Crucible Base (E14:365 / 1464)
Fig. 142
Black vitreous material visible internally, red vitreous material externally. E area, south of river.
D 20mm. H 30mm. Ext. Diam. 19mm. Base T 10mm. Rim T 3mm.

86. Crucible Base (E14:269)
Fig. 142
Fragment. Approx. 40% of the exterior surface is covered by a light layer of grey porous slag. E area, north of river.
D 25mm. H 34mm. Base T 9mm. Rim T 6mm.

87. Crucible Base (E14:274)
Fig. 142
Fragment. The exterior is covered with grey porous slag and red vitreous material. E area, north of river.
D 20mm. H 29mm. Base T 9mm. Rim T 9mm.

88. Crucible Base (E14:275)
Fig. 142
Fragment. Traces of red vitreous material on the exterior. E area, north of river.
D 27mm. H 39mm. Base T 12mm. Rim T 6mm.

89. Crucible Base (E14:276 A)
Fig. 142
No visible residues. Flat bottom allows it to stand independent of any support. E area, north of river.
D 14mm. H 21mm. Ext. Diam. 30mm. Base T 7mm. Rim T 5mm.

90. Crucible Base (E14:291)
Fig. 142
No visible residues. E area, north of river.
D 6mm. H 13mm. Ext. Diam. 28mm. Base T 7mm. Rim T 6mm.

91. Crucible Base (E14:293)
Fig. 142
Fragment. Red vitreous material visible externally. E area, north of river.
D 27mm. H 35mm. Base T 8mm. Rim T 4mm.

92. Crucible Base (E14:455)
Fig. 142
The exterior is encrusted with grey porous slag and red vitreous material. SE area.
D 11mm. H 19mm. Ext. Diam. 18mm. Base T 8mm. Rim T 5mm.

93. Crucible Base (E14:460)
Fig. 143
Fragment. Grey porous slag and occasional red vitreous material visible externally. SE area.
D 16mm. H 29mm. Base T 13mm. Rim T 7mm.

94. Crucible Base (E14:463)
Fig. 143
Faint traces of red vitreous material visible externally. SE area.
D 12mm. H 19mm. Ext. Diam. 35mm. Base T 7mm. Rim T 7mm.

95. Crucible Base (E14:465)
Fig. 143
Miniscule traces of green metallic residue visible on the interior. The exterior is heavily coated with grey porous slag and approx. 5% red vitreous material. SE area.
D 7mm. H 21mm. Ext. Diam. 32mm. Base T 14mm. Rim T 10mm.

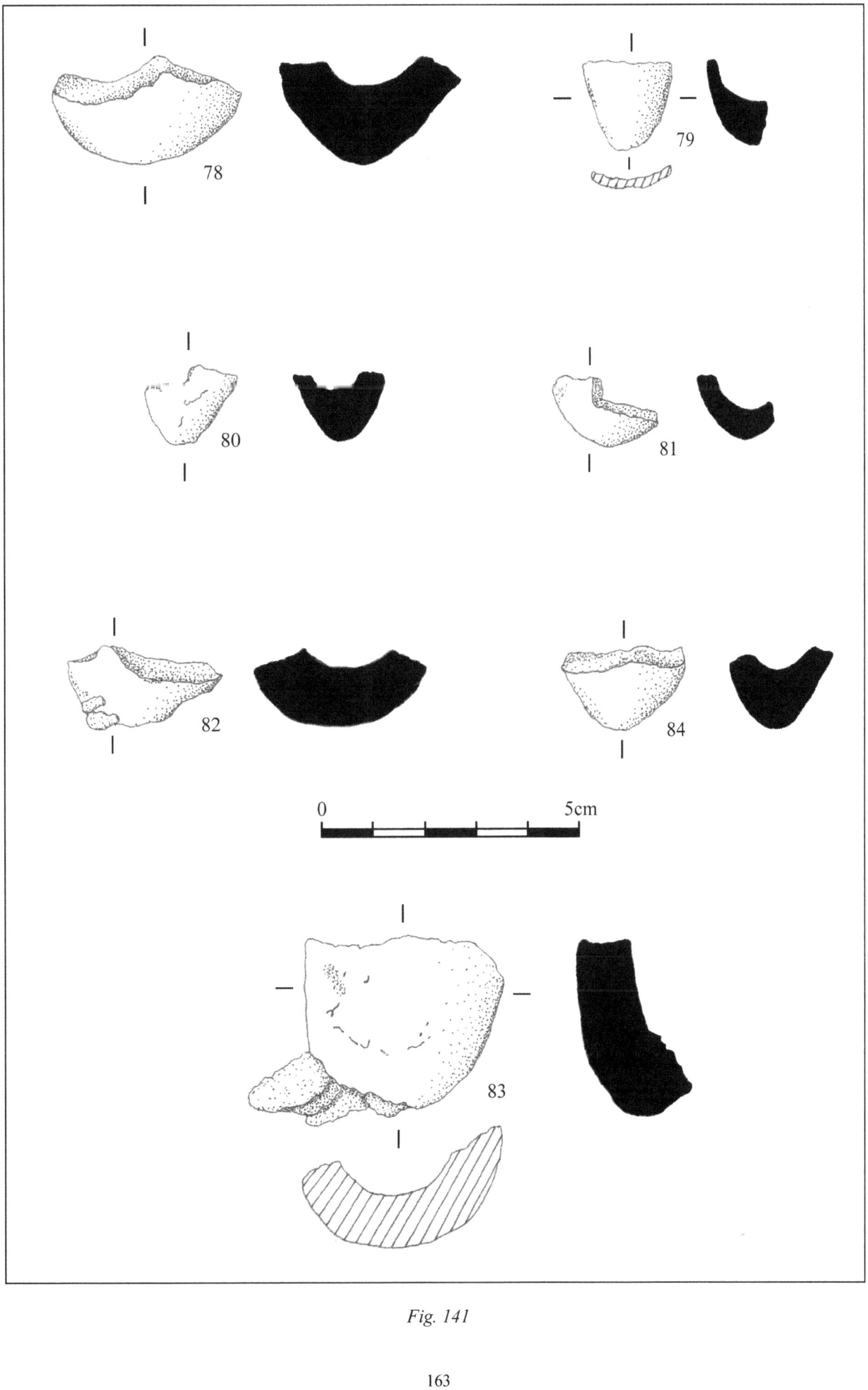

Fig. 141

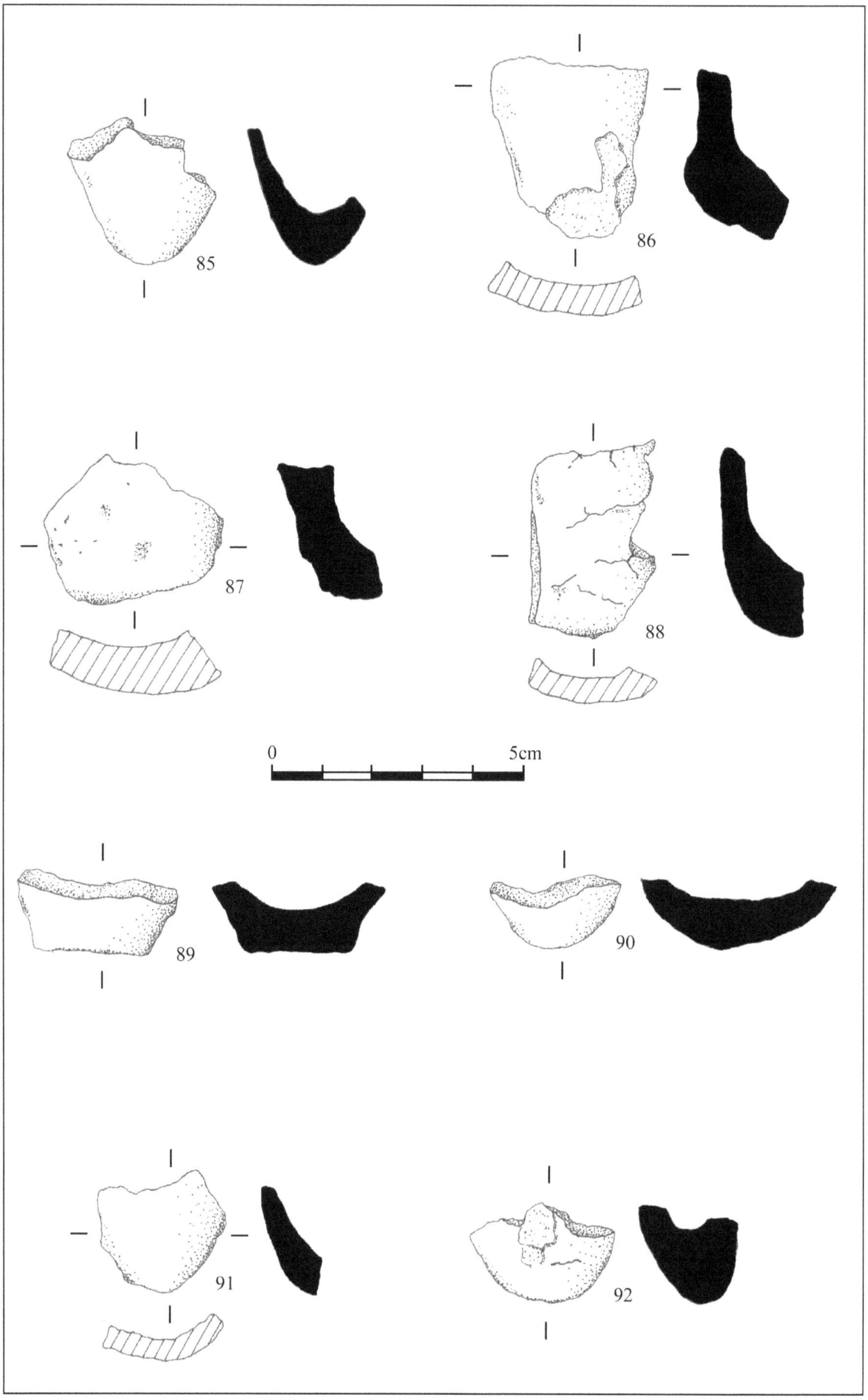

Fig. 142

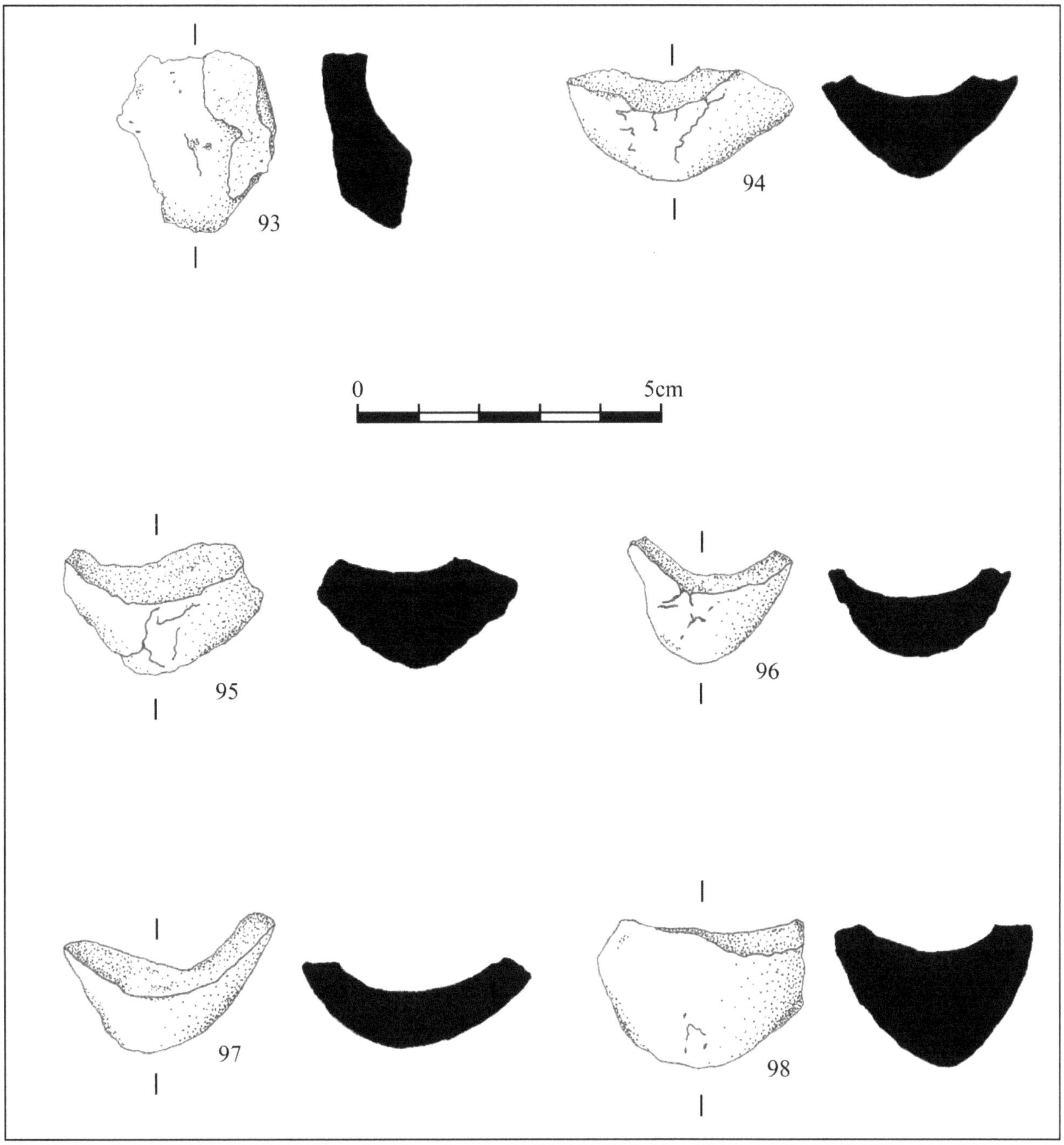

Fig. 143

96. Crucible Base (E14:466)
Fig. 143
Approx. 10% of the exterior is covered with grey porous slag and red vitreous material. SE area.
D 11mm. H 22mm. Ext. Diam. 30mm. Base T 11mm. Rim T 4mm.

97. Crucible Base (E14:418)
Fig. 143
The interior is blackened, the exterior glazed. NW area.
D 12mm. H 22mm. Ext. Diam. 35mm. Base T 10mm. Rim T 7mm.

98. Crucible Base (E14:434)
Fig. 143
The exterior is covered with grey porous slag and faint red vitreous material. NW area.
D 3mm. H 23mm. Ext. Diam. 35mm. Base T 20mm. Rim T 7mm.

Crucible Rims

99. Crucible Rim (E14:196)
Fig. 144
Sheet bears red vitreous material both internally and externally.
L 15mm. B 9mm. H 7mm.

100. Crucible Rim (E14:524)
Fig. 144
The interior is heat-reddened, the exterior covered with a layer of grey slag and trace red vitreous material. Disturbed central area. L 26mm. B 36mm. T 6mm.

101. Crucible Rim (E14:525)
Fig. 144
The interior is blackened, the exterior glazed. Disturbed central area. L 24mm. B 35mm. T 5mm.

102. Crucible Rim (E14:248)
Fig. 144
No visible residues. L 22mm. B 24mm. T 4mm.

103. Crucible Rim (E14:249)
Fig. 144
The exterior bears grey porous slag.
L 40mm. B 34mm. T 4mm.

104. Crucible Rim (E14:250 / 1298)
Fig. 144
The rim and approx. 30% of the exterior are with red vitreous material. E area, north of river. Found inside inner palisade in peaty material above crannóg.
L 32mm. B 37mm. T 5mm.

105. Crucible Rim (E14:262 / 1343)
Fig. 144
The interior is heat-reddened, the rim covered with red vitreous material. E area, north of river in brown gravelly soil. L 27mm. B 32mm. T 7mm.

106. Crucible Rim (E14:263 / 1300)
Fig. 144
The rim and exterior are covered with red vitreous material. E area, north of river. Found in hearth A, layer IV, inside inner palisade.
L 21mm. B 36mm. T 7mm.

107. Crucible Rim (E14:376)
Fig. 145
The exterior is glazed, with a small area of red vitreous material. NE area. L 28mm. B 32mm. T 3mm.

108. Crucible Rim (E14:380)
Fig. 145
The rim and exterior are covered with red vitreous material. NE area. L 31mm. B 28mm. T 6mm.

109. Crucible Rim (E14:381)
Fig. 145
The rim and exterior are encrusted with grey porous slag and red vitreous material. NE area.
L 29mm. B 26mm. T 5mm.

110. Crucible Rim (E14:385)
Fig. 145
Orange vitreous material is visible on the interior, a thin layer of grey slag and red vitreous material on the rim and exterior. NE area. 54mm x 41mm x 10mm.

111. Crucible Rim (E14:390)
Fig. 145
Dark green-black vitreous material visible externally, red vitreous material internally. NE area.
L 45mm. B 32mm. T 8mm.

112. Crucible Rim (E14:394)
Fig. 145
The interior is heat-reddened, the exterior covered with red vitreous material. NE area.
L 14mm.. B 26mm T 4mm.

113. Crucible Rim (E14:404)
Fig. 145
The exterior is covered with a layer of light grey slag. NE area.
L 23mm. B 23mm. T 6mm; slag 1mm, crucible 5mm.

114. Crucible Rim (E14:406)
Fig. 145
The exterior is covered with grey porous slag and red vitreous material. NE area. L 24mm. B 30mm. T 4mm.

115. Crucible Rim (E14:407)
Fig. 145
The exterior is covered with grey porous slag and red vitreous material. NE area. L 25mm. B 28mm. T 7mm.

116. Crucible Rim (E14:410)
Fig. 145
The exterior and the rim are covered with red vitreous material. NE area. L 26mm. B 40mm. T 4mm.

117. Crucible Rim (E14:411)
Fig. 146
The interior is blackened, the exterior covered with grey porous slag and red vitreous material. NE area.
L 41mm. B 39mm. T 5mm.

118. Crucible Rim (E14:372 B)
Fig. 146
The interior is heat-reddened, the exterior lightly slagged. NE area. L 14mm. B 18mm. T 5mm.

119. Crucible Rim (E14:372 C)
Fig. 146
The rim and exterior are covered with red vitreous material and grey porous slag. NE area.
L 7mm. B 15mm. T 4mm.

120. Crucible Rim (E14:372 D)
Fig. 146
Completely covered with grey porous slag. NE area.
L 10mm. B 9mm. T 4mm.

121. Crucible Rim (1929:1324)
Fig. 146
The exterior and rim are covered with red vitreous material. L 26mm. B 30mm. T 6mm.

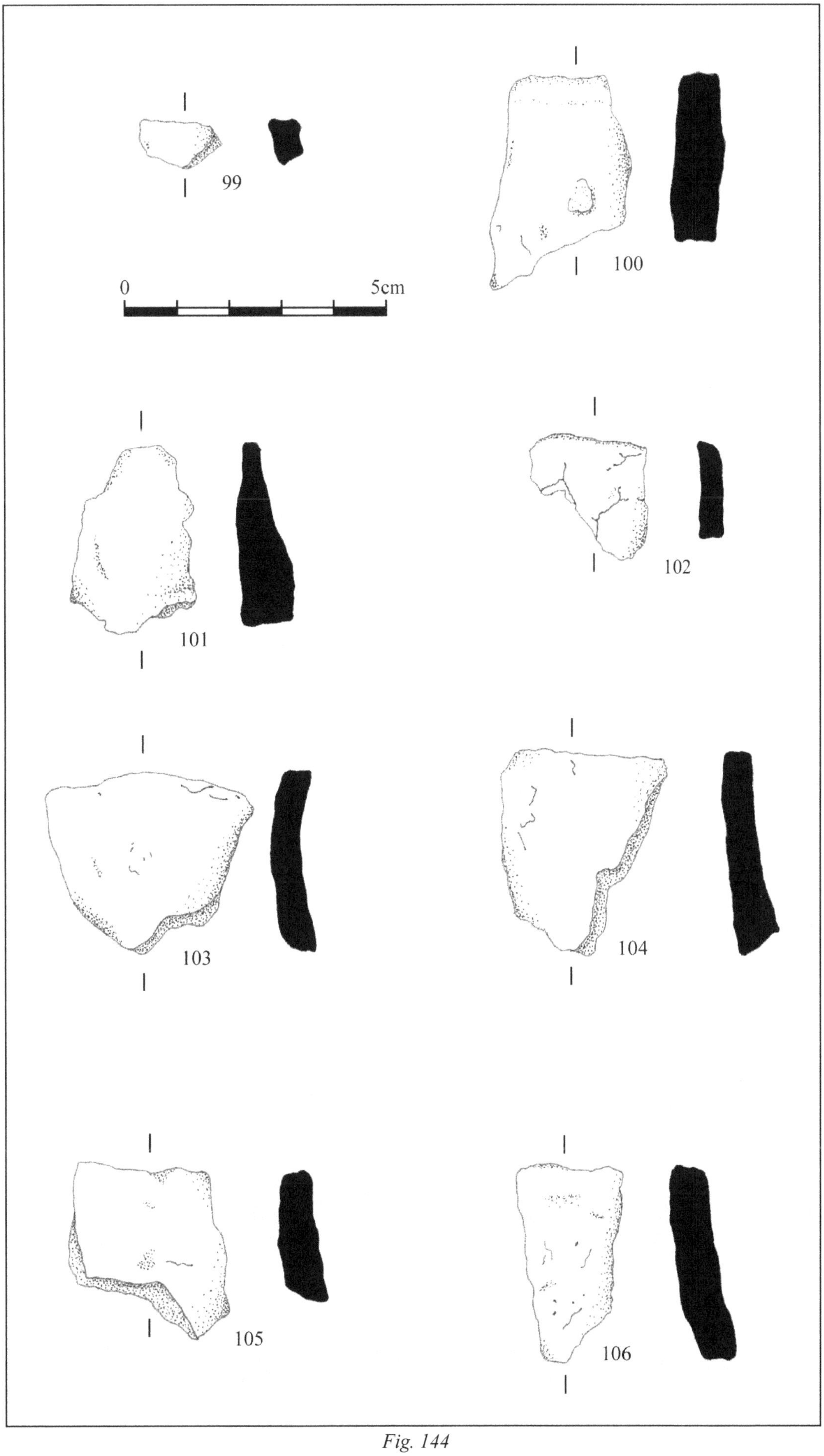

Fig. 144

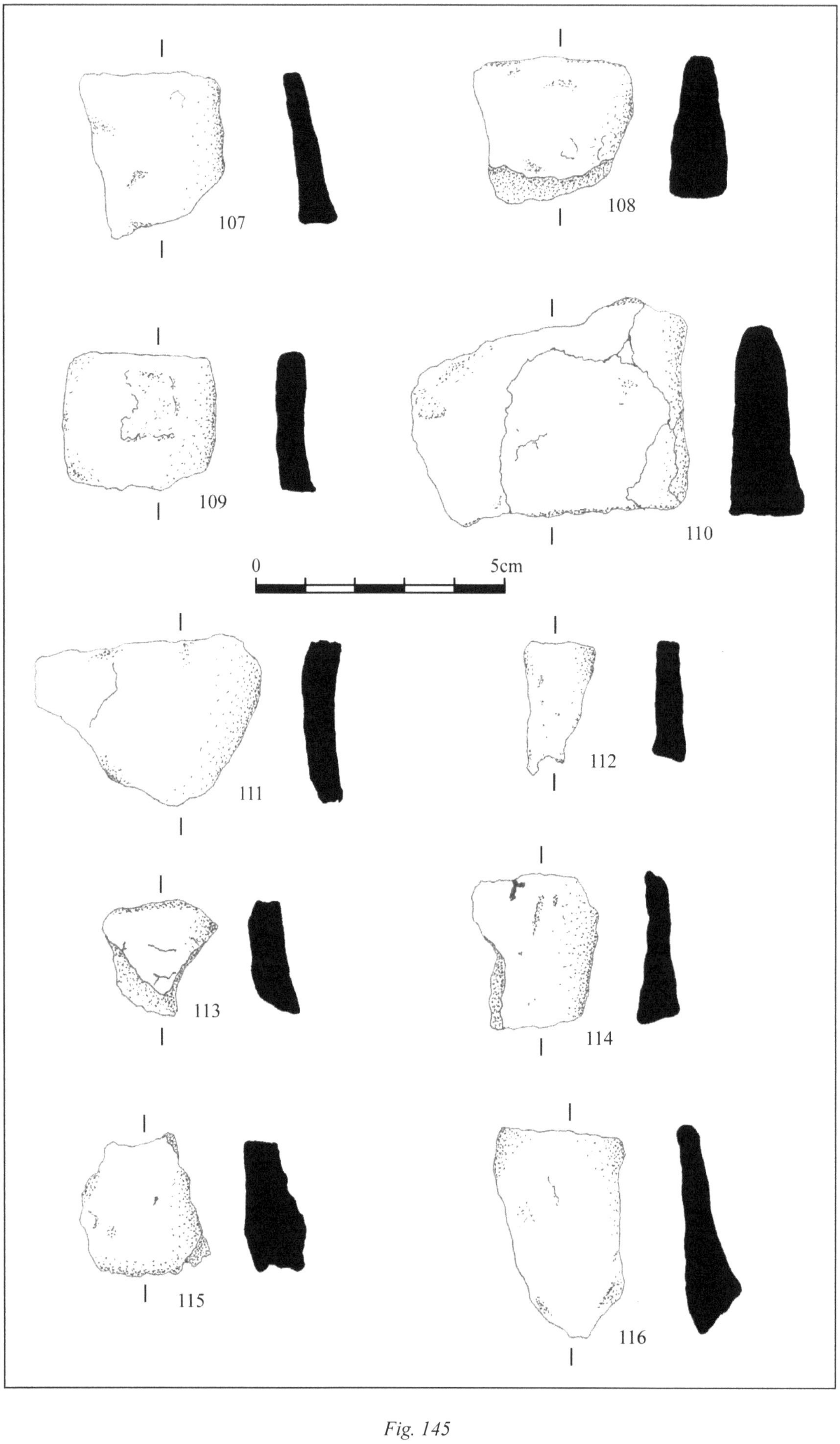

Fig. 145

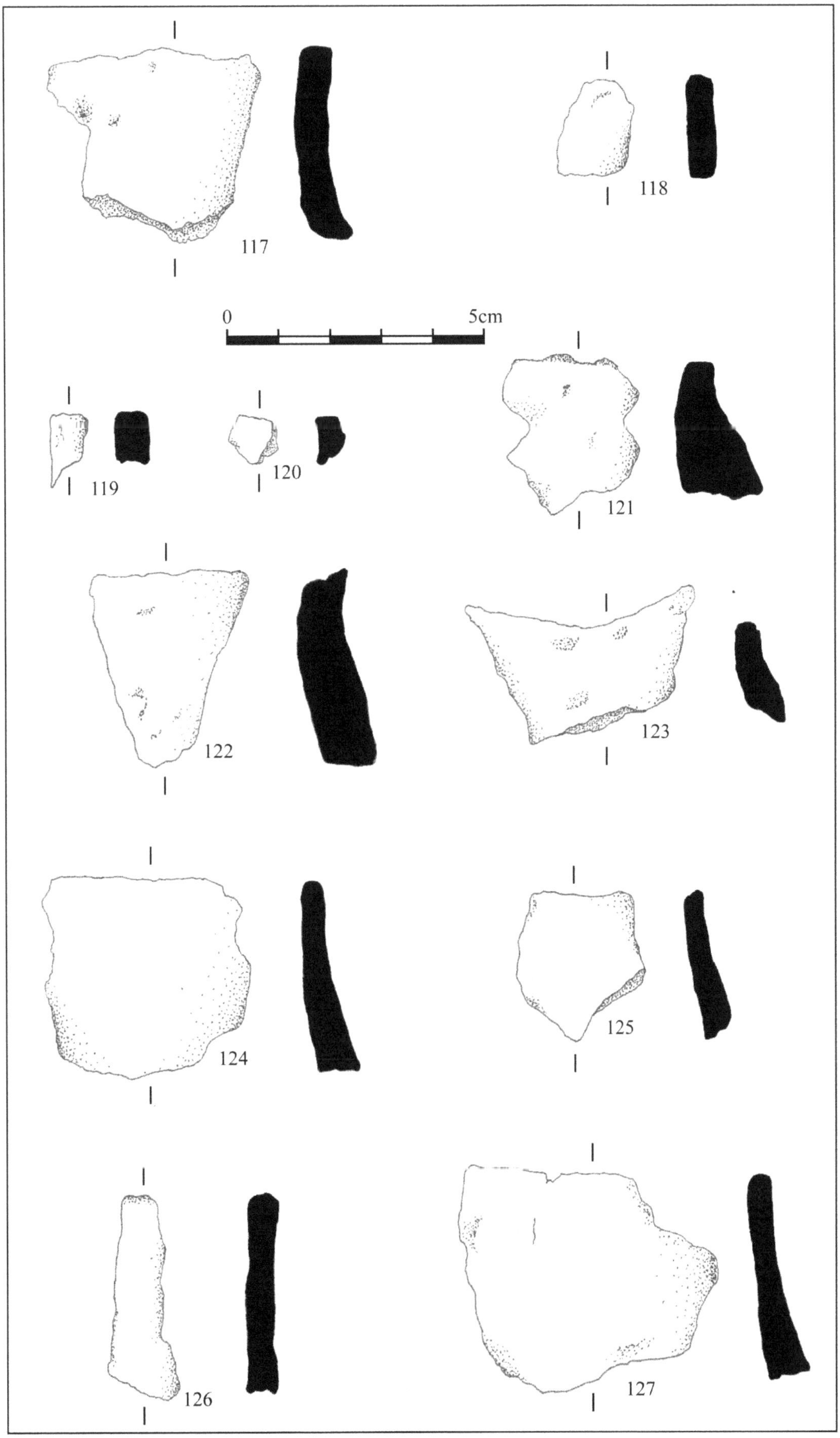

Fig. 146

122. Crucible Rim (E14:511)
Fig. 146
The rim, both internally and externally, is covered with red vitreous material.
L 31mm. B 37mm. T 8mm.

123. Crucible Rim (E14:512)
Fig. 146
The exterior is glazed.
L 45mm. B 27mm. T 4mm.

124. Crucible Rim (E14:513)
Fig. 146
The exterior bears a little red vitreous material.
L 37mm. B 38mm. T 4mm.

125. Crucible Rim (E14:514)
Fig. 146
No visible residues.
L 25mm. B 28mm. T 4mm.

126. Crucible Rim (E14:515)
Fig. 146
Approx. 20% of the exterior is covered with red vitreous material. L 38mm. B 13mm. T 6mm.

127. Crucible Rim (E14:516)
Fig. 146
No visible residues. L 49mm. B 42mm. T 5mm.

128. Crucible Rim (E14:478)
Fig. 147
The exterior and rim are covered with grey porous slag and red vitreous material. SW area.
L 16mm. B 27mm. T 6mm.

129. Crucible Rim (E14:469 and E14:470 F)
Fig. 147
The rim and approx. 30% of the exterior are covered with red vitreous material. SE area.
L 26mm. B 18mm. T 2mm.

130. Crucible Rim (E14:470 D)
Fig. 147
The exterior is glazed. SE area.
L 20mm. B 17mm. T 5mm.

131. Crucible Rim (E14:483)
Fig. 147
One surface bears a thin layer of black slag. SW area.
L 24mm. B 33mm. T 5mm.

132. Crucible Rim (E14:484)
Fig. 147
The exterior is covered with red vitreous material. SW area. L 30mm. B 42mm. T 8mm.

133. Crucible Rim (E14:504 / 1065)
Fig. 147
Completely covered with red vitreous material. WNW area. L 42mm. B 31mm. T 12mm.

134. Crucible Rim (E14:322 A)
Fig. 147
The interior is covered with red vitreous material. E area, north of river.
L 16mm. B 28mm. T 6mm.

135. Crucible Rim (E14:326)
Fig. 147
The rim and interior are covered with red vitreous material. E area, north of river.
L 27mm. B 34mm. T 6mm.

136. Crucible Rim (E14:328)
Fig. 147
The exterior is glazed. E area, north of river.
L 32mm. B 31mm. T 4mm.

137. Crucible Rim (E14:329)
Fig. 147
The exterior is covered with grey porous slag and green metallic residue. E area, north of river.
L 23mm. B 30mm. T 5mm.

138. Crucible Rim (E14:330 / 1574)
Fig. 147
The interior is slightly heat-reddened, the exterior and rim covered with red vitreous material and grey porous slag. E area, north of river in brushwood layer 1.
L 21mm. B 45mm. T 10mm.

139. Crucible Rim (E14:332 / 1299)
Fig. 148
The interior and rim are covered with red vitreous material. E area, north of river. Found inside inner palisade in brown peat layer 3.
L 16mm. B 27mm. T 5mm.

140. Crucible Rim (E14:334 / 1297)
Fig. 148
No visible residues. E area, north of river. Found inside inner palisade in black peaty material above crannóg. L 24mm. B 29mm. T 8mm.

141. Crucible Rim (E14:338 / 1391)
Fig. 148
The interior of the rim is covered with red vitreous material. E area, north of river. Found between palisades in brown peat layer 3.
L 15mm. B 25mm. T 3mm.

142. Crucible Rim (E14:344)
Fig. 148
The rim is encrusted with grey porous slag. E area, north of river.
L 23mm. B 31mm. T 9mm.

143. Crucible Rim (E14:349)
Fig. 148
The exterior is thickly coated with grey porous slag. E area, north of river. L 19mm. B 20mm. T 6mm.

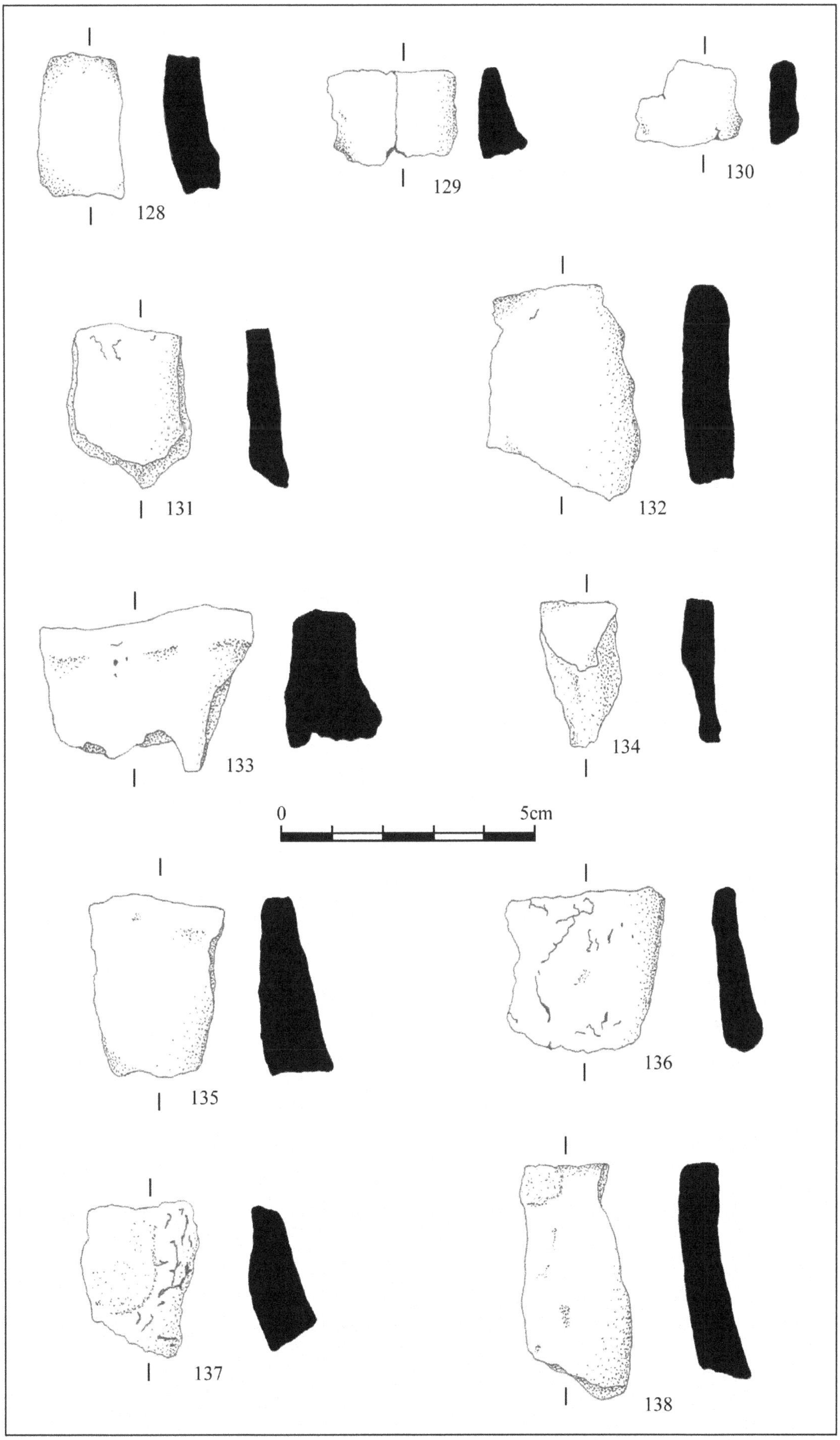

Fig. 147

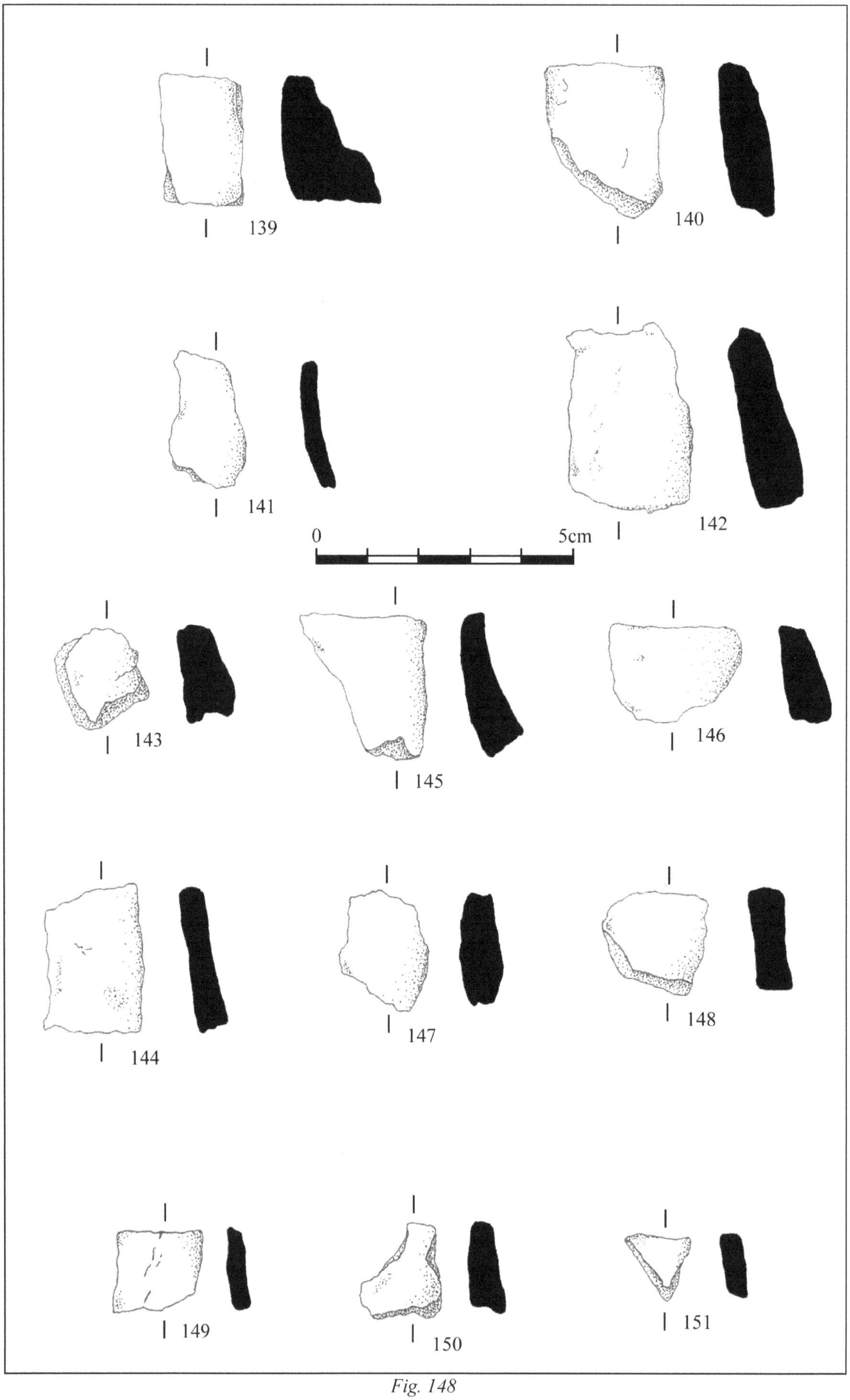

Fig. 148

144. Crucible Rim (E14:287)
Fig. 148
The rim and exterior are covered with grey porous slag and red vitreous material. E area, north of river.
L 20mm. B 29mm. T 5mm.

145. Crucible Rim (E14:285)
Fig. 148
The interior is heat-reddened, the exterior covered with red vitreous material. E area, north of river.
L 25mm. B 27mm. T 5mm.

146. Crucible Rim (E14:286)
Fig. 148
Completely covered with red vitreous material. E area, north of river. L 25mm. B 18mm. T 5mm.

147. Crucible Rim (E14:353 / 1350)
Fig. 148
A thin layer of grey porous slag covers the rim and interior. E area, north of river. Found just inside inner palisade in black peaty material.
L 17mm. B 24mm. T 6mm.

148. Crucible Rim (E14:355)
Fig. 148
The rim and exterior are covered with grey porous slag and red vitreous material. E area, north of river.
L 22mm. B 20mm. T 6mm.

149. Crucible Rim (E14:356 B / 1285)
Fig. 148
The interior is heat-reddened, the exterior glazed. E area, north of river. Found in sand layer 2 between palisades. L 17mm. B 16mm. T 3mm.

150. Crucible Rim (E14:357 / 1334)
Fig. 148
The interior is heat-reddened, the exterior bears trace red vitreous material. E area, north of river. Found inside inner palisade in black peaty material.
L 16mm. B 19mm. T 6mm.

151. Crucible Rim (E14:362 / 1450)
Fig. 148
Completely covered with red vitreous material. E area, north of river. Found between palisades in brown peat layer 3. L 13mm. B 12mm. T 4mm.

152. Crucible Rim (E14:266)
Fig. 149
The rim and exterior are covered with red vitreous material. E area, north of river.
L 30mm. B 33mm. T 6mm.

153. Crucible Rim (E14:265)
Fig. 149
The interior is blackened, the exterior showing faint red vitreous material. E area, north of river.
L 47mm. B 54mm. T 8mm.

154. Crucible Rim (E14:268)
Fig. 149
The interior is heat-reddened, the rim covered with red vitreous material. E area, north of river.
L 37mm. B 30mm. T 4mm.

155. Crucible Rim (E14:271)
Fig. 149
The exterior is covered with red vitreous material. E area, north of river. L 29mm. B 30mm. T 5mm.

156. Crucible Rim (E14:272 A)
Fig. 149
The exterior bears faint red vitreous material. E area, north of river. L 25mm. B 35mm. T 3mm.

157. Crucible Rim (E14:273)
Fig. 149
The exterior and rim are covered with grey porous slag and red vitreous material. E area, north of river.
L 17mm. B 39mm. T 4mm.

158. Crucible Rim (E14:278)
Fig. 149
The rim is covered with red vitreous material. E area, north of river. L 29mm. B 24mm. T 4mm.

159. Crucible Rim (E14:294)
Fig. 149
One surface is heavily coated with grey porous slag. E area, north of river. L 16mm. B 29mm. T 9mm.

160. Crucible Rim (E14:295)
Fig. 149
The rim and interior are covered with red vitreous material. E area, north of river.
L 18mm. B 20mm. T 3mm.

161. Crucible Rim (E14:299)
Fig. 149
The exterior is covered with red vitreous material. E area, north of river.
L 32mm. B 26mm. T 4mm.

162. Crucible Rim (E14:305)
Fig. 149
The rim bears trace red vitreous material, the exterior is glazed. E area, north of river.
L 25mm. B 28mm. T 3mm.

163. Crucible Rim (E14:306)
Fig. 149
The exterior is covered with grey porous slag. E area, north of river. L 18mm. B 20mm. T 8mm.

164. Crucible Rim (E14:308)
Fig. 149
One surface is lightly slagged and glazed. E area, north of river.
L 13mm. B 33mm. T 4mm.

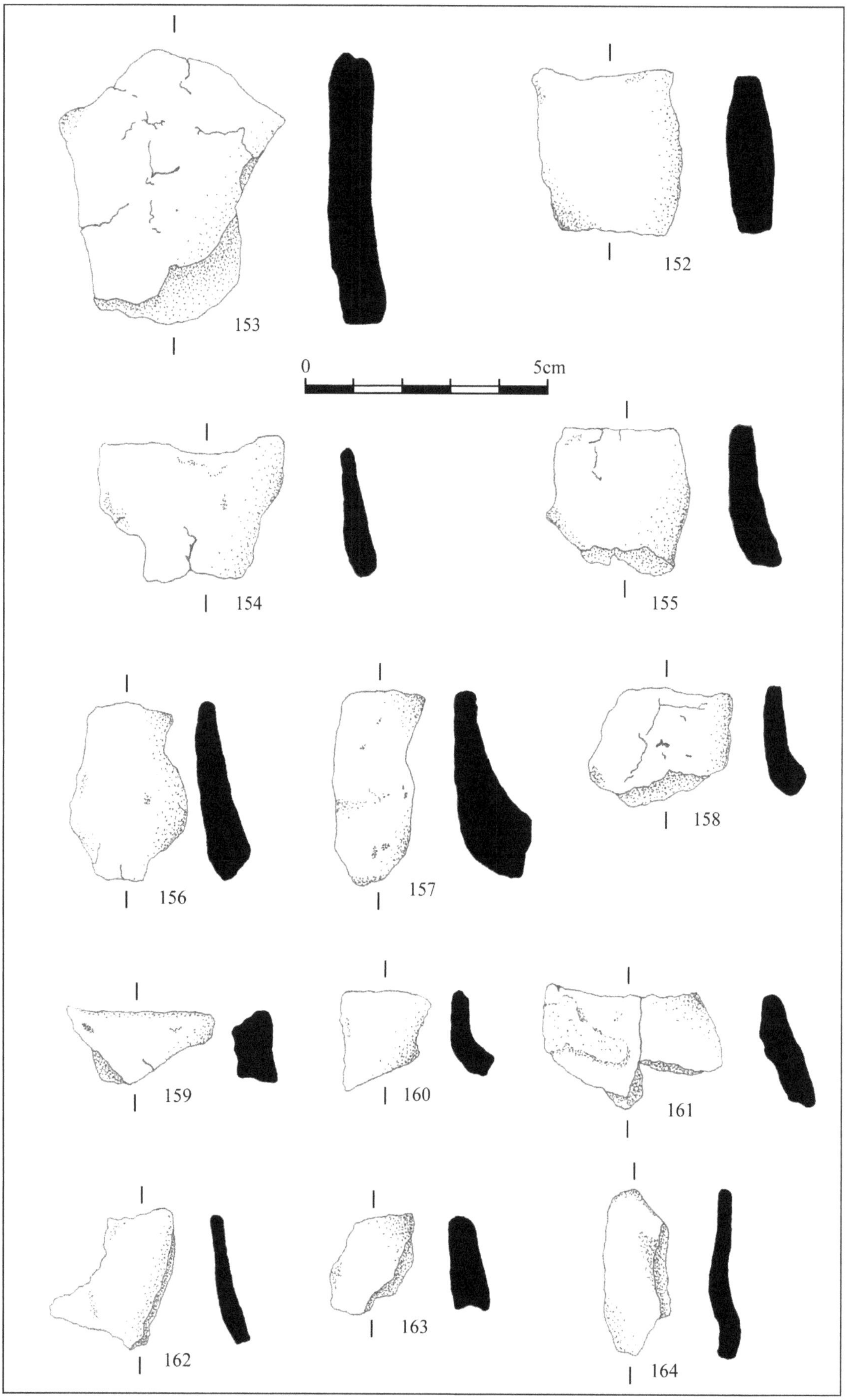

Fig. 149

165. Crucible Rim (E14:312)
Fig. 150
The interior is covered with grey porous slag, the exterior and rim with grey porous slag and red vitreous material. E area, north of river.
L 20mm. B 19mm. T 5mm.

166. Crucible Rim (E14:316)
Fig. 150
The exterior is covered with grey porous slag, with trace red vitreous material on the rim. E area, north of river. L 12mm. B 22mm. T 4mm.

167. Crucible Rim (E14:320)
Fig. 150
No visible residues. E area, north of river.
L 11mm. B 13mm. T 2mm.

168. Crucible Rim (E14:446)
Fig. 150
The exterior is covered with grey porous slag and trace red vitreous material. SE area.
L 23mm. B 32mm. T 5mm.

169. Crucible Rim (E14:447)
Fig. 150
The exterior is glazed. SE area.
L 35mm. B 31mm. T 5mm.

170. Crucible Rim (E14:451)
Fig. 150
No visible residues. SE area.
L 18mm. B 13mm. T 4mm.

171. Crucible Rim (E14:454)
Fig. 150
This sherd is made of orange clay. The exterior is covered with a patchy film of green vitreous material. SE area. L 32mm. B 39mm. T 7mm.

172. Crucible Rim (E14:462)
Fig. 150
Completely covered with dark grey porous slag and trace red vitreous material on the rim interior. SE area.
L 28mm. B 26mm. T 10mm.

173. Crucible Rim (E14:467)
Fig. 150
The exterior is covered with grey porous slag. SE area.
L 22mm. B 29mm. T 4mm.

174. Crucible Rim (E14:468)
Fig. 150
The interior is heat-reddened, the exterior covered with grey porous slag. SE area. L 13mm. B 26mm. T 8mm.

175. Crucible Rim (E14:417)
Fig. 150
The exterior is glazed, with a trace of red vitreous material also visible. NW area.
L 29mm. B 43mm. T 6mm.

176. Crucible Rim (E14:441 B)
Fig. 150
The interior is slightly blackened. NW area.
L 21mm. B 34mm. T 8mm.

177. Crucible Rim (E14:335 A)
Fig. 150
The interior is heat-reddened, the exterior covered with grey porous slag which slightly overhangs the rim. E area, north of river. Found in brown peat layer 3.
L 32mm. B 28mm. T 4mm.

178. Crucible Rim (E14:400 A, B)
Fig. 151
The rim and interior are covered with dark grey and red vitreous material. NE area. L 39mm. B 33mm. T 8mm.

179. Crucible Rim (E14:497)
Fig. 151
The rim, exterior and interior are covered with red vitreous material. WSW area.
L 21mm. B 29mm. T 5mm.

180. Crucible Rim (E14:502)
Fig. 151
Approx. 40% of the exterior is covered with grey porous slag, with trace red vitreous material near the rim. WSW area. L 42mm. B 40mm. T 7mm.

181. Crucible Rim (E14:499)
Fig. 151
The rim, interior and exterior are covered with red vitreous material. WSW area.
L 41mm. B 42mm. T 5mm.

182. Crucible Rim (E14:500)
Fig. 151
The rim and exterior are covered with grey porous slag and red vitreous material. WSW area.
L 29mm. B 34mm. T 7mm.

183. Crucible Rim (E14:269 B)
Fig. 151
The rim bears red vitreous material. E area, north of river. L 32mm. B 20mm. T 6mm.

Crucible Sherds

184. Crucible Sherd (E14:492)
Fig. 152
The interior is heat-reddened, the exterior covered with grey porous slag and red vitreous material. W area.
L 32mm. B 43mm. T 5mm.

185. Crucible Sherd (E14:495)
Fig. 152
No visible residues. W area.
L 21mm. B 18mm. T 6mm.

186. Crucible Sherd (E14:489)
Fig. 152
No visible residues. W area. L 25mm. B 28mm. T 3mm.

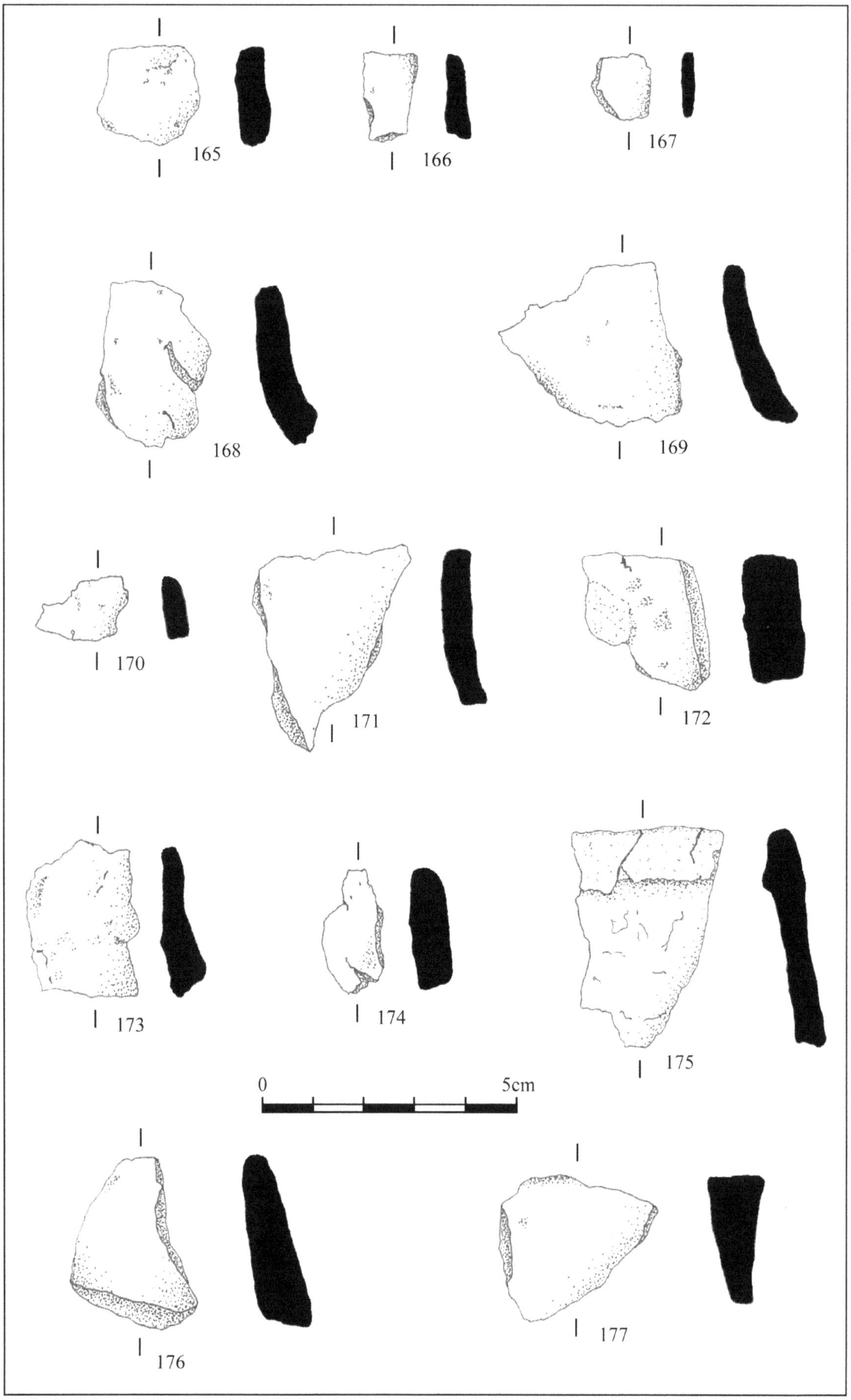

Fig. 150

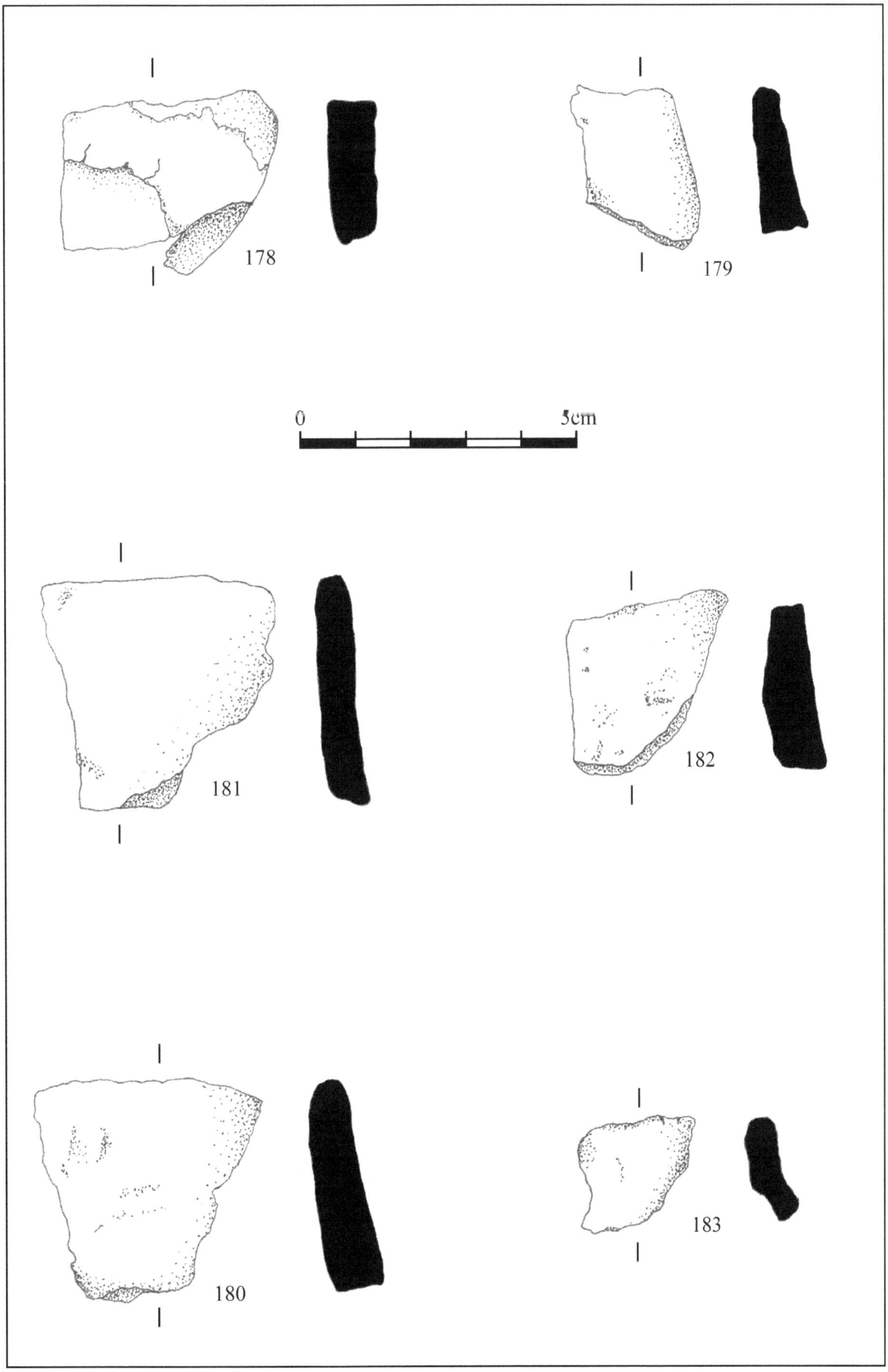

Fig. 151

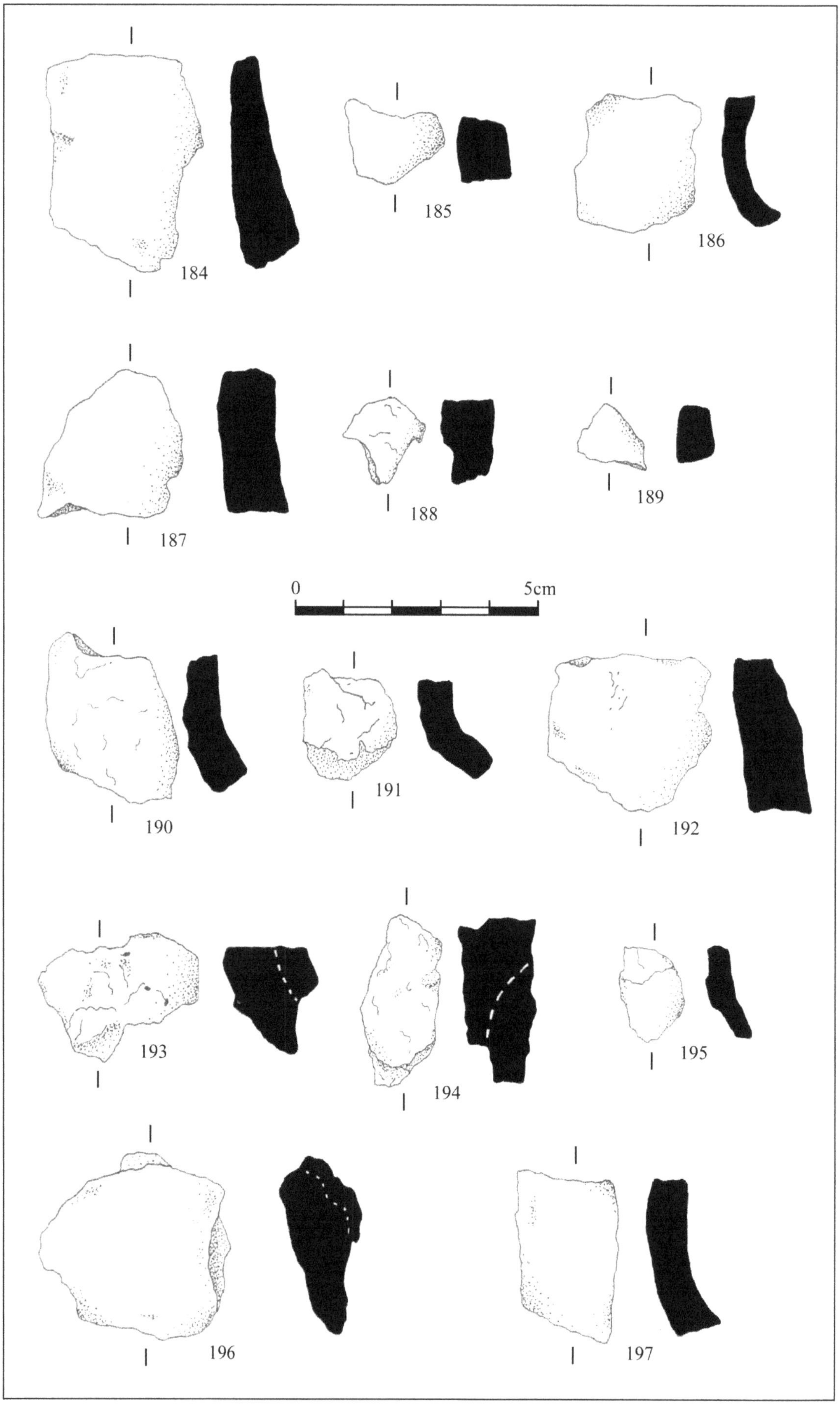

Fig. 152

187. Crucible Sherd (E14:496 A)
Fig. 152
Rust-coloured residue visible internally. W area.
L 29mm. B 33mm. T 10mm.

188. Crucible Sherd (E14:496 C)
Fig. 152
The exterior is covered with grey porous slag and red vitreous material. W area. L 17mm. B 18mm. T 12mm.

189. Crucible Sherd (E14:496 D)
Fig. 152
No visible residues. W area. L 15mm. B 14mm. T 6mm.

190. Crucible Sherd (E14:519)
Fig. 152
The exterior is encrusted with grey porous slag and red vitreous material. Disturbed central area.
L 23mm. B 34mm. T 3mm.

191. Crucible Sherd (E14:247)
Fig. 152
The exterior is lightly slagged.
L 20mm. B 23mm. T 10mm.

192. Crucible Sherd (E14:254)
Fig. 152
No visible residues. L 34mm. B 31mm. T 8mm.

193. Crucible Sherd (E14:255)
Fig. 152
Heavily coated with grey porous slag and trace red vitreous material. L 33mm B 25mm T 6mm (crucible); 21mm (including slag).

194. Crucible Sherd (E14:256)
Fig. 152
Heavily coated with grey porous slag.
L 35mm. B 16mm. T 7mm (crucible); 16mm (including slag).

195. Crucible Sherd (E14:257)
Fig. 152
No visible residues. L 19mm. B 13mm. T 5mm.

196. Crucible Sherd (E14:260 / 1341)
Fig. 152
One surface bears occasional encrustations of grey porous slag. E area, north of river. Found in brown gravelly soil. L 40mm. B 39mm. T 9mm.

197. Crucible Sherd (E14:261 / 1340)
Fig. 152
No visible residues. E area, north of river. Found in hearth A inside inner palisade.
L 21mm. B 34mm. T 7mm.

198. Crucible Sherd (E14:378)
Fig. 153
The exterior is glazed. NE area.
L 30mm. B 36mm. T 6mm.

199. Crucible Sherd (E14:382)
Fig. 153
The exterior is heavily slagged and also bears red vitreous material. NE area.
L 37mm. B 34mm T 7mm (crucible); 18mm (including slag).

200. Crucible Sherd (E14:387)
Fig. 153
The exterior is covered with grey porous slag and red vitreous material. NE area. L 22mm. B 32mm. T 6mm.

201. Crucible Sherd (E14:391)
Fig. 153
The exterior bears red vitreous material and is glazed. NE area. L 27mm. B 30mm. T 7mm.

202. Crucible Sherd (E14:393)
Fig. 153
The exterior is glazed. NE area.
L 21mm. B 23mm. T 6mm.

203. Crucible Sherd (E14:395)
Fig. 153
The exterior bears red vitreous material and is glazed. NE area. L 19mm. B 18mm. T 3mm.

204. Crucible Sherd (E14:397)
Fig. 153
The exterior bears red and black vitreous material. NE area.
L 27mm. B 36mm. T 5mm.

205. Crucible Sherd (E14:405)
Fig. 153
The exterior bears red vitreous material. NE area.
L 24mm. B 34mm. T 8mm.

206. Crucible Sherd (E14:372 A)
Fig. 153
The interior is heat-reddened, the exterior bears trace red vitreous material and grey porous slag. NE area.
L 22mm. B 21mm. T 7mm.

207. Crucible Sherd (E14:471)
Fig. 153
No visible residues. SW area.
L 26mm. B 17mm. T 10mm.

208. Crucible Sherd (E14:472)
Fig. 153
The exterior is glazed. SW area.
L 17mm. B 14mm. T 4mm.

209. Crucible Sherd (E14:473)
Fig. 153
The exterior bears red vitreous material and is glazed. SW area. L 21mm. B 9mm. T 5mm.

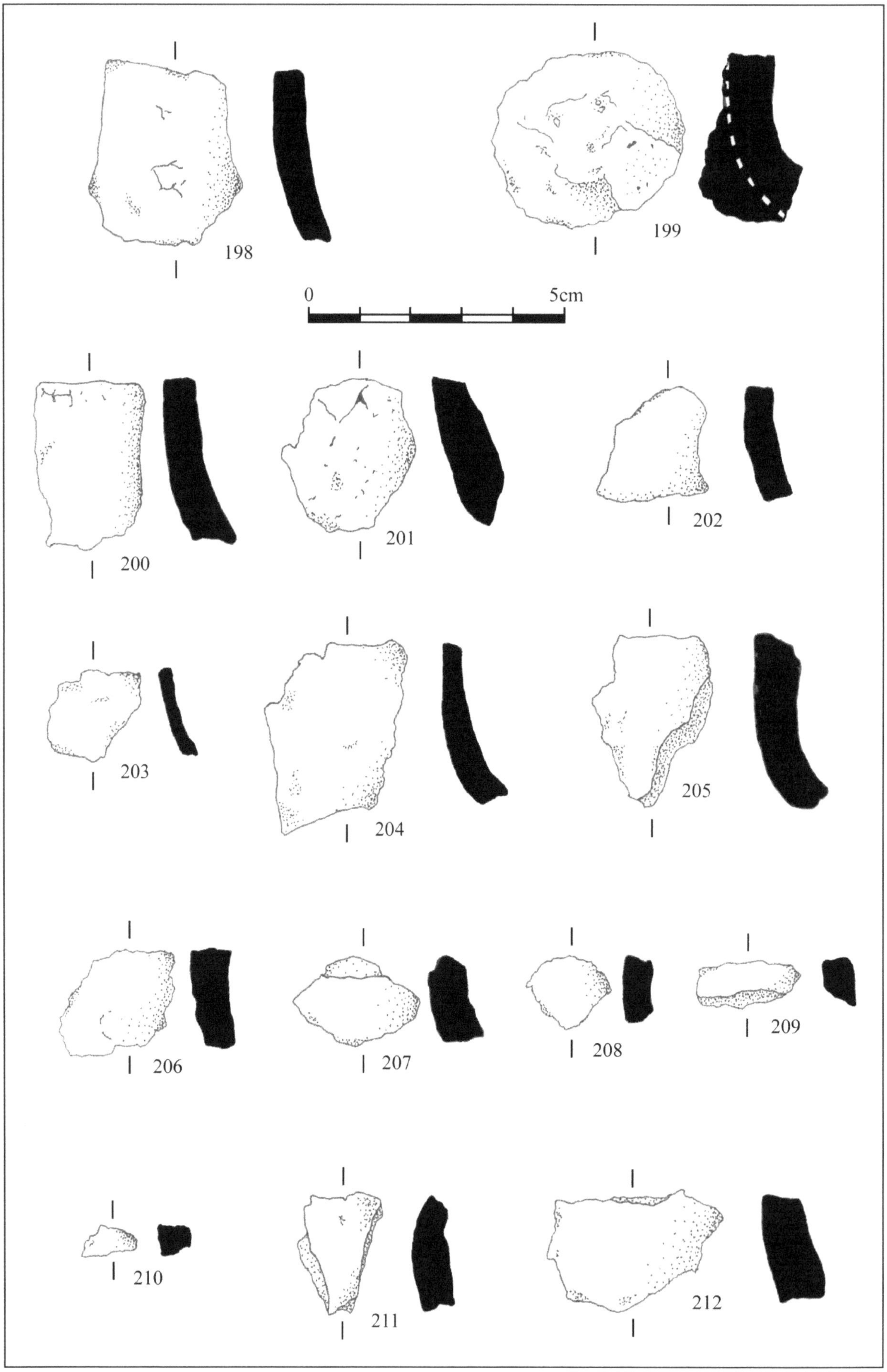

Fig. 153

210. Crucible Sherd (E14:474)
Fig. 153
The exterior bears red vitreous material and is glazed. SW area. L 12mm. B 6mm. T 7mm.

211. Crucible Sherd (E14:475)
Fig. 153
The interior is covered with a thin layer of grey porous slag and some green metallic residue. The exterior is also covered with a layer of grey porous slag. SW area. L 17mm. B 24mm. T 7mm.

212. Crucible Sherd (E14:476)
Fig. 153
The exterior bears red vitreous material and is glazed. SW area. L 33mm. B 22mm. T 7mm.

213. Crucible Sherd (E14:470 B)
Fig. 154
No visible residues. SE area. L 12mm. B 6mm. T 7mm.

214. Crucible Sherd (E14:482)
Fig. 154
A trace of green metallic residue is visible internally. The exterior is very heavily slagged and also bears red vitreous material. SW area. L 19mm. B 31mm. T 5mm (crucible); 13mm (including slag).

215. Crucible Sherd (E14:470 G)
Fig. 154
The interior is blackened, the exterior glazed. SE area. L 17mm. B 13mm. T 5mm.

216. Crucible Sherd (E14:505 / 1060)
Fig. 154
No visible residues. WNW disturbed area.
L 18mm. B 27mm. T 6mm.

217. Crucible Sherd (E14:333 / 1296)
Fig. 154
The exterior is covered with a layer of grey porous slag and red vitreous material. E area, north of river. Found in brown peat layer 3. L 25mm. B 3mm. T 10mm (crucible); 15mm (including slag).

218. Crucible Sherd (E14:335 B / 1303)
Fig. 154
The exterior is covered with a very thin layer of grey porous slag. E area, north of river. Found in brown peat layer 3. L 27mm. B 24mm. T 8mm.

219. Crucible Sherd (E14:336 / 1348)
Fig. 154
A thin layer of grey porous slag is visible on one surface, a rust-coloured residue on the other. E area, north of river. Found inside inner palisade in black peaty material. L 29mm. B 37mm. T 4mm.

220. Crucible Sherd (E14:337 / 1336)
Fig. 154
The exterior is covered with grey porous slag and red vitreous material. E area, north of river. Found inside inner palisade in black peaty material.
L 40mm. B 35mm. T 8mm.

221. Crucible Sherd (E14:339)
Fig. 154
The exterior is covered with a thin layer of grey porous slag. E area, north of river. L 23mm. B 34mm. T 6mm.

222. Crucible Sherd (E14:340)
Fig. 154
The exterior is covered with a layer of grey porous slag. E area, north of river. L 22mm. B 29mm. T 6mm (crucible); 10mm (including slag).

223. Crucible Sherd (E14:342)
Fig. 154
No visible residues. E area, north of river.
L 23mm. B 28mm. T 7mm.

224. Crucible Sherd (E14:343)
Fig. 154
Largely grey porous slag with a tiny crucible sherd attached. E area, north of river.
L 15mm. B 37mm. T 2mm (crucible); 11mm (including slag).

225. Crucible Sherd (E14:345)
Fig. 154
The exterior bears grey porous slag. E area, north of river. L 18mm. B 27mm. T 9mm.

226. Crucible Sherd (E14:346)
Fig. 154
The exterior bears grey porous slag. E area, north of river. L 29mm. B 25mm. T 9mm.

227. Crucible Sherd (E14:350 / 1389)
Fig. 154
No visible residues. E area, north of river. Found between palisades in sand layer 2.
L 16mm. B 24mm. T 6mm.

228. Crucible Sherd (E14:283)
Fig. 155
The exterior bears red vitreous material. E area, north of river. L 27mm. B 24mm. T 8mm.

229. Crucible Sherd (E14:354 / 1396)
Fig. 155
No visible residues. E area, north of river. Found between palisades in brown peat layer 3.
L 16mm. B 16mm. T 6mm.

230. Crucible Sherd (E14:356 A)
Fig. 155
The interior bears trace green metallic residue, the exterior red vitreous material and black slag. E area, north of river.
L 28mm. B 21mm. T 5mm.

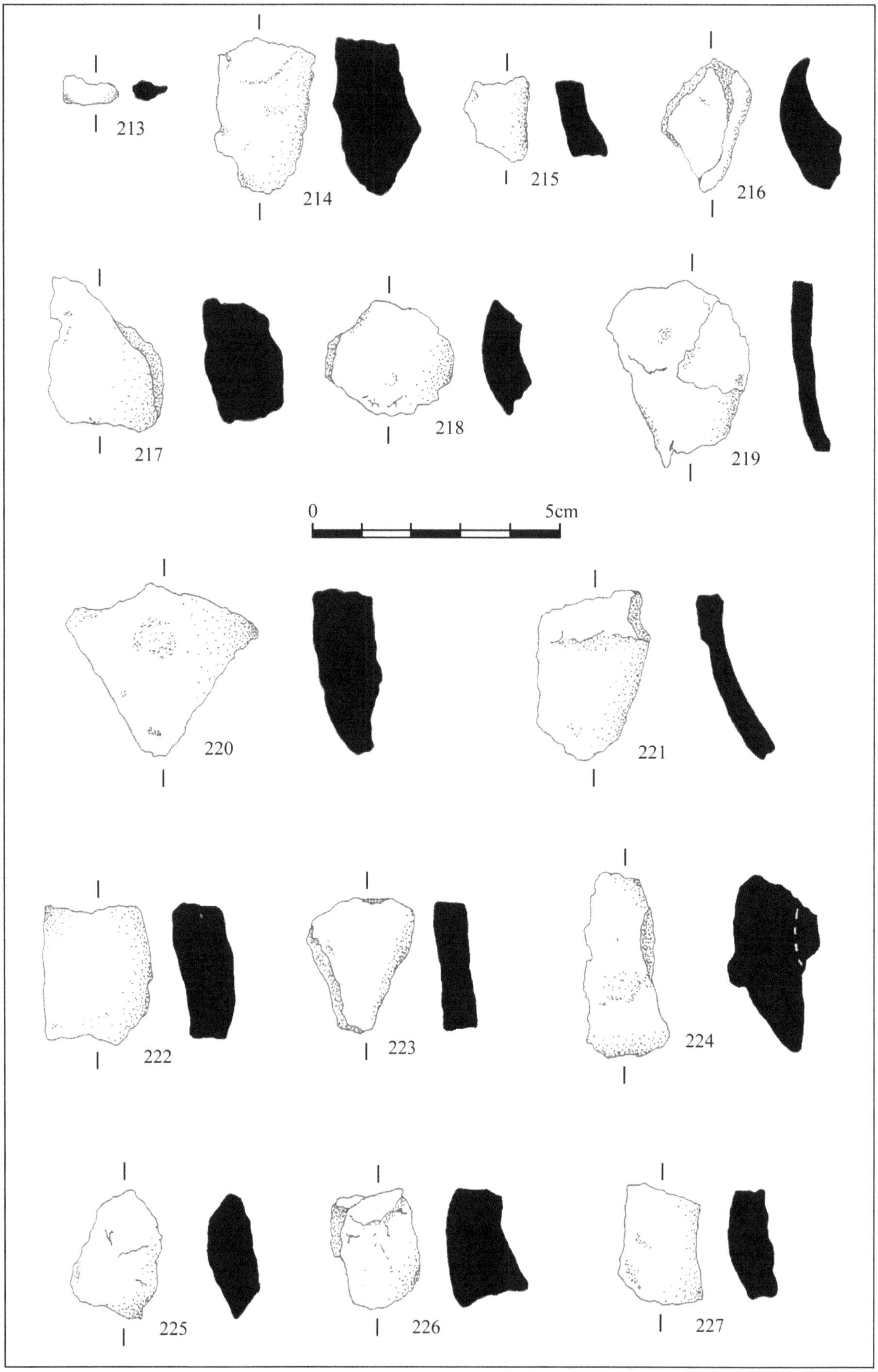

Fig. 154

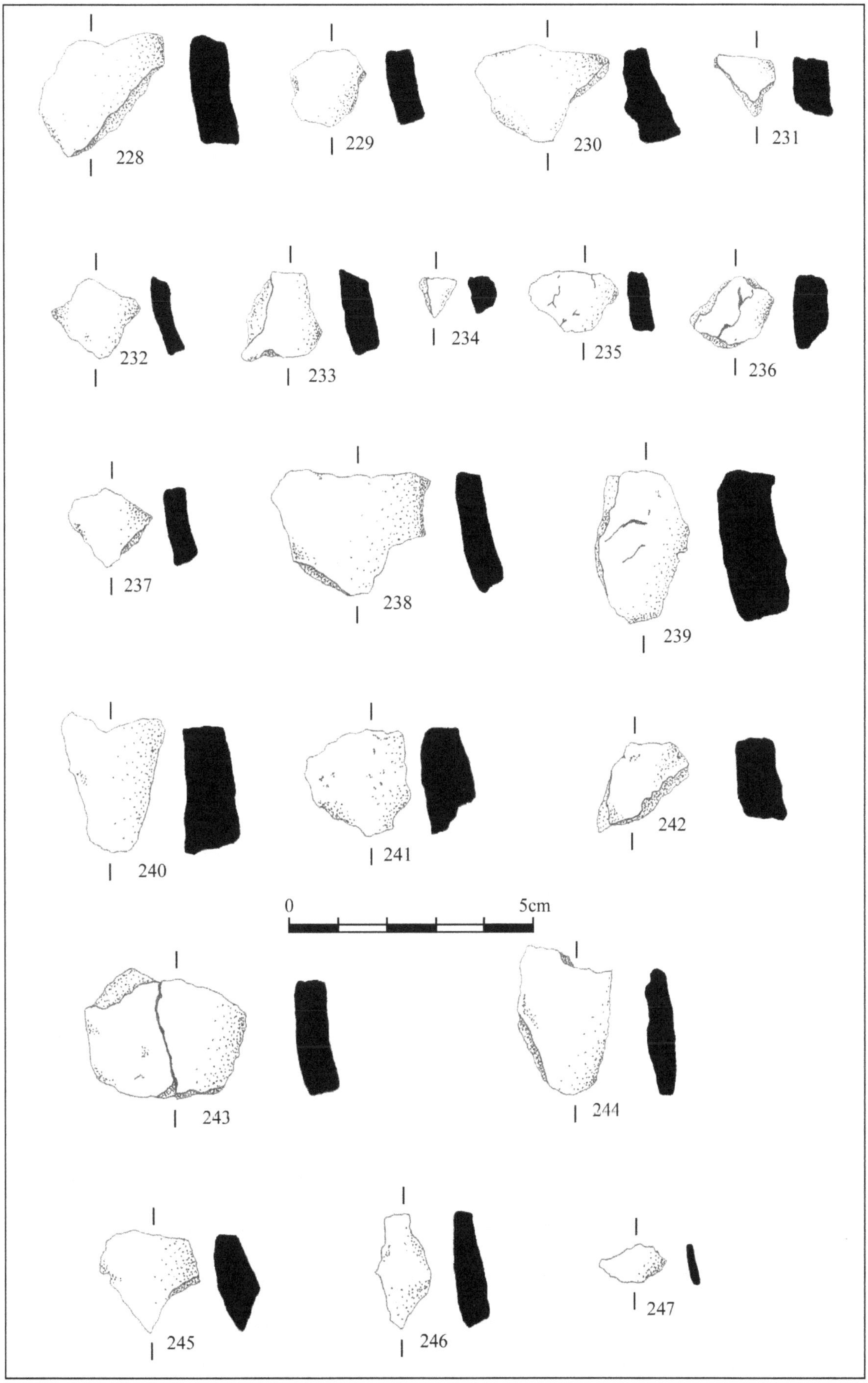
228
229
230
231
232
233
234
235
236
237
238
239
240
241
242
0
5cm
243
244
245
246
247

Fig. 155

231. Crucible Sherd (E14:359)
Fig. 155
The exterior is covered with grey porous slag and red vitreous material. E area, north of river.
L 13mm. B 12mm. T 8mm.

232. Crucible Sherd (E14:360 / 1342)
Fig. 155
The interior is heat-reddened. E area, north of river. Found in brown gravelly soil.
L 19mm. B 17mm. T 4mm.

233. Crucible Sherd (E14:361 / 1295)
Fig. 155
Approx. 50% of the exterior is covered with red vitreous material. E area, north of river. Found in hearth A, inside palisade. L 17mm. B 17mm. T 7mm.

234. Crucible Sherd (E14:364 / 1422)
Fig. 155
No visible residues. E area, north of river. Found between palisades in brown peat layer 3.
L 8mm. B 8mm. T 5mm.

235. Crucible Sherd (E14:470 H)
Fig. 155
The exterior is glazed. SE area.
L 19mm. B 13mm. T 4mm.

236. Crucible Sherd (E14:272 B)
Fig. 155
No visible residues. E area, north of river.
L 18mm. B 15mm. T 6mm.

237. Crucible Sherd (E14:276 B)
Fig. 155
The exterior bears red vitreous material. E area, north of river. L 17mm. B 17mm. T 5mm.

238. Crucible Sherd (E14:277)
Fig. 155
The exterior is glazed. E area, north of river.
L 23mm. B 26mm. T 5mm.

239. Crucible Sherd (E14:289)
Fig. 155
A thick layer of grey porous slag and red vitreous material covers the exterior. E area, north of river.
L 20mm. B 31mm. T 3mm (crucible); 11mm (including slag).

240. Crucible Sherd (E14:290)
Fig. 155
The exterior is glazed. E area, north of river.
L 22mm. B 29mm. T 9mm.

241. Crucible Sherd (E14:296)
Fig. 155
The exterior is covered with grey porous slag. E area, north of river. L 21mm. B 22mm. T 10mm.

242. Crucible Sherd (E14:297)
Fig. 155
The interior is heat-reddened, the exterior covered with grey porous slag and red vitreous material. E area, north of river. L 20mm. B 19mm. T 8mm.

243. Crucible Sherd (E14:300 and E14:309)
Fig. 155
This sherd bears a clay projection on one side – possibly a 'handle' or the result of a collapse. The exterior is covered with a layer of grey porous slag and red vitreous material. E area, north of river. L 34mm. B 26mm. T 6mm (crucible); 9mm (including slag).

244. Crucible Sherd (E14:307)
Fig. 155
No visible residues. E area, north of river.
L 20mm. B 30mm. T 4mm.

245. Crucible Sherd (E14:315)
Fig. 155
The interior is heat-reddened, the exterior glazed. E area, north of river. L 21mm. B 22mm. T 5mm.

246. Crucible Sherd (E14:317)
Fig. 155
A thin layer of dark grey slag is visible internally. E area, north of river. L 23mm. B 23mm. T 4mm.

247. Crucible Sherd (E14:321)
Fig. 155
No visible residues. E area, north of river.
L 14mm. B 8mm. T 1mm.

248. Crucible Sherd (E14:461)
Fig. 156
The exterior is covered with grey porous slag and red vitreous material. SE area.
L 23mm. B 33mm. T 10mm.

249. Crucible Sherd (E14:464)
Fig. 156
The exterior is covered with grey porous slag and red vitreous material. SE area. L 24mm. B 31mm. T 8mm.

250. Crucible Sherd (E14:419)
Fig. 156
The interior is heat-reddened, the exterior covered with grey porous slag and red vitreous material. NW area. L 42mm. B 49mm. T 7mm.

251. Crucible Sherd (E14:425 A)
Fig. 156
The exterior bears red vitreous material and is glazed. SE area. L 18mm. B 17mm. T 3mm.

252. Crucible Sherd (E14:425 B)
Fig. 156
No visible residues. SE area.
L 11mm. B 17mm. T 5mm.

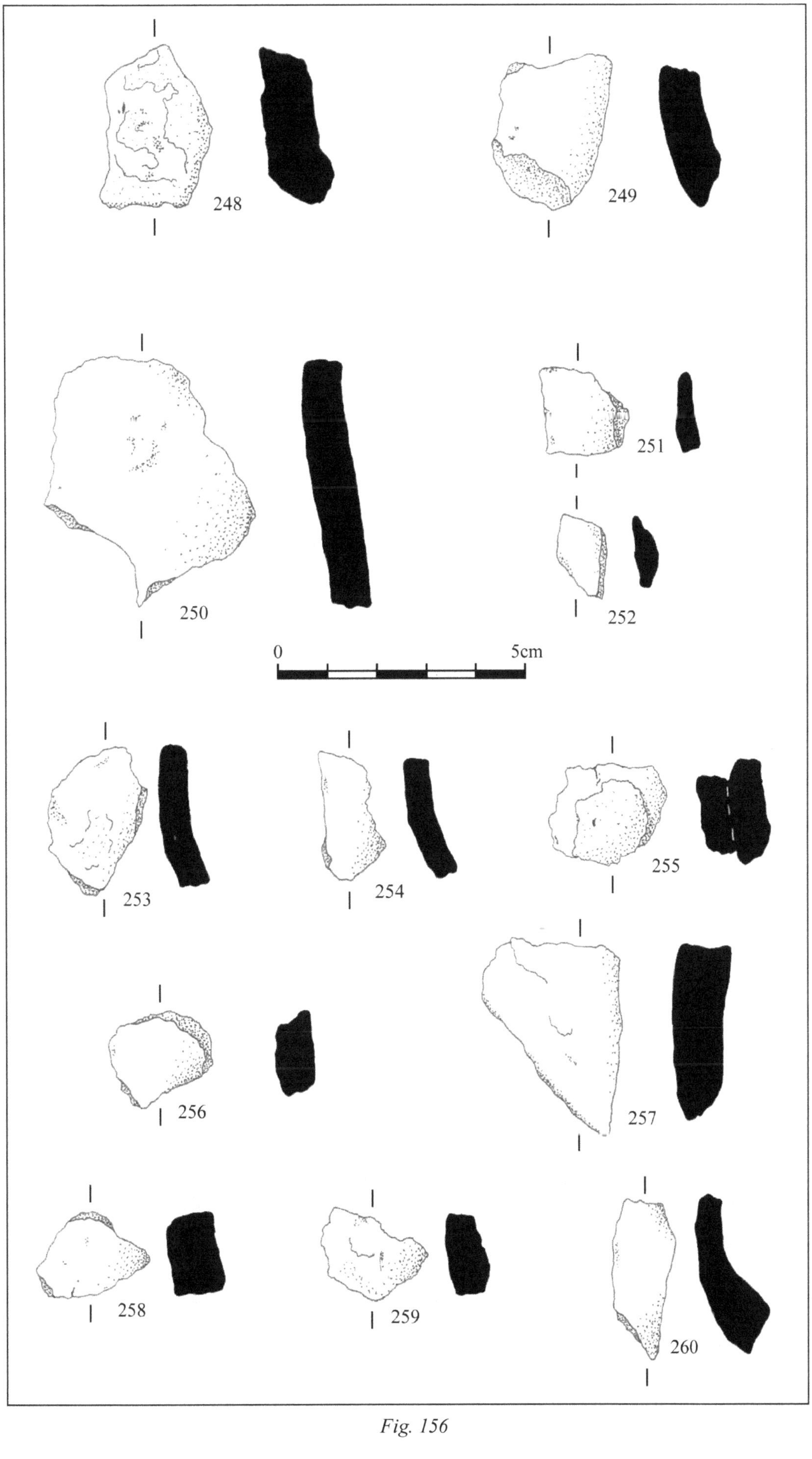

Fig. 156

253. Crucible Sherd (E14:429)
Fig. 156
The exterior bears red vitreous material. SE area.
L 20mm. B 30mm. T 5mm.

254. Crucible Sherd (E14:436 / 486)
Fig. 156
The interior is heat-reddened. NW area.
L 13mm. B 37mm. T 5mm.

255. Crucible Sherd (E14:438)
Fig. 156
The exterior bears large encrustations of grey porous slag. NW area.
L 25mm. B 21mm. T 13mm.

256. Crucible Sherd (E14:441 D)
Fig. 156
No visible residues. NW area.
L 23mm. B 20mm. T 8mm.

257. Crucible Sherd (E14:252 A)
Fig. 156
No visible residues.
L 29mm. B 40mm. T 11mm.

258. Crucible Sherd (E14:252 B)
Fig. 156
No visible residues.
L 23mm. B 18mm. T 10mm.

259. Crucible Sherd (E14:501)
Fig. 156
The exterior is covered with grey porous slag and red vitreous material. WSW area.
L 21mm. B 19mm. T 8mm.

260. Crucible Sherd (E14:503)
Fig. 156
No visible residues. WSW area.
L 13mm. B 31mm. T 5mm.

Crucible Fragments

261. Crucible Fragment (E14:490 A, B)
Fig. 157
Consists of two sherds, one of which is a rim sherd. The interior is blackened. W area.
H 69mm. B 33mm. Rim T 5mm.

262. Crucible Fragment (E14:253 A, B, C / 1374)
Fig. 157
Consists of three sherds, two of which are adjoining. No rim present. No visible residues. E area, north of river. Found between palisades in brown peat layer 3.
L 55mm (A, B). B 41mm (A, B). T 6mm (A, B).
L 20mm(C). B 26mm(C). T 5mm (C).

263. Crucible Fragment (E14:269 A and E14:319)
Fig. 157
Consists of two sherds including rim. The rim and exterior are covered with black and red vitreous material. E area, north of river.
L 52mm. B 32mm. T 4mm.

264. Crucible Fragment (E14:396)
Fig. 157
Rim included. A layer of grey slag is clearly visible on the exterior. NE area.
L 49mm. B 38mm. T 3mm (crucible); 6mm (including slag).

265. Crucible Fragment (E14:398 and E14:409)
Fig. 157
No rim present. Consists of two sherds. The exterior is very heavily slagged, with red and green vitreous material also present. NE area.
L 70mm. B 51mm. T 7mm (crucible); 12mm (including slag).

266. Crucible Fragment (E14:510)
Fig. 158
Rim included. The exterior bears red vitreous material.
L 62mm. B 42mm. T 5mm.

267. Crucible Fragment (E14:301)
Fig. 158
No rim present. A trace of green residue is visible internally, the exterior is covered with grey porous slag and glaze. E area, north of river.
L 21mm. B 35mm. T 7mm.

268. Crucible Fragment (E14:304 and E14:258)
Fig. 158
Consists of two sherds, including rim. No visible residues. E area, north of river.
L 40mm. B 36mm. T 4mm.

269. Crucible Fragment (E14:444 and E14:450)
Fig. 158
Consists of two sherds. No rim present. Approx. 45% of the exterior is covered with red vitreous material. SE area. L 53mm. B 36mm. T 6mm.

270. Crucible Fragment (E14:459)
Fig. 158
Rim included. The rim and exterior are covered with grey porous slag and red vitreous material. SE area.
L 47mm. B 37mm. T 7mm.

Miscellaneous crucibles

271. Polypod Crucible (E14:364)
Fig. 159
A complete, three-legged or 'polypod' vessel. There is also the hint of a projection at the side. Perfectly balanced and very well made. Approx. 5% of the exterior is covered with black vitreous material. E area, north of river. Found between palisades in brown peat layer 3. D: 36mm. Diam. 35mm (internal); 56mm (external). T 5mm, 13mm including projection.

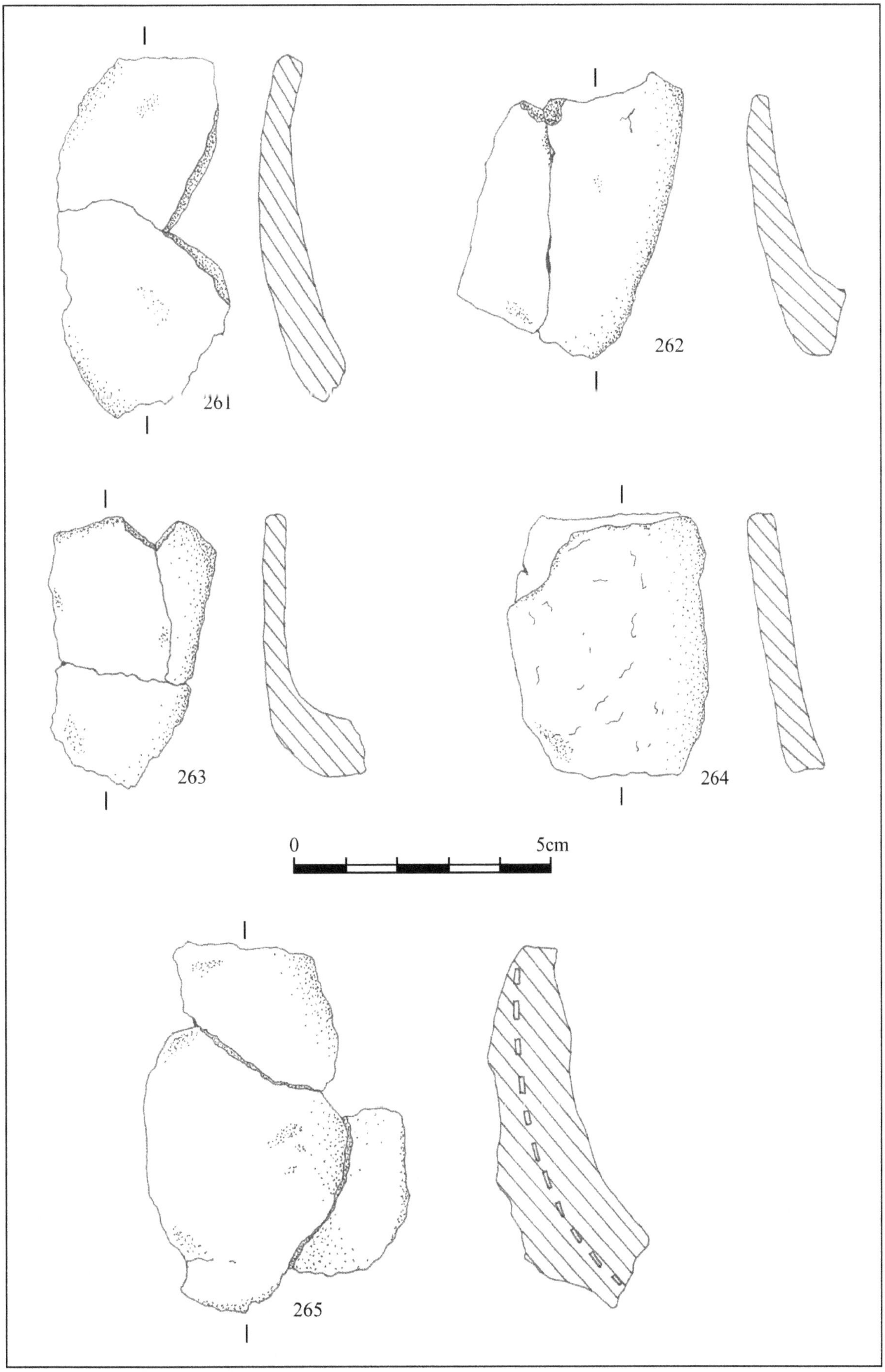

Fig. 157

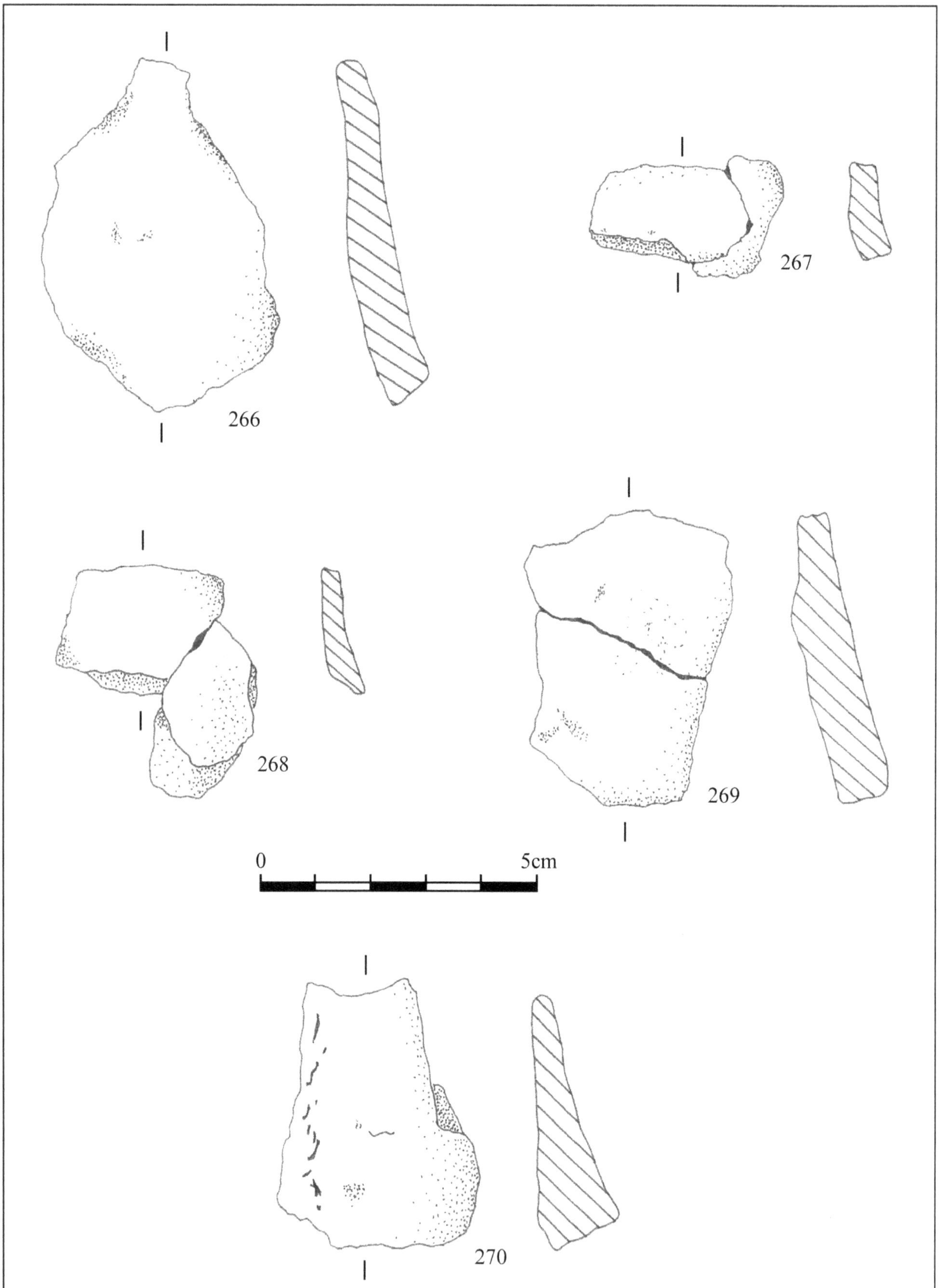

Fig. 158

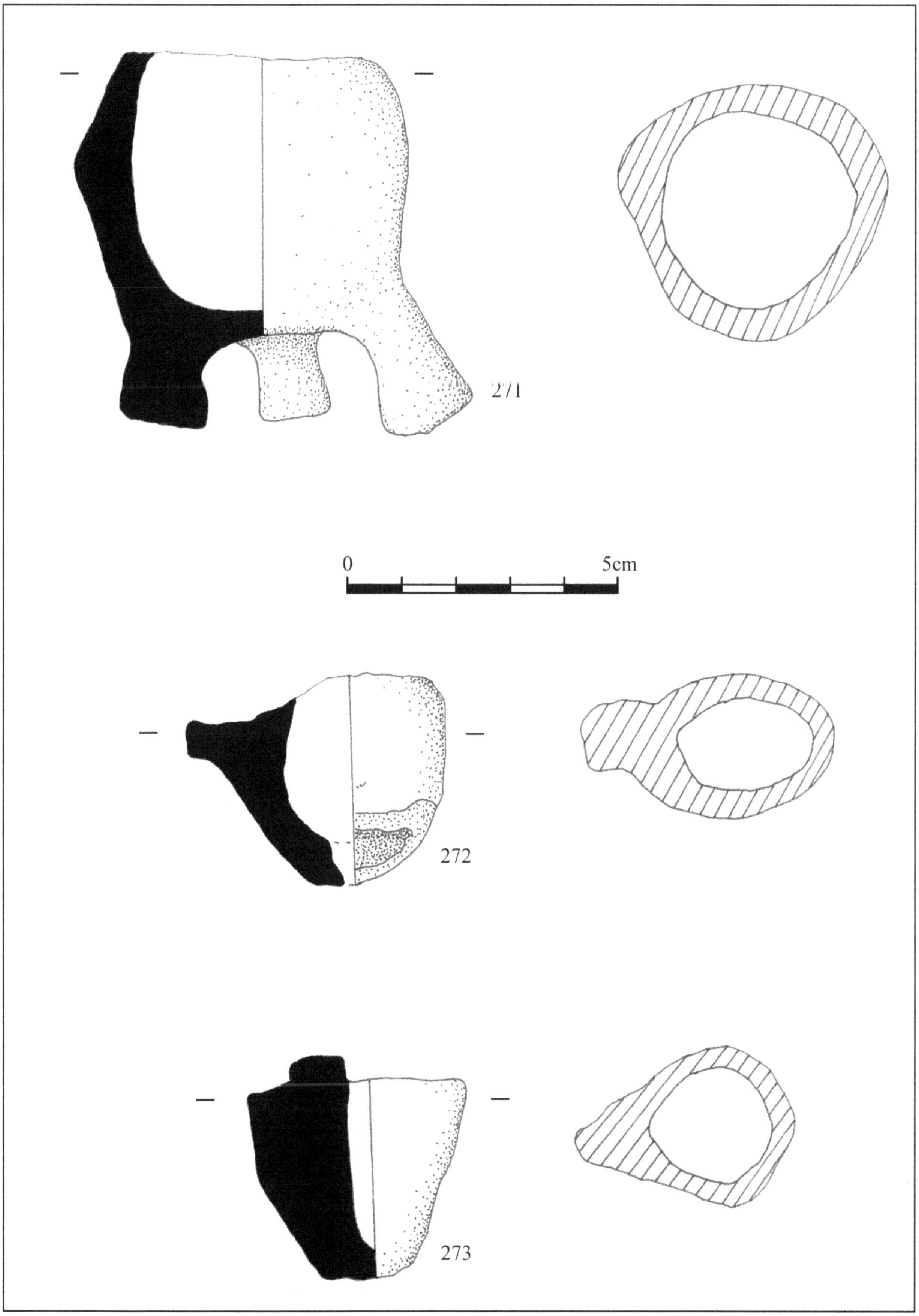

Fig. 159

272. Handled Crucible (E14:458)
Fig. 159
Oval-shaped vessel with handle. Base missing. Trace of green residue visible internally, externally some red and black vitreous material. SE area.
'Handle' L 18mm. B 13mm. D 34mm.
Diam. 22mm (internal). T 4mm.

273. Handled Crucible ('Pot D')
Fig. 159
A clay crucible with two projections, one at right angles to the other. The rim bears light red vitreous material, the exterior grey porous slag and red, black and mustard-colour vitreous material.
Projection 11mm x 11mm. D 38mm. Diam. 23mm (internal); 31mm (external). T 4mm.

274. Clay Projection (E14:310)
Fig. 160
No visible residues. E area, north of river. T 4mm.

275. Clay Projection (E14:313)
Fig. 160
The interior is heat-reddened. E area, north of river.
T 7mm.

276. Clay Projection (E14:314)
Fig. 160
No visible residues. E area, north of river.

277. Clay Projection (E14:413 A, B)
Fig. 160
Clay projection and rim sherd. No visible residues. NE area. T 3mm.

278. Clay Projection (E14:449)
Fig. 160
Approx. 90% of the exterior is covered with grey porous slag. SE area.

279. Crucible Lid (1278)
Fig. 160
Clay object which is most probably a crucible lid. Bears a projection to facilitate gripping with tongs. No visible residues on the interior, approx. 20% of the exterior is covered with grey porous slag.
B of projection 19mm. Diam. 33mm (internal) 63mm (external).

MISCELLANEOUS

This section includes slag from crucibles (12 pieces), possible fragments of furnace lining (6 fragments), and possible tuyère fragments (4 fragments). The measurements of these miscellaneous sherds and fragments consist of max. height; breadth; length; diameter and thickness.

280. Possible Furnace Lining Fragment (E14:412)
Fig. 161
Made of orange clay with a dark green glaze on the exterior. Possibly a fragment of furnace rim. NE area.
H 30mm. W 34mm. T 5mm–14mm.

281. Possible Furnace Lining Fragment (E14:374)
Fig. 161
Made of orange clay. Bears grey porous slag on one side. NE area.
H 32mm. W 42mm. T 9mm.

282. Possible Furnace Lining Fragment (E14:281)
Fig. 161
Made of orange clay. Possible furnace lining rim. E area, north of river.
H 28mm. W 37mm. T 9mm.

283. Possible Furnace Lining Fragment (E14:282)
Fig. 161
Made of orange clay. The exterior and top are covered with grey porous slag. Possible furnace lining rim. E area, north of river.
H 27mm. W 40mm. T 14mm.

284. Possible Furnace Lining Fragment (E14:284)
Fig. 161
Made of orange clay. One side is covered with a layer of grey porous slag and red vitreous material. Possible furnace lining rim. E area, north of river.
H 44mm. W 21mm. T 6mm (crucible); 17mm (including slag).

285. Possible Furnace Lining Fragment (1681)
Fig. 161
Made of orange clay with one vitrified black surface.
H 25mm. W 28mm. T 10mm.

286. Possible Furnace/Tuyère Fragment (E14:480)
Fig. 162
Made of orange clay. Dark grey slag is visible around the rim, the exterior encrusted with grey porous slag. SW area.
H 64mm. W 25mm. T 12mm.

287. Possible Tuyère Fragment (E14:479)
Fig. 162
Glazed externally. SW area.
H 19mm. W 38mm. T 15mm.

288. Tuyère Fragment (E14:399 and E14:375)
Fig. 162
Tuyère fragment consisting of two pieces. Trace red vitreous material and grey porous slag are visible around the rim exterior. NE area.
Diam. of opening 10mm. T 5mm–22mm.

289. Possible Tuyère Fragment (E14:367)
Fig. 162
Made of orange clay. No visible residues. E area, south of river. H 24mm. W 24mm. T 11mm.

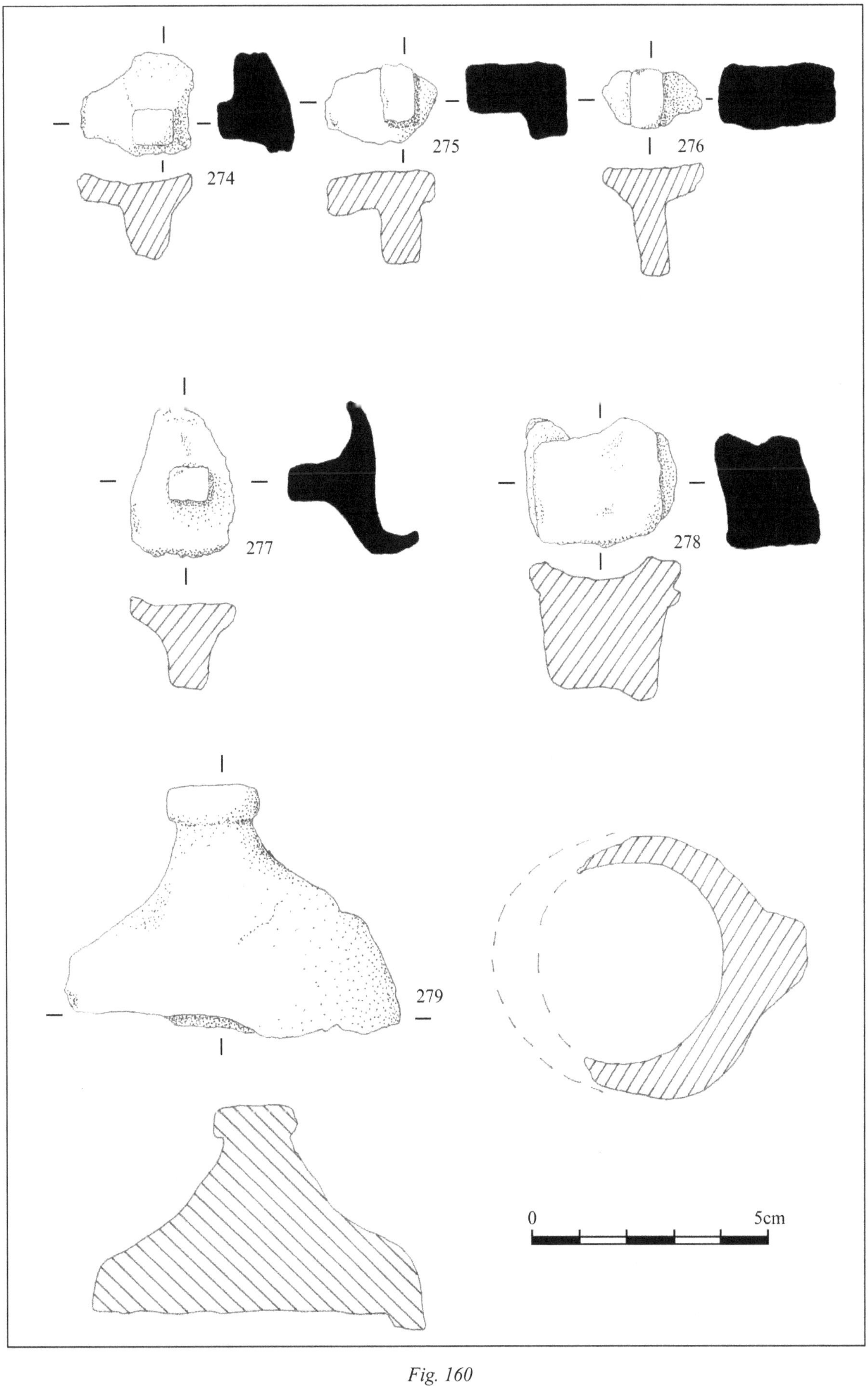

Fig. 160

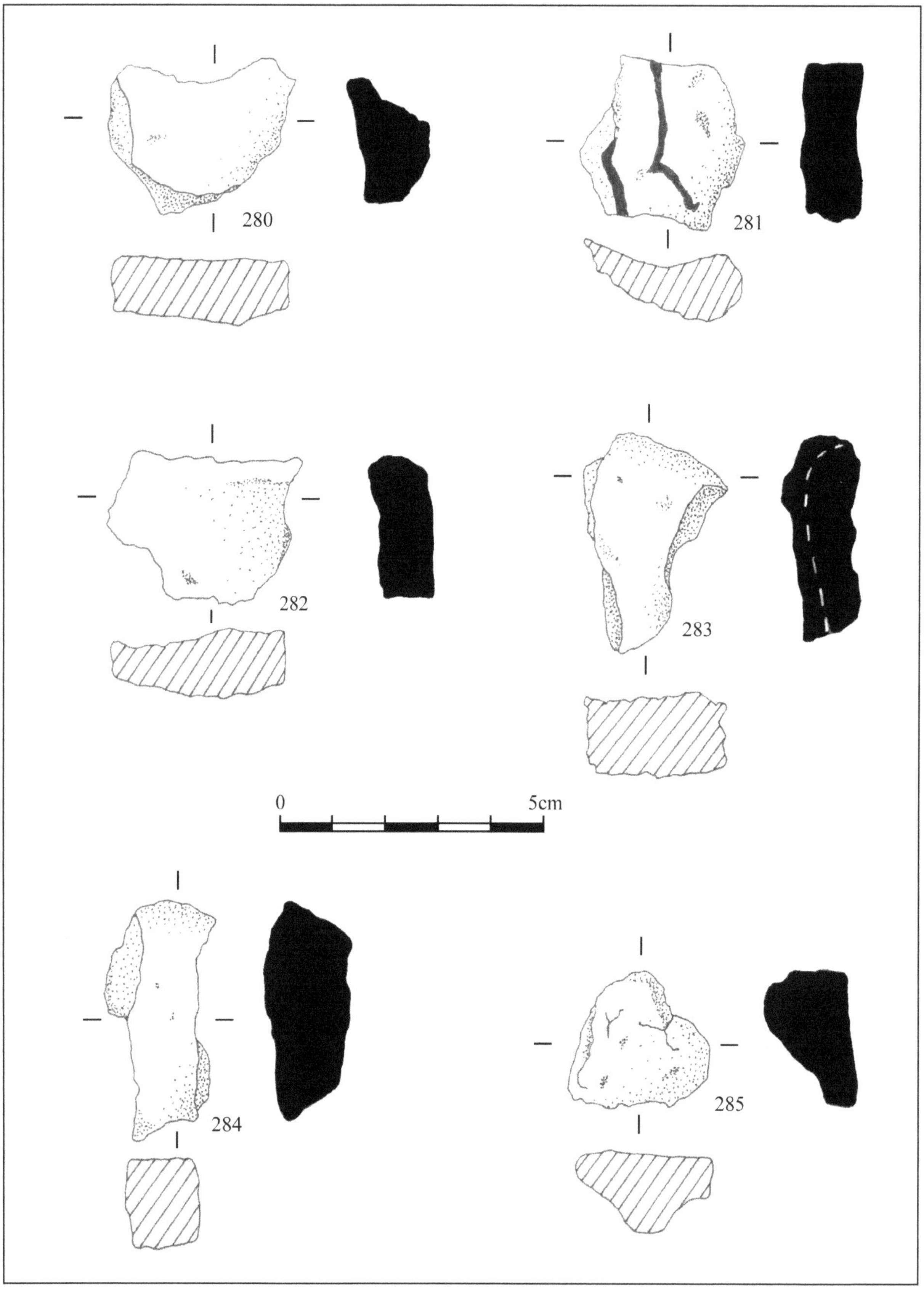

Fig. 161

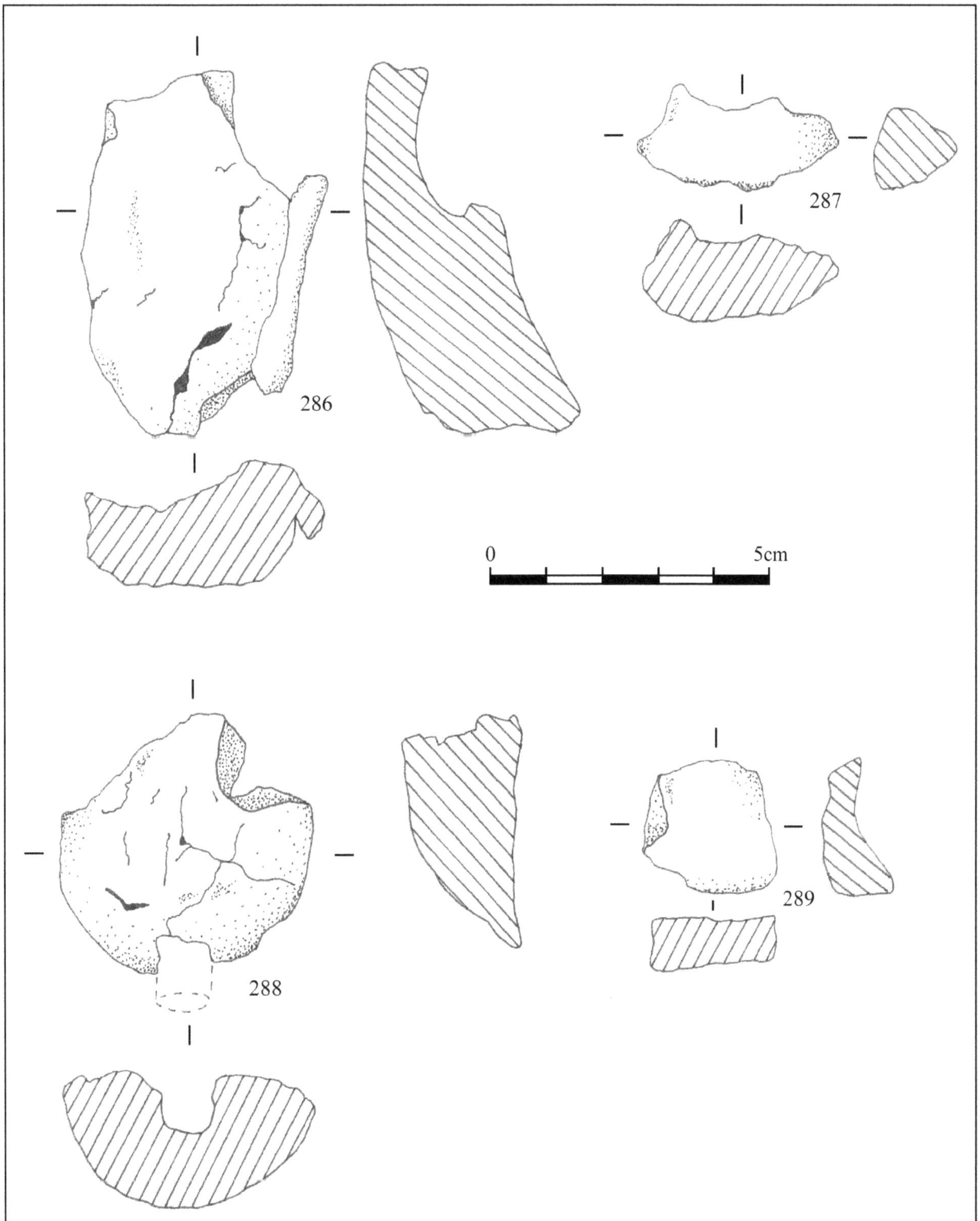

Fig. 162

290. Slag (E14:521)
Fig. 163
Grey porous slag with red vitreous surface. Probably dislodged from around a crucible. Disturbed central area. H 39mm. W 36mm. T 20mm.

291. Slag (E14:251)
Fig. 163
A lump of grey porous slag with a curved interior. Most likely dislodged from the exterior of a crucible. H 31mm. W 50mm. T 18mm.

292. Slag (E14:389)
Fig. 163
A small lump of grey porous slag. NE area.
H 29mm. W 18mm. T 7mm.

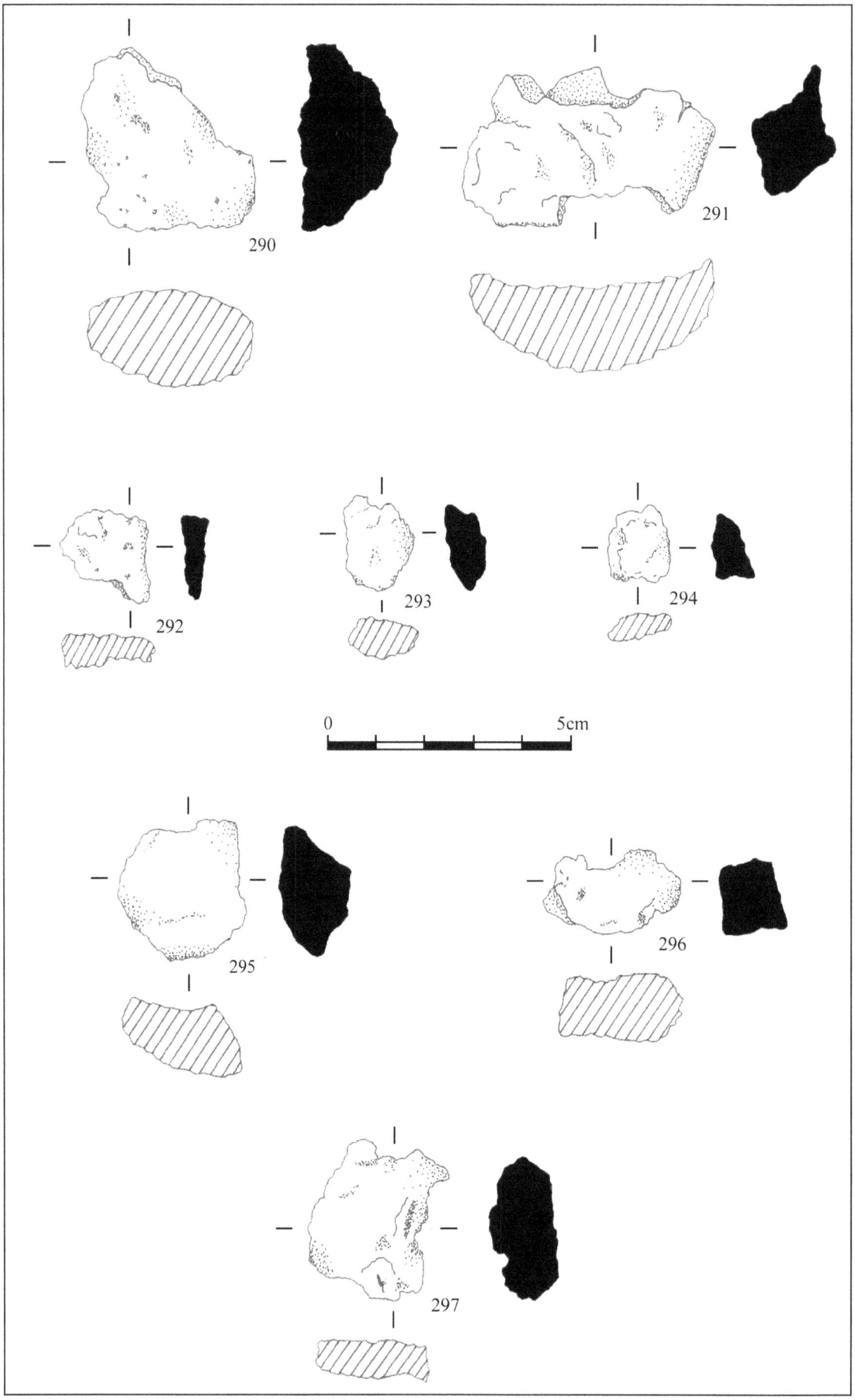

Fig. 163

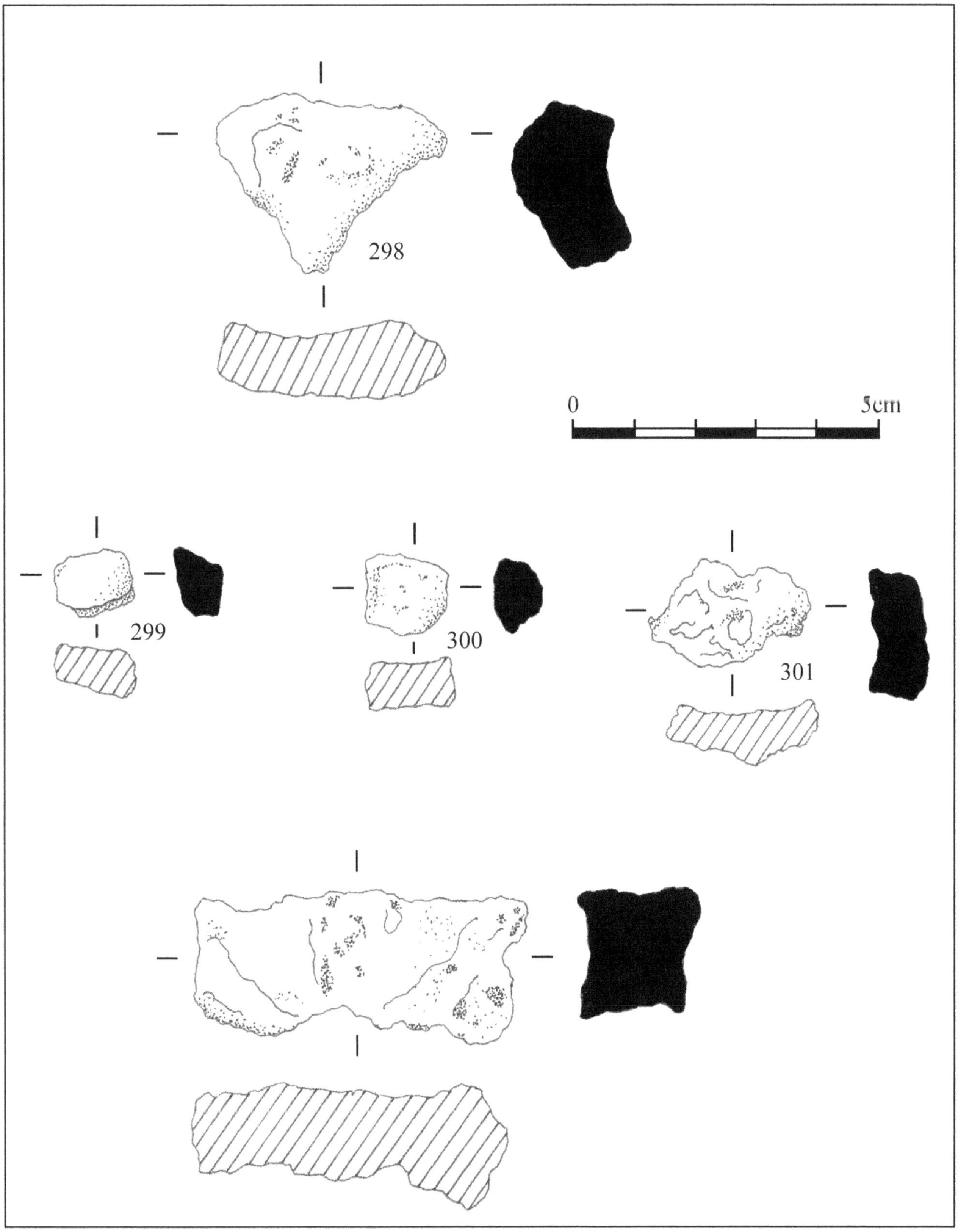

Fig. 164

293. Slag (E14:403)
Fig. 163
Grey porous slag. NE area.
H 20mm. W 15mm. T 8mm.

294. Slag (E14:457)
Fig. 163
Small lump of grey porous slag. SE area.
H 21mm. W 23mm. T 7mm.

295. Slag (E14:481)
Fig. 163
Dark grey porous slag. SW area.
H 24mm. W 53mm. T 12mm–20mm.

296. Slag (E14:506 / 1063)
Fig. 163
Grey porous slag with red vitreous material on the surface. WNW area. H 18mm. W 29mm. T 13mm.

297. Slag (E14:348)
Fig. 163
Grey porous slag with traces of white clay on one surface. Either from crucible or furnace lining. E area, north of river. H 33mm. W 30mm. T 6mm.

298. Slag (E14:368)
Fig. 164
Grey porous slag from a crucible exterior with traces of clay on the inner surface. E area, south of river.
H 24mm. W 37mm. T 11mm.

299. Slag (E14:298)
Fig. 164
Grey porous slag with red vitreous material from a crucible rim. E area, north of river. H 11mm. W 13mm. T 7mm.

300. Slag (E14:318)
Fig. 164
Small lump of grey porous slag. E area, north of river.
H 13mm. W 14mm. T 89mm.

301. Slag (E14:433)
Fig. 164
Grey slag. NW area. H 18mm. W 26mm. T 7mm.

MOTIF - PIECES

Twelve artefacts are included in this category, with one dubious (Six stone, five bone, and one wood). All are in good condition and have been described in detail by Ó Meadhra (1979; 1987). The wooden artefact, W181, may also have functioned as a mould. It differs from other motif-pieces in that its overall form was shaped as opposed to the incision of a shape on a larger surface. The records begin by stating the artefact type and the National Museum of Ireland registration and excavation numbers. These are followed by a description, measurements and illustration references. The majority of measurements consist of length/height; breadth and thickness. The individual habitat of each artefact is also recorded where such information is available.

302. Bone motif-piece (W29) (Old Find)
Fig. 165
This highly polished end of a long bone is probably the most well known of all the Lagore motif-pieces. The decoration consists of five main panels and a variety of smaller, less prominent incisions. The five main panels are chip-carved, two of which contain a single interlaced animal, another contains two very similar animals intertwined, the fourth and fifth a simple interlace ornament. The two panels of interlace ornament have billeted or hatched borders. The less prominent motifs include some smaller panels of chip-carved interlace and incised triquetras. L 217mm. Diam. 44mm (at broken end), 70mm (at knuckle end). C89, CD36:9.

303. Bone motif-piece (Wk. 98 G.310) (Old Find)
Fig. 169
Sub-triangular portion of bone, the surface of which is in good condition. Both surfaces of the piece bear a six-petalled marigold motif within a circle. These motifs were unevenly incised with the aid of a compass.
L 74mm. B 66mm. T 9mm. B26:23.

304. Bone motif-piece (E14:526 / 1086)
Fig. 166
This piece could not be located in the Museum, therefore the following description is based on Ó Meadhra's 1979 catalogue (p.90). The piece consists of the end fragment of blade bone. The bone is incomplete though the motif piece is complete. The bleached surface has a slight handling polish and is in good condition. The patterns lie centrally on an area of soft bone on one face only and the main motif consists of badly chip-carved interlace. The other motifs on the piece are lightly incised arcs and a plait-like motif. WSW area, in sand above crannóg.
L 153mm. B 50mm. T 18mm (where carved).

305. Bone Motif-Piece (E14:527 / 589)
Fig. 166
This piece could not be located in the Museum, therefore the description which follows is based on Ó Meadhra's 1979 catalogue (p.90). Fragment of a blade bone. The patterns are scattered over the softest bone of one face only. There are four motifs on the piece and of these only one is well defined. This consists of a small chip-carved pointed triquetra. NW area, outside palisade.
L 133mm. B 63mm. T 15mm (where carved).

306. Bone motif-piece (E14:531 / 324)
Fig. 167
This end portion of a cow radius bears interlace panels which are very similar to those of the bone motif-piece W29. The worked area is complete and in good condition with the whole surface having been highly polished. Approx. 30 motifs consisting of main panels of chip-carved interlace (two of which have hatched borders), smaller panels or strips of chip-carved interlace and, finally, lightly incised motifs such as triquetras. NW area, outside palisade.
L 165mm. B 42mm. T 59mm. Viking Exhibition.

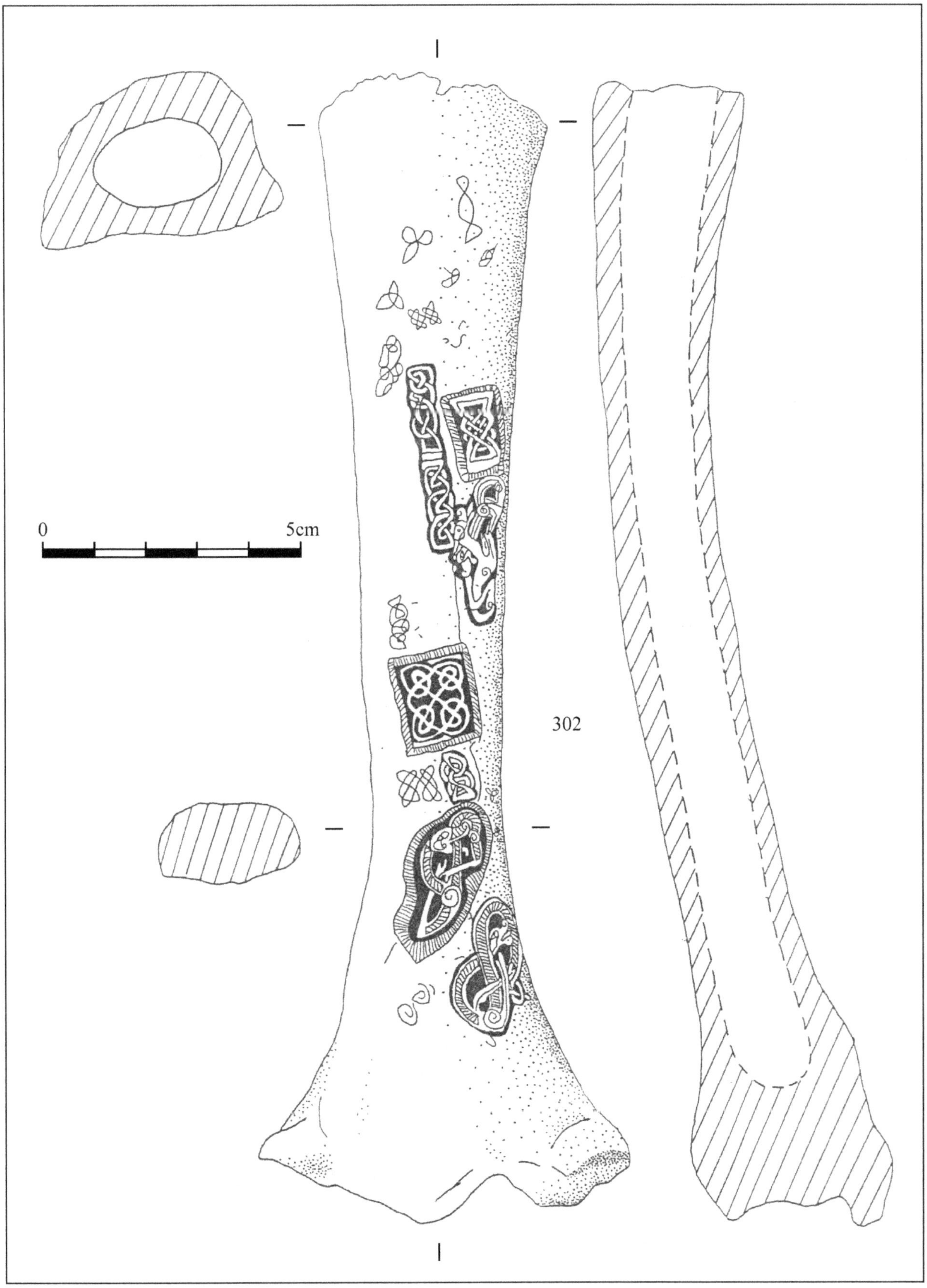

Fig. 165

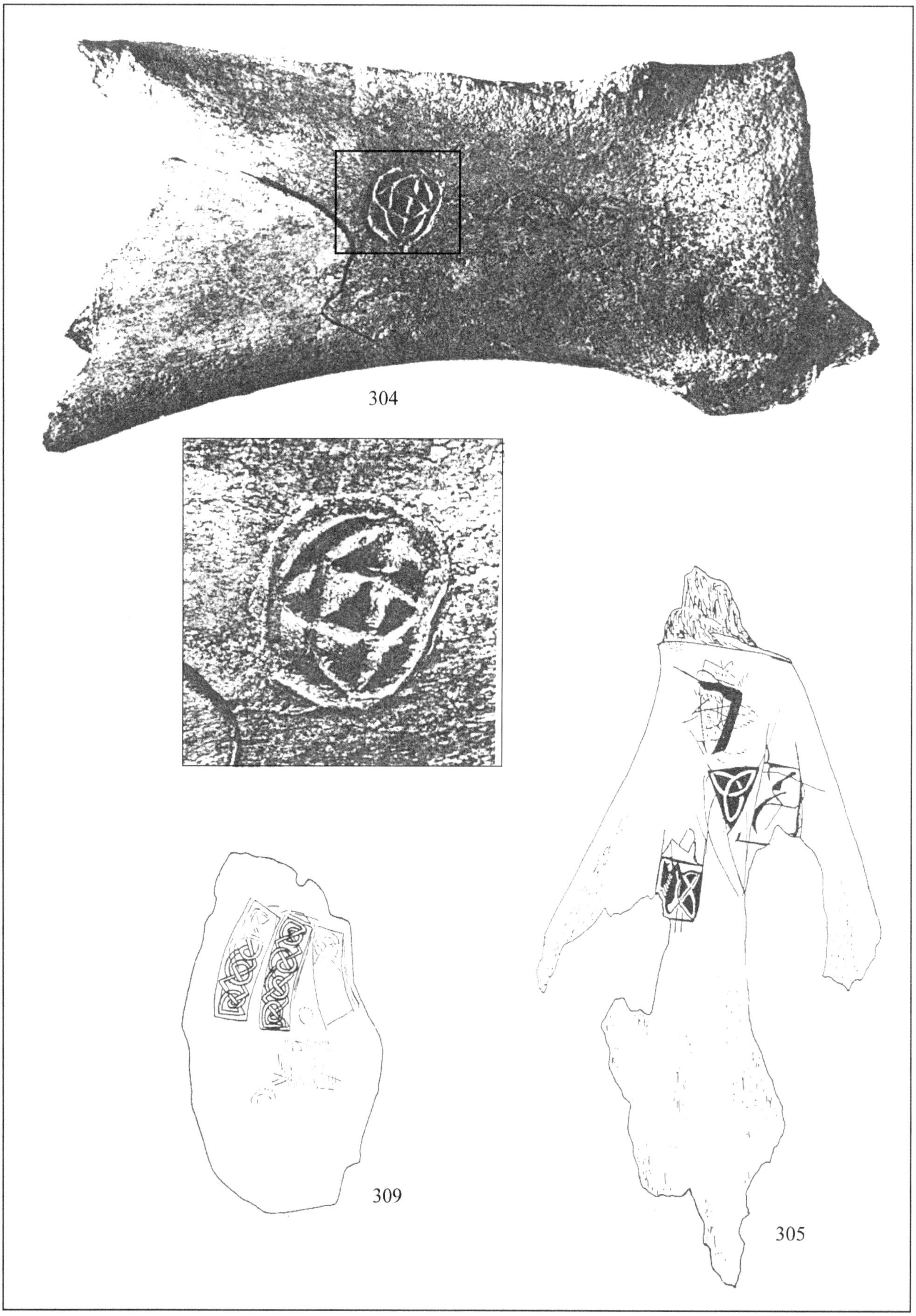

Fig. 166

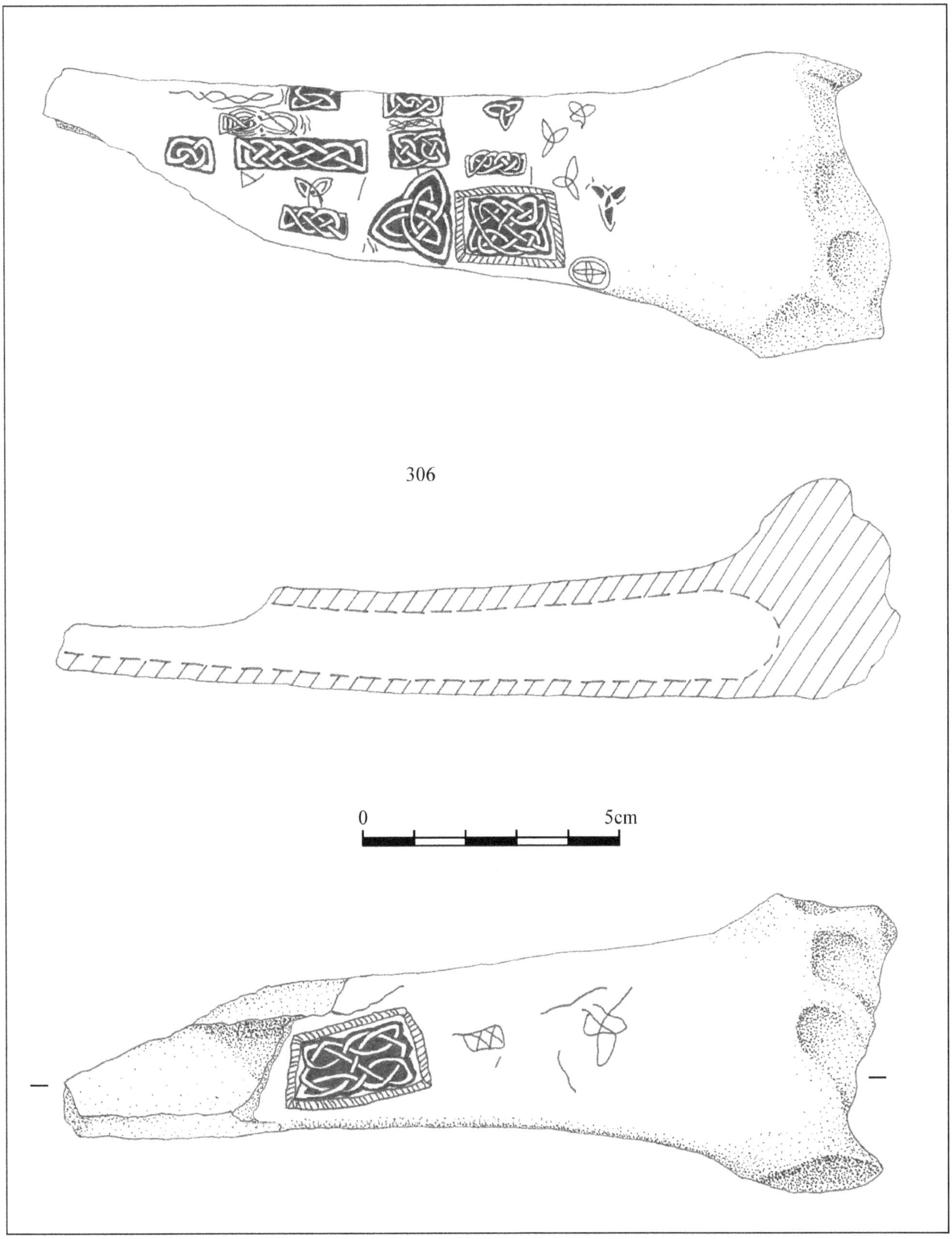

Fig. 167

307. Stone Motif-Piece (P.962)
Fig. 168
This artefact consists of a flat, triangular piece of shale. It is complete with only one of its faces decorated. this decoration consists of interlaced bands which covers practically all of one surface. The voids formed by the interlace are, in some cases, hatched or cross-hatched. The design is incised and not of a very good quality. L 140mm. B 110mm. T 42mm. D1:13.

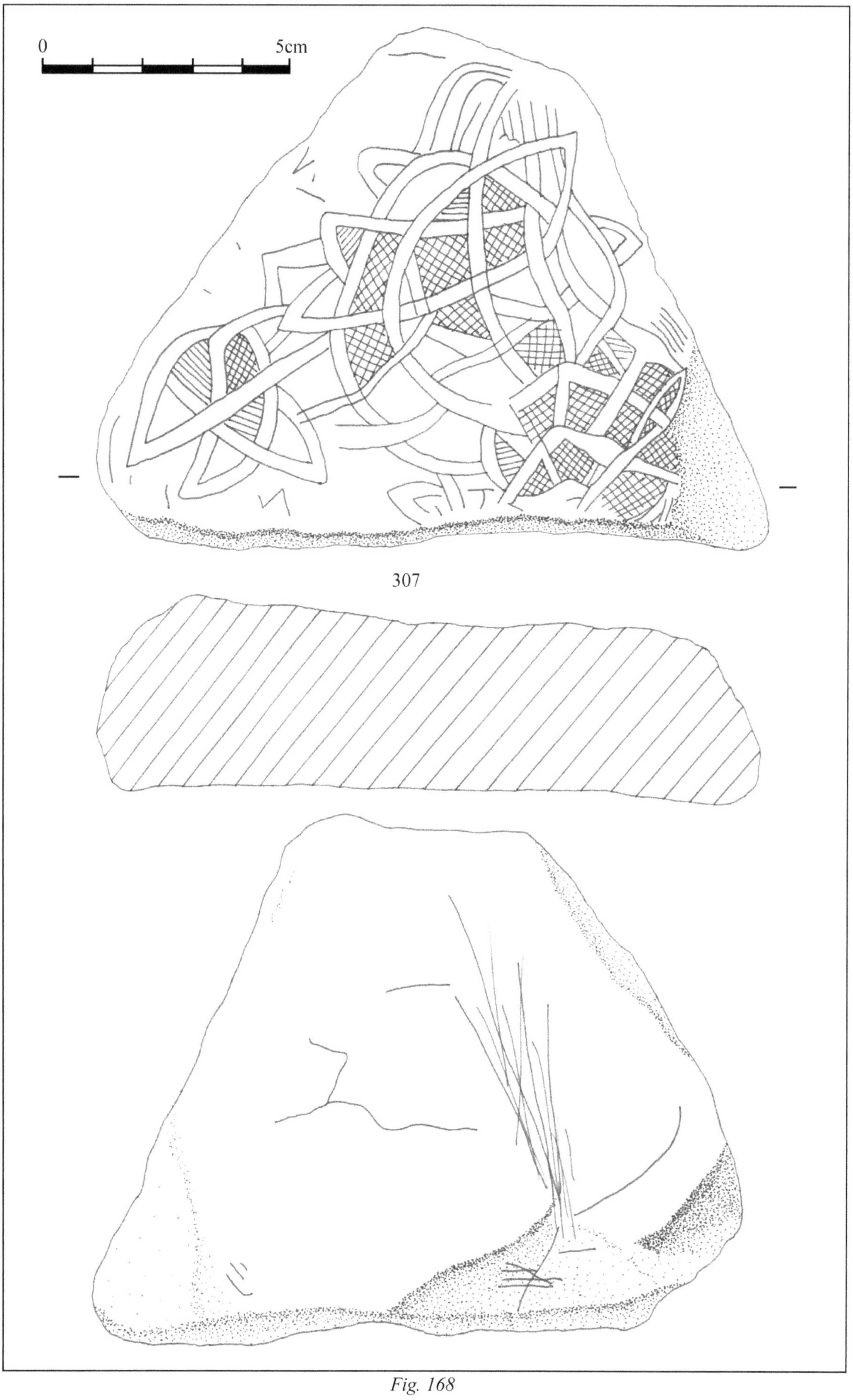

Fig. 168

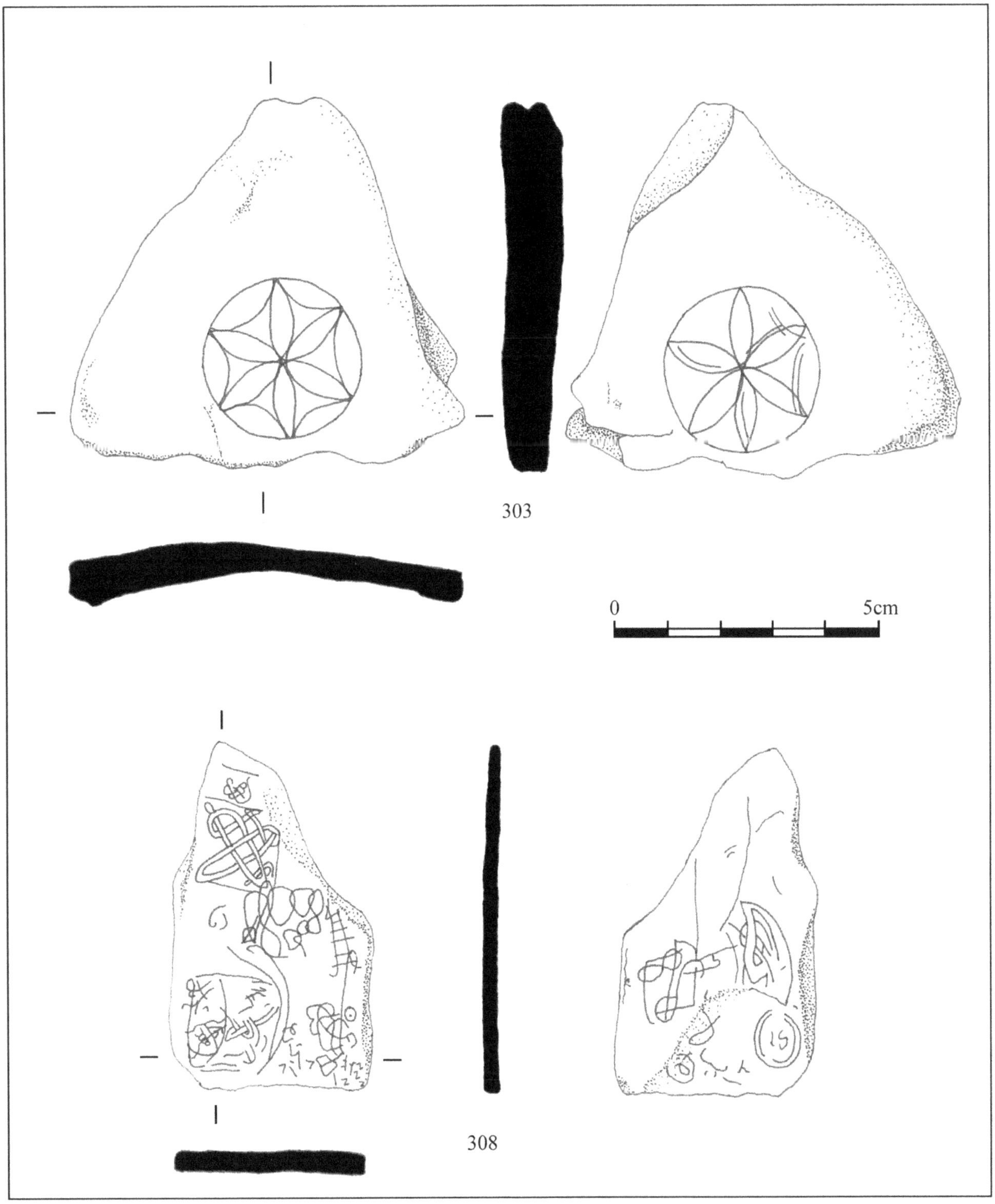

Fig. 169

308. Stone Motif-Piece (72)
Fig. 169
A thin sub-triangular sliver of green slate. Both surfaces are worked although the motifs are not very clear. Some panels of very rudimentary incised interlace are identifiable however. In many cases the motifs actually overlie each other. H 63mm. B 34mm. T 2mm–4mm. B19:25.

309. Stone Motif-Piece (73)
Fig. 166
This piece could not be located in the Museum so, therefore, the description which follows is based on Ó Meadhra's 1979 catalogue. Roughly oval sliver of soft shale which appears to be complete. This piece bears three main motifs all on the one face. The motifs consist of arced panels of incised interlace, one of which is unfinished. L 68mm. B 40mm. T 3mm.

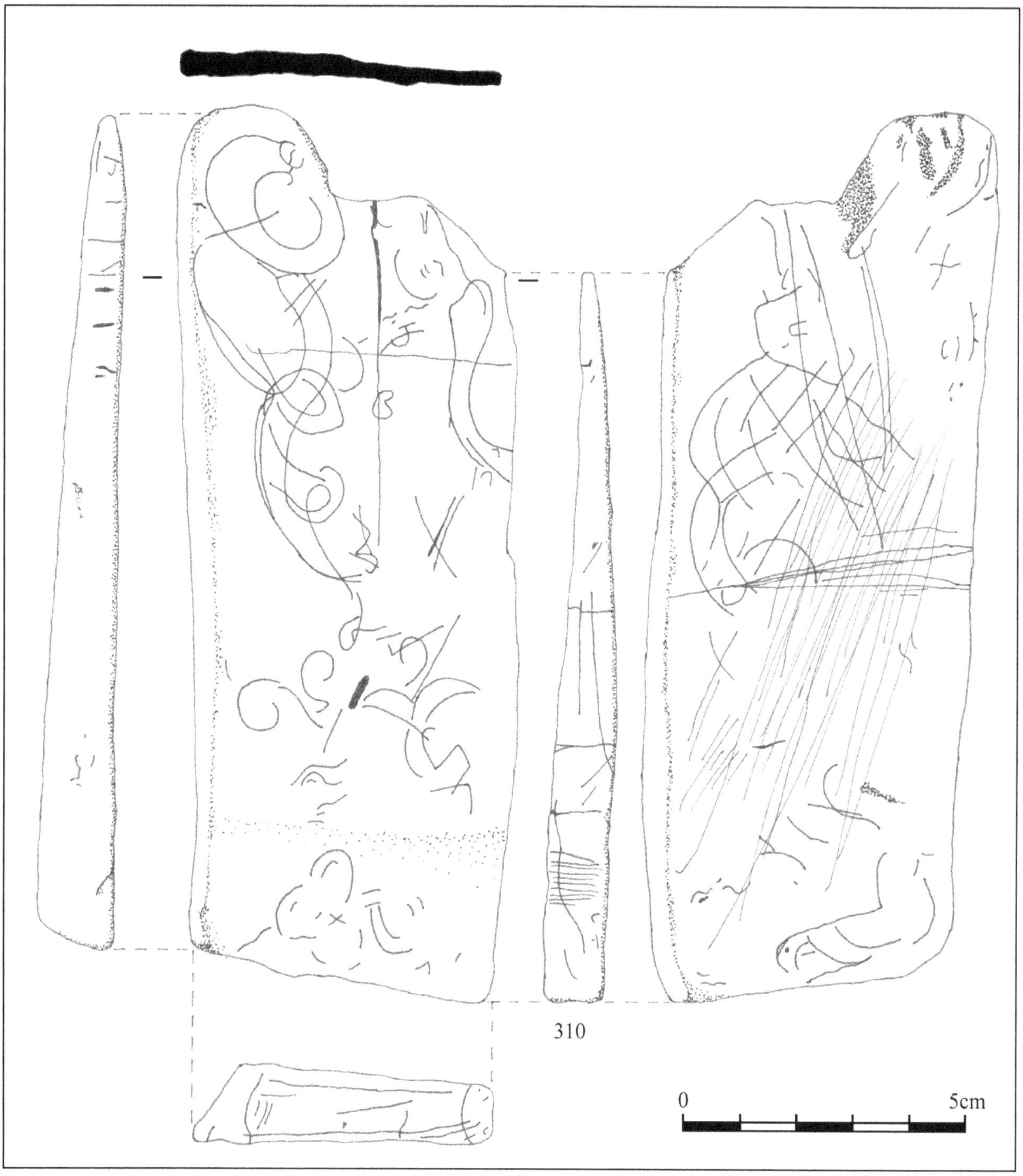

Fig. 170

310. Stone Motif-Piece (240)
Fig. 170
This shale motif-piece is somewhat larger than the others and is also very different motif-wise. Both surfaces bear unfinished and extremely unclear motifs which stretch from the top to the bottom of the piece. The coarse incisions are very worn and it would also appear that one face was used as a whetstone. H 156mm. B 69mm. T 4mm–13mm. B18:14.

311. Stone Motif-Piece (958)
Fig. 171
This piece consists of a flat piece of green slate, bearing motifs on both faces. The piece is in relatively good condition with two main panels of design present, one on either face. These panels contain two incised interlace animals, one intertwined with the other. The panels differ only in minor detail. Other less prominent and much smaller incisions are visible elsewhere on the piece. L 70mm. B 27mm. T 3mm. B19:25.

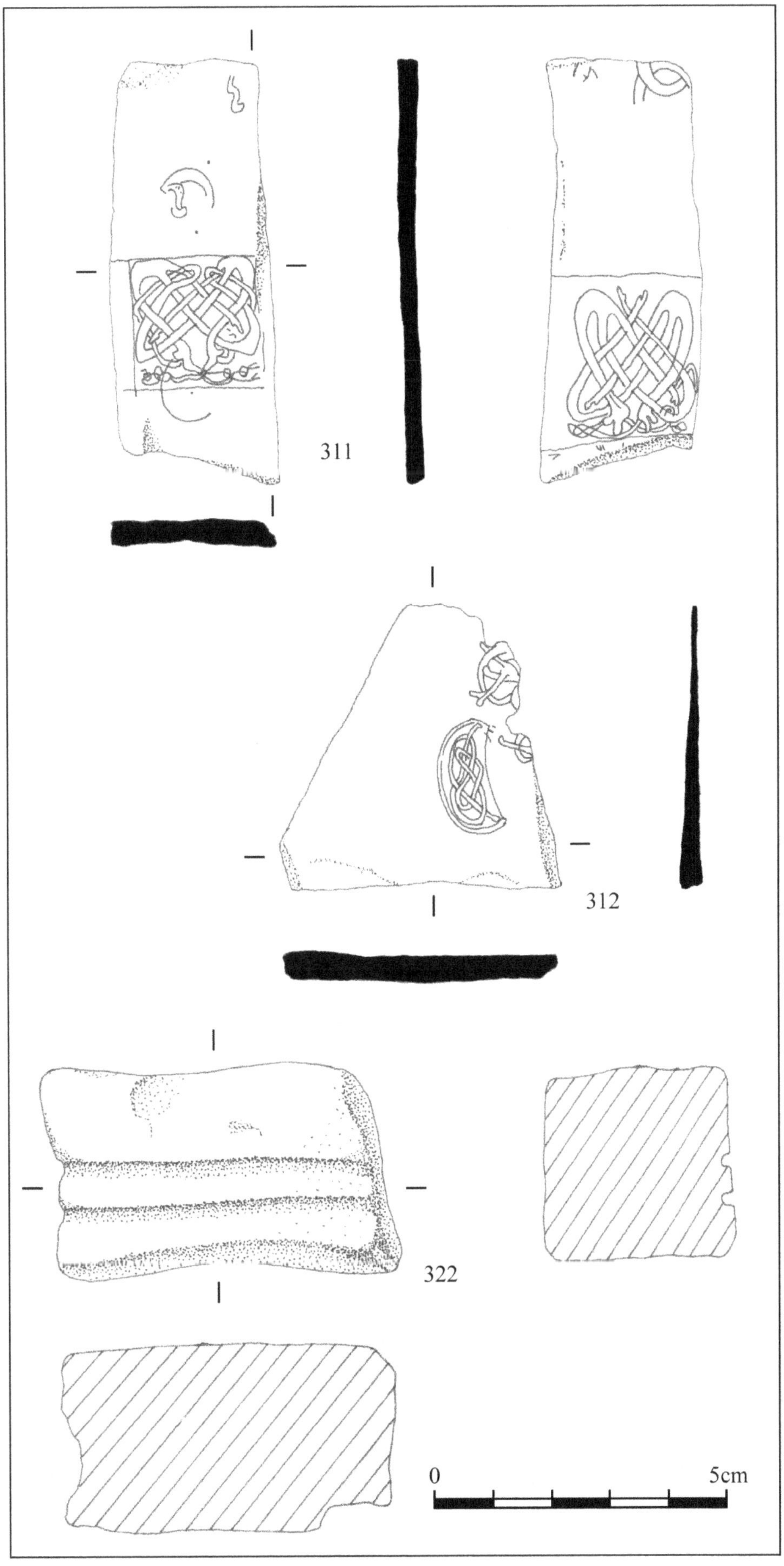

Fig. 171

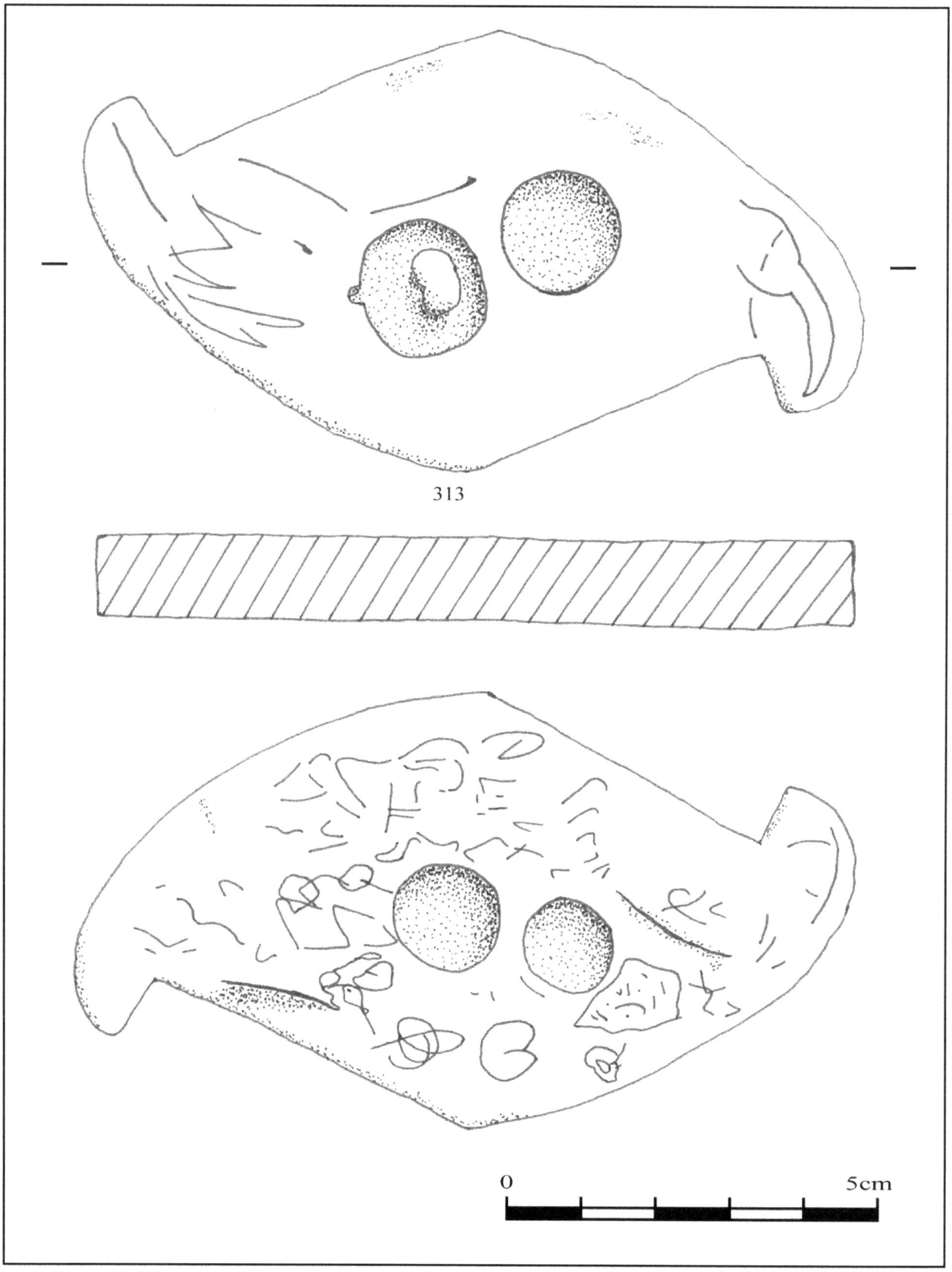

Fig. 172

312. Stone Motif-Piece (1358)
Fig. 171
This piece consists of a sub-triangular sliver of soft green slate. The surface of the piece has been rubbed and scraped thus erasing some of the original motifs, traces of which remain. The most prominent remaining panel is semi-circular in shape and contains a lightly incised ribbon interlace. H 43mm. B 47mm. T 1mm–4mm. B19:25.

313. Wooden Motif-Piece or Mould? (W181)
Fig. 172
This piece is made of yew and is lozenge shaped with alternatively curled ends. As a motif-piece the artefact is of very poor quality. One surface bears a very basic, incised outline of a bird's head, the other a series of incised, badly worn nondescript scratches. It is highly possible that this piece may also have been used as a mould for casting wax models. Both surfaces bear two circular depressions, one of which appears to be a mould for the ring of a ring-headed pin or for a small pseudo-penannular brooch.
H 103mm. B 57mm. T 8mm.
Circular depression 1: Diam. 15mm. 2: Diam. 17mm. 3: Diam. 12mm. 4: Diam. 13mm. 1-4: D 2mm.

MOULDS

Nine moulds are included here, two of clay, and seven of stone. The wooden artefact, W181, can also be referred to as a possible mould. Of the two clay moulds, one is a bivalve fragment, the other a complete valve. Six of the seven stone moulds bear ingot matrices, ranging in size from 38mm in length (matrix 4. of mould no. 28) to 94mm long (mould with no recorded number). One mould is possibly that for casting pin shanks while three others (including W181) bear circular matrices. Of the seven stone moulds, six are of sandstone, one of schist, and four of the moulds are complete valves. The moulds are listed according to type, with their National Museum of Ireland registration and excavation numbers, description, measurements, illustration references, and habitat also recorded. The measurements differ for each mould and are explained in individual entries.

314. Clay Mould (872)
Fig. 173
Fragment of one valve of a bivalve clay mould for casting a large ring, possibly that of a penannular brooch. The pouring gate is present on the fragment. There are no residues visible on the mould.
Mould; L 58mm. B 48mm. T 12mm.
Matrix; D approx. 3mm. C17:12.

315. Clay Mould (1517)
Fig. 173
One complete valve of a bivalve clay mould for casting the ring of a ring-headed pin. It is apparent from the mould that this ring would have had two head-like bosses on opposing sides. As the valve is complete, both the pouring gate and three holes for receiving dowels are present.
Mould; L 37mm. B 25mm B. T 8mm.
Matrix; external Diam. 12mm, internal Diam. 6mm. D 2–3mm. B18:8.

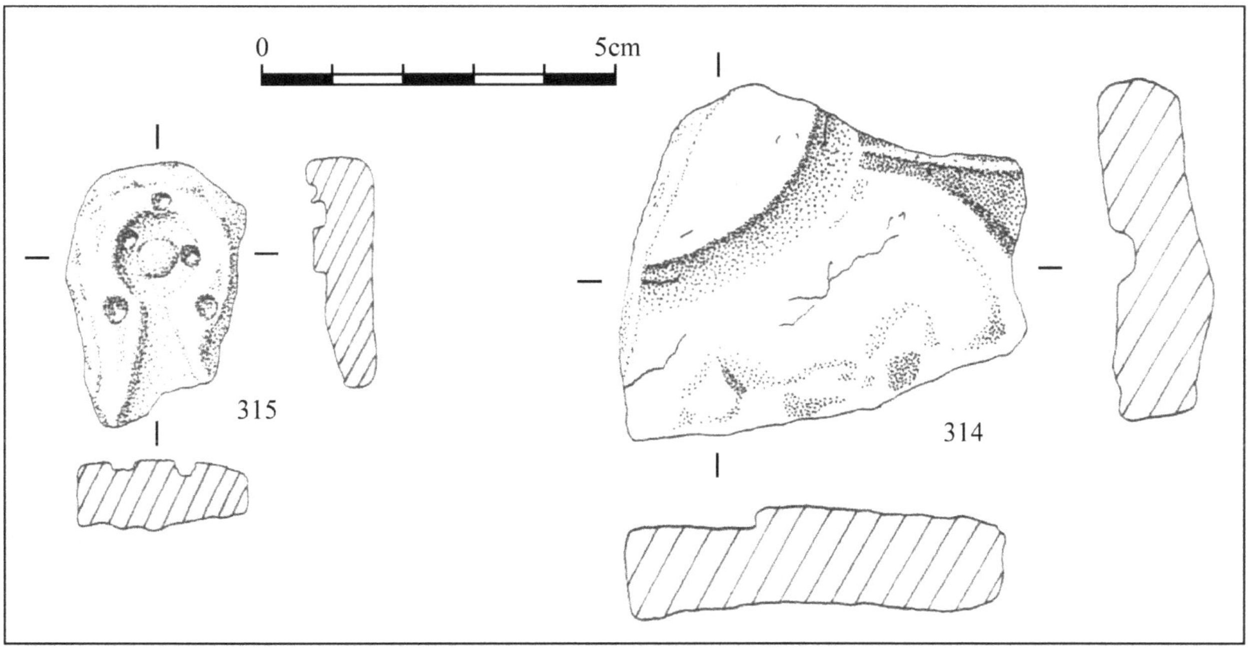

Fig. 173

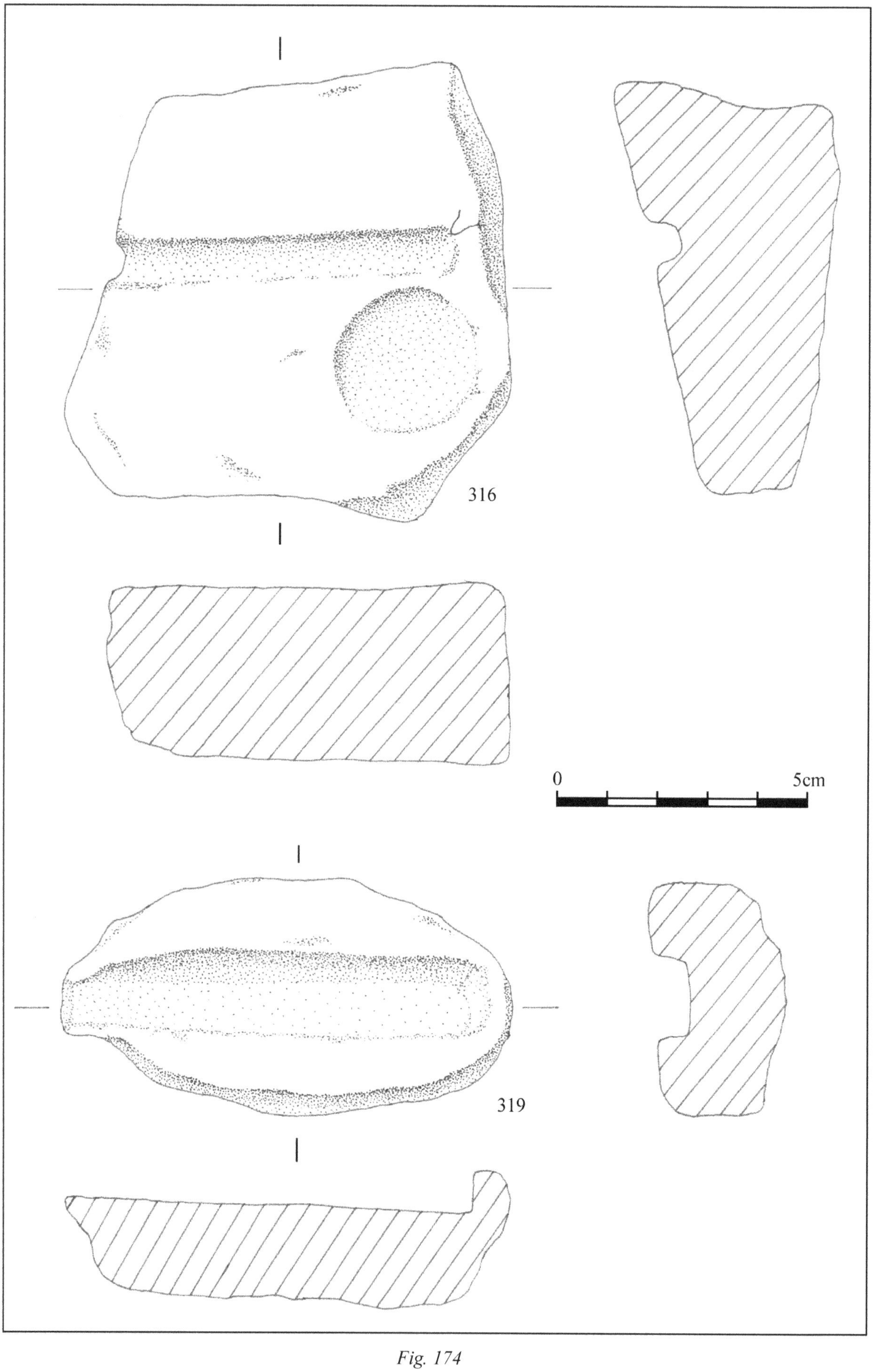

Fig. 174

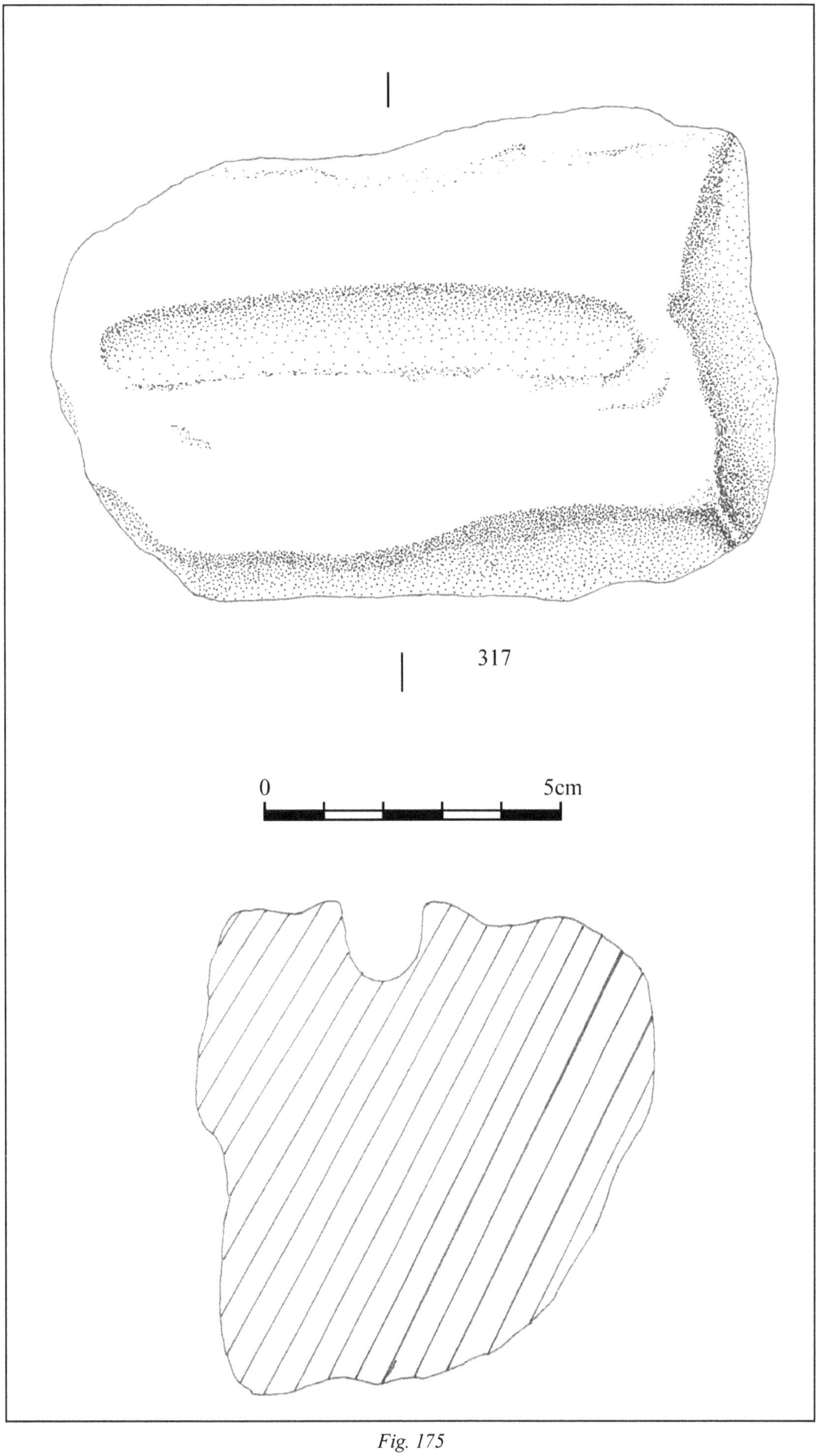

Fig. 175

316. Stone Mould (Wk.22, W97) (Old Find)
Fig. 174
Practically complete sandstone mould - one end broken away. Only one surface is worked, it bears two matrices - one circular depression and one ingot matrix, of which one end is missing. The surface of the stone around these matrices has been deliberately flattened and made smooth. This upper surface is blackened while the matrices contain a rust-coloured residue. This mould has a roughly triangular cross-section and so had to be supported in a horizontal position for casting, possibly using clay or sand, or perhaps this is the top valve of a bivalve mould?
Mould; L 90mm. B 88mm. T 46mm.
Ingot matrix; L 70mm. B 9mm. D 5mm.
Circular matrix; Diam. 30mm. D 4mm. D1:13.

317. Stone Mould
Fig. 175
Sandstone mould containing one ingot matrix. There are no residues visible on the mould. The piece appears to be complete. The upper surface of the mould is very uneven while the cross-section is flat enough to allow it to stand upright.
Mould; L 121mm. B 80mm. T 80mm.
Matrix; L 94mm. B 13mm. D 12mm.

318. Stone Mould (28)
Fig. 176
A sandstone mould with four ingot matrices of varying sizes on one face and one ingot matrix on the opposing face. The mould is incomplete although, judging by the present size of the ingot matrices, it is not likely that the complete mould was very much larger than the existing piece. Neither surface of the mould is very flat while the face bearing the four matrices is partially damaged. There are no visible residues on the mould.
Mould; L 95mm. B 110mm. T 50mm.
Matrix 1; L 48mm. B 13mm. D 7mm.
Matrix 2; L 56mm x B 7mm x D 6mm.
Matrix 3; L 70mm x B 19mm x D 17mm.
Matrix 4; L 38mm x B 10mm x D 7mm.
Matrix 5; L 55mm x B 10mm x D 8mm.

319. Stone Mould (115)
Fig. 174
A virtually complete schist mould with a single ingot matrix. There are no residues visible on the mould. The upper surface of the mould is relatively flat.
Mould; L 90mm. B 47mm. T 23mm.
Matrix; L 80mm. B 16mm. D 8mm.

320. Stone Mould (E14:1571 / 499)
Fig. 177
Sandstone mould, complete, with a single ingot matrix on one surface and a pseudo- cylindrical cavity on the other. Relatively flat-bottomed with no visible residues. The surface of the mould is not flat enough to have been successfully used with a lid or covering stone. Disturbed SE area.
Mould; L 140mm. B 98mm B. T 52mm.
Ingot matrix; L 70mm. B 15mm. D 8mm.
Cavity; D 40mm. Diam. at surface 25mm. Diam. at base of cavity 10mm. B18:14.

321. Stone Mould (1027)
Fig. 178
Sandstone mould, one surface of which bears a small damaged ingot matrix. The opposing surface bears two circular depressions, possibly moulds, one of which is incomplete (due to incomplete nature of the mould itself). Also on this surface is a very small groove, perhaps part of a pin matrix? A relatively large hollow has been worn in the surface bearing the ingot matrix, suggesting that the piece was also used as a whetstone. The surface containing the two circular depressions and groove is completely flat. Whether or not the other surface was in such condition is impossible to say due to damage and its use as a whetstone.
Mould; L 118mm. B 94mm. T 42mm.
Ingot matrix; L 32mm. B 8mm. D 10mm.
Circular matrix 1; external Diam. 33mm, internal Diam. 20mm, D 5mm.
Circular matrix 2; external Diam. 26mm, internal Diam. 12mm, D 5mm.
Groove ; L 16mm. B 2mm. D 2mm.

322. Stone Mould (923)
Fig. 171
Fragment of one valve of a bivalve mould for casting pin shanks. One surface bears two groove matrices. The mould is of sandstone and shows signs of burning.
Mould; L 60mm. B 33mm. T 31mm.
Grooves; L 50mm. B 3mm. D 3mm.

METALLIC OBJECTS RELEVANT TO NON-FERROUS METALWORKING

This section of the catalogue includes iron, bronze, lead and silver artefacts. The iron artefacts are tools which may have been used in non-ferrous working, tools such as awls and gouges. The bronze objects consist mainly of waste and scrap, while silver is represented by a cut fragment of a Viking silver bracelet, and lead by a model for the ring of a ringed pin. The condition of the artefacts vary.

Artefact type, National Museum of Ireland registration and excavation numbers, description, measurements, illustration references, and habitat are recorded. Again, the measurements are explained in the individual entries.

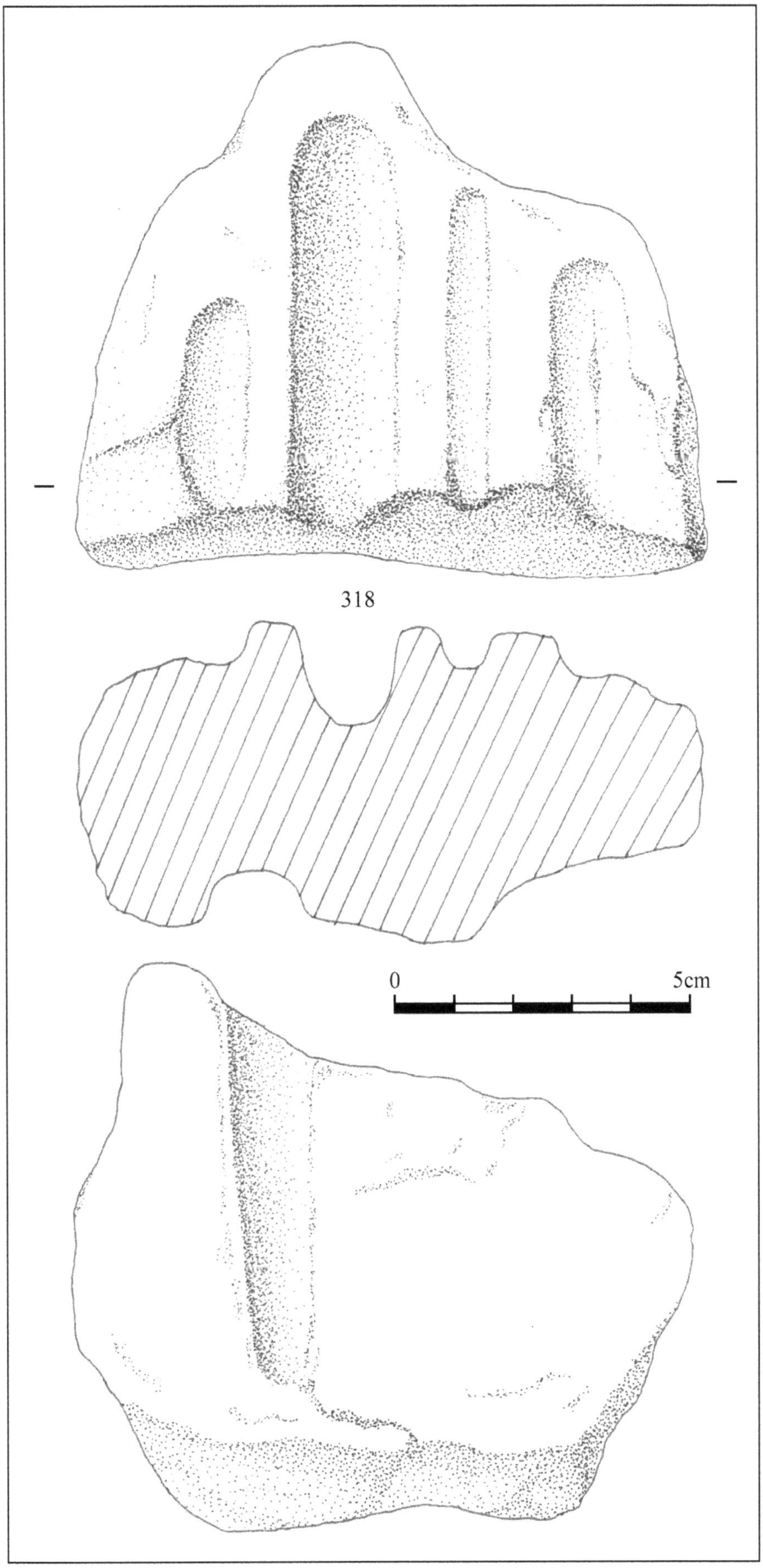

Fig. 176

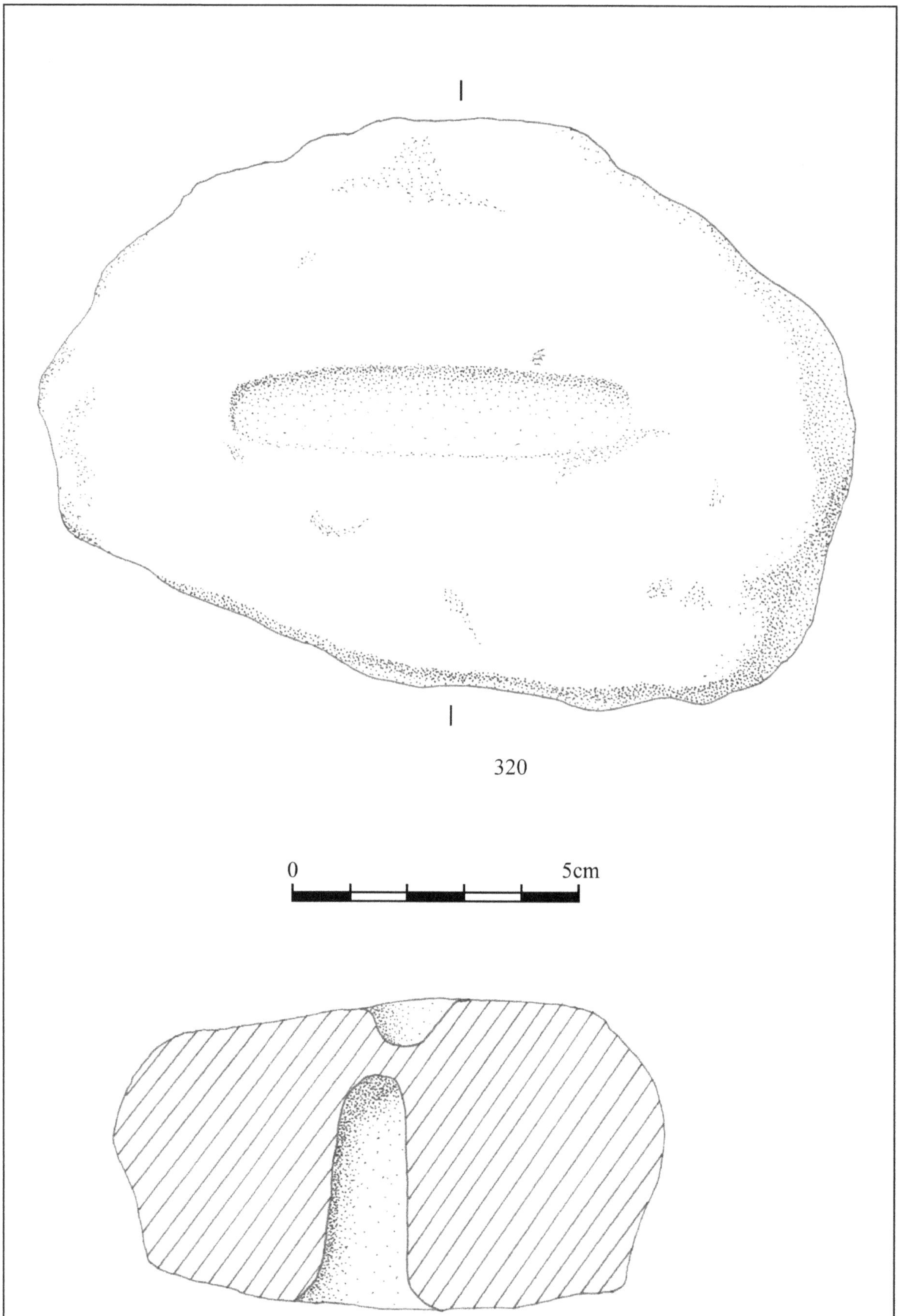

Fig. 177

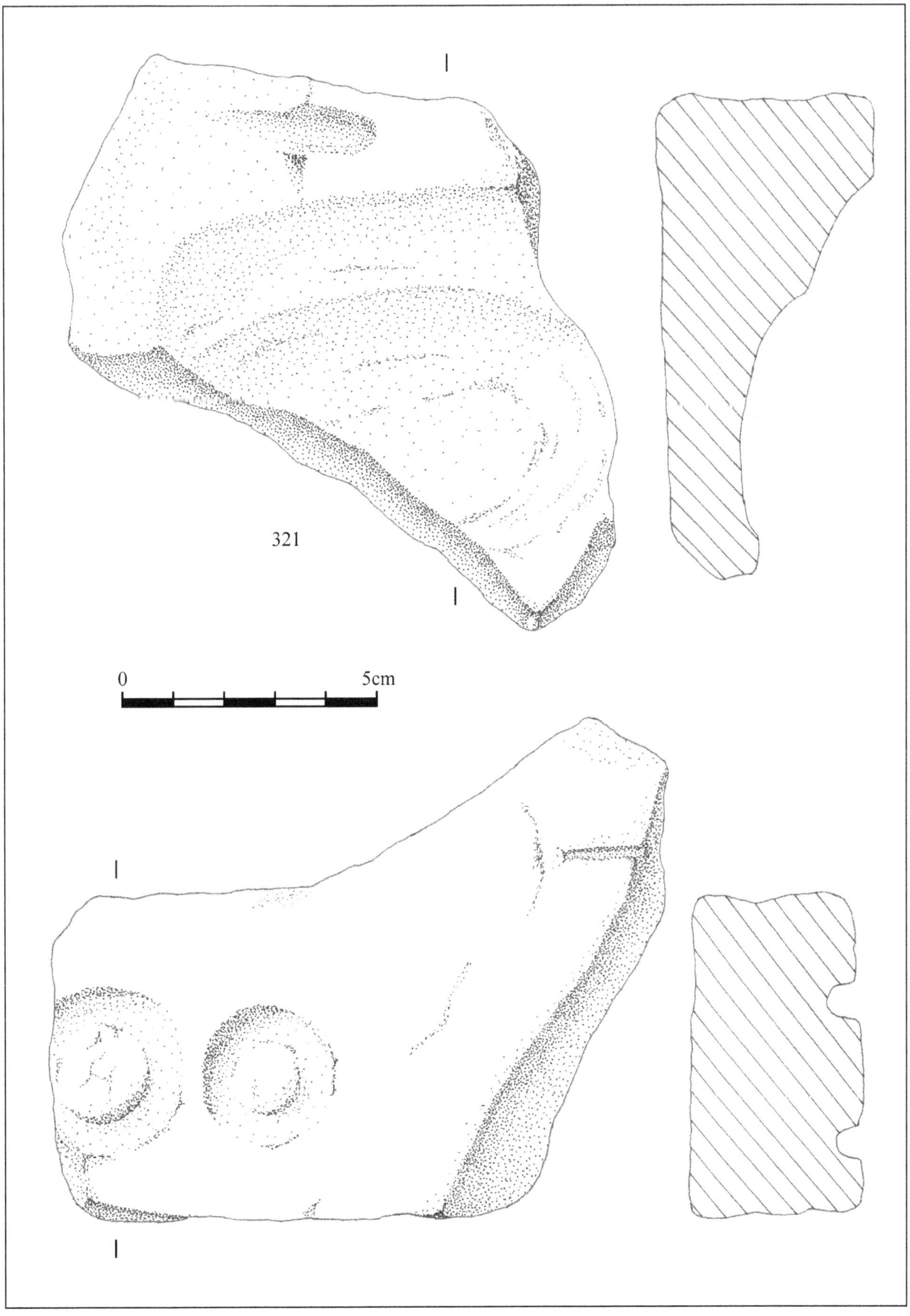

Fig. 178

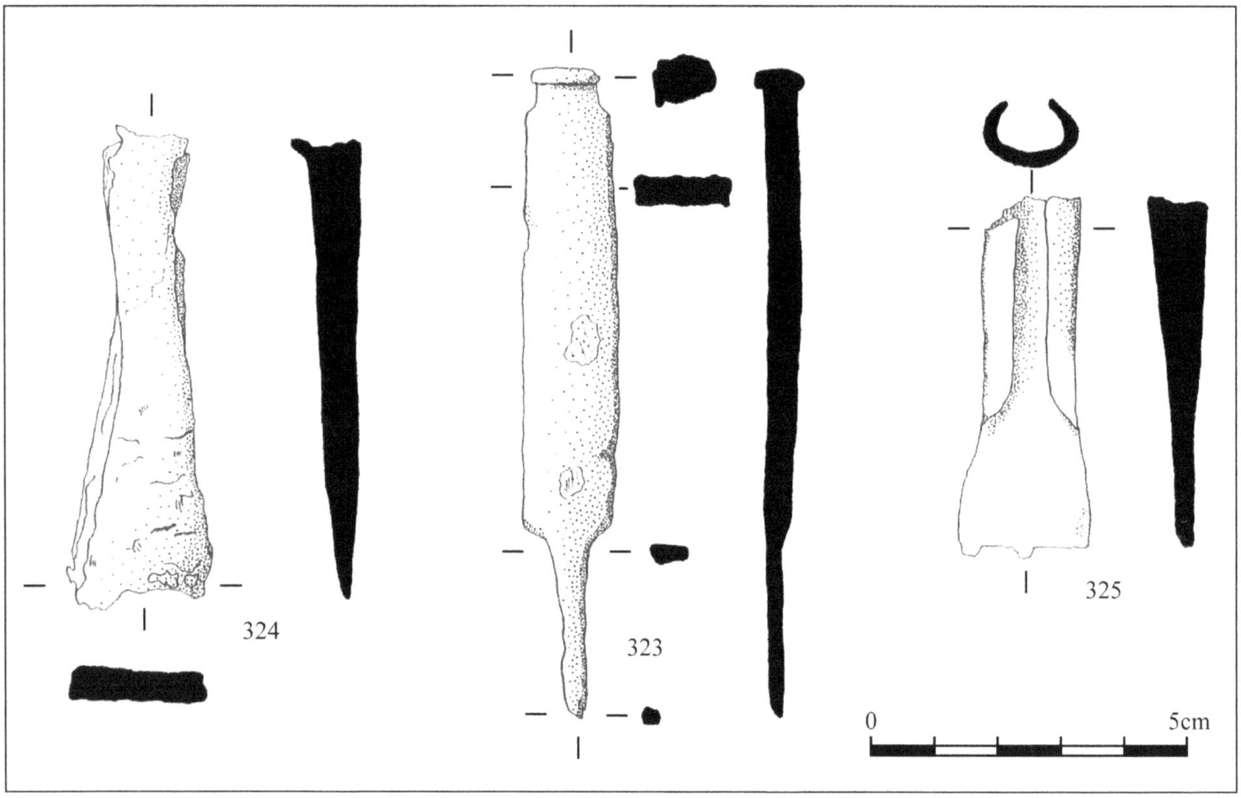

Fig. 179

323. Iron Object (E14:1108 / 58)
Fig. 179
An iron punch, slightly corroded with an intact striking platform. NW area, in wood of palisade.
L 99mm. Point L 27mm. Point Diam. 2mm. Striking platform; L 10mm x B 7mm x T 2mm. Body of tool; B 15mm. T 4mm. C17:15.

324. Iron Object (E14:1104 / 1678)
Fig. 179
A badly corroded iron wedge, tapering to a very thin blade. Disturbed, WNW area.
L 73mm. B 24mm. T at 'top' 8mm. T of 'blade' 2mm. C17:15.

325. Iron Object (E14:1346 / 1425)
Fig. 179
A small socketed iron tool, originally having three small prongs one of which has since been broken off. The surface of the tool is badly corroded. E area, south of river. Found between palisades in sand layer 2.
L 54mm. B at working end 20mm. Internal Diam. of socket 10mm. C17:6.

326. Iron Tool (E14:1105 / 171)
Fig. 180
A large socketed iron gouge, obviously meant to be hafted. The tool is in very good condition. NW area, outside palisade in brushwood.
L 207mm. Socket; L 27mm. external Diam. 24mm, T 2mm. B 11mm–27mm. T 6mm–24mm. T of metal at working end 1mm. C17:12.

327. Iron Tool (E14:1106 / 183)
Fig. 181
A well preserved iron gouge, smaller than E14:1105. The tool is well preserved but was not intended for hafting. Its shaft is looped forwards to form a handle. NE area, outside palisade.
L 188mm. B 8mm. T 3mm.
T of metal at working end 1mm. C17:15.

328. Iron Tool (E14:1107 / 1187)
Fig. 181
A well preserved iron gouge roughly the same size as E14:1106. In this case, however, the shaft is looped backwards to form a handle. E area, south of river. Found outside palisade in brushwood.
L 189mm. B 8mm. T 2mm. T at working end 1mm. C17:15.

329. Iron Tool (E14:1110 / 239)
Fig. 182
Iron awl of rectangular cross-section, the surface of which has been corroded. From the rectangular cross-section the tool tapers to a point. NE area. Found between palisades in brushwood.
L 102mm. B 2-6mm. T 2–6mm. C17:15.

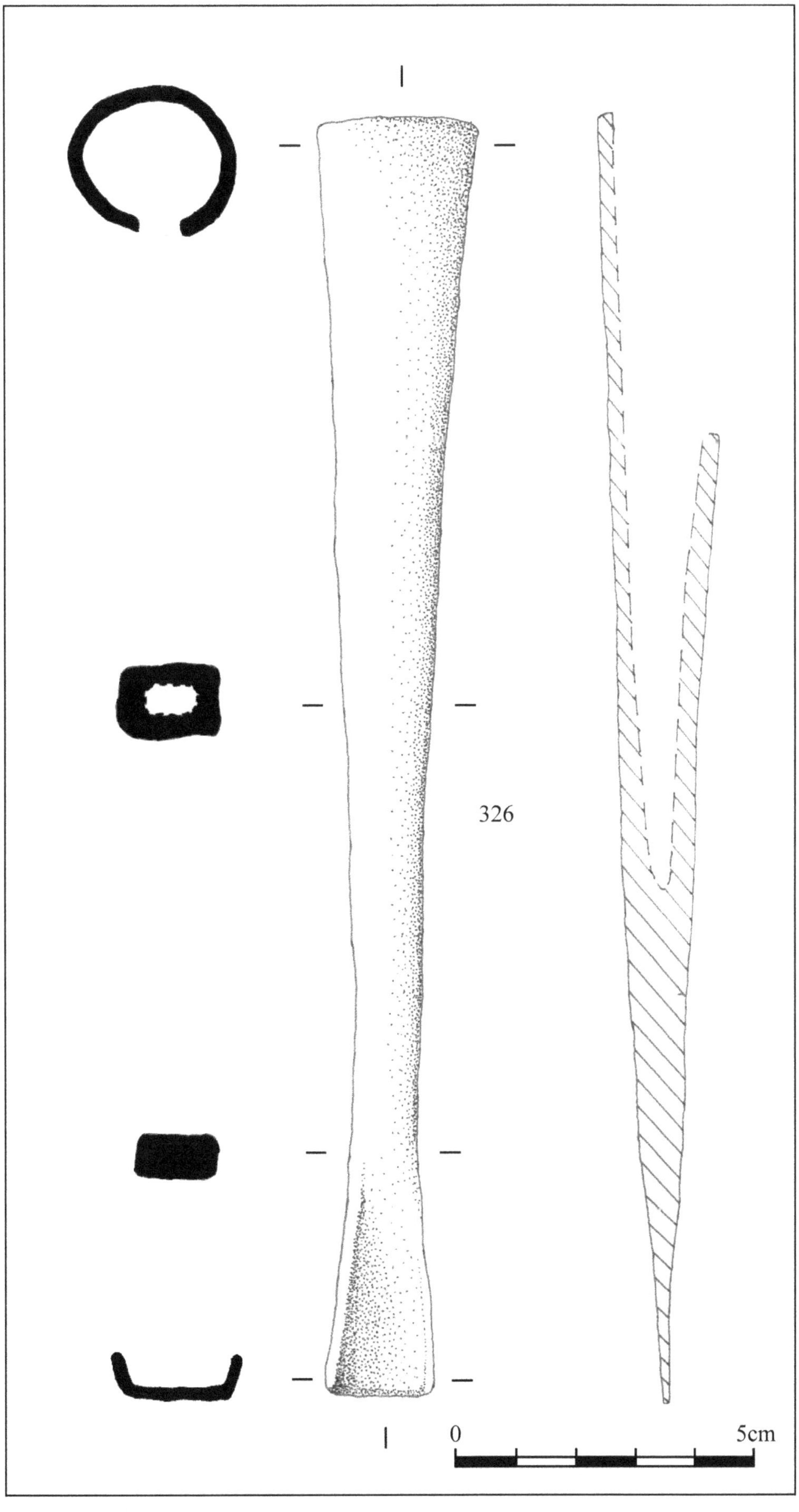

Fig. 180

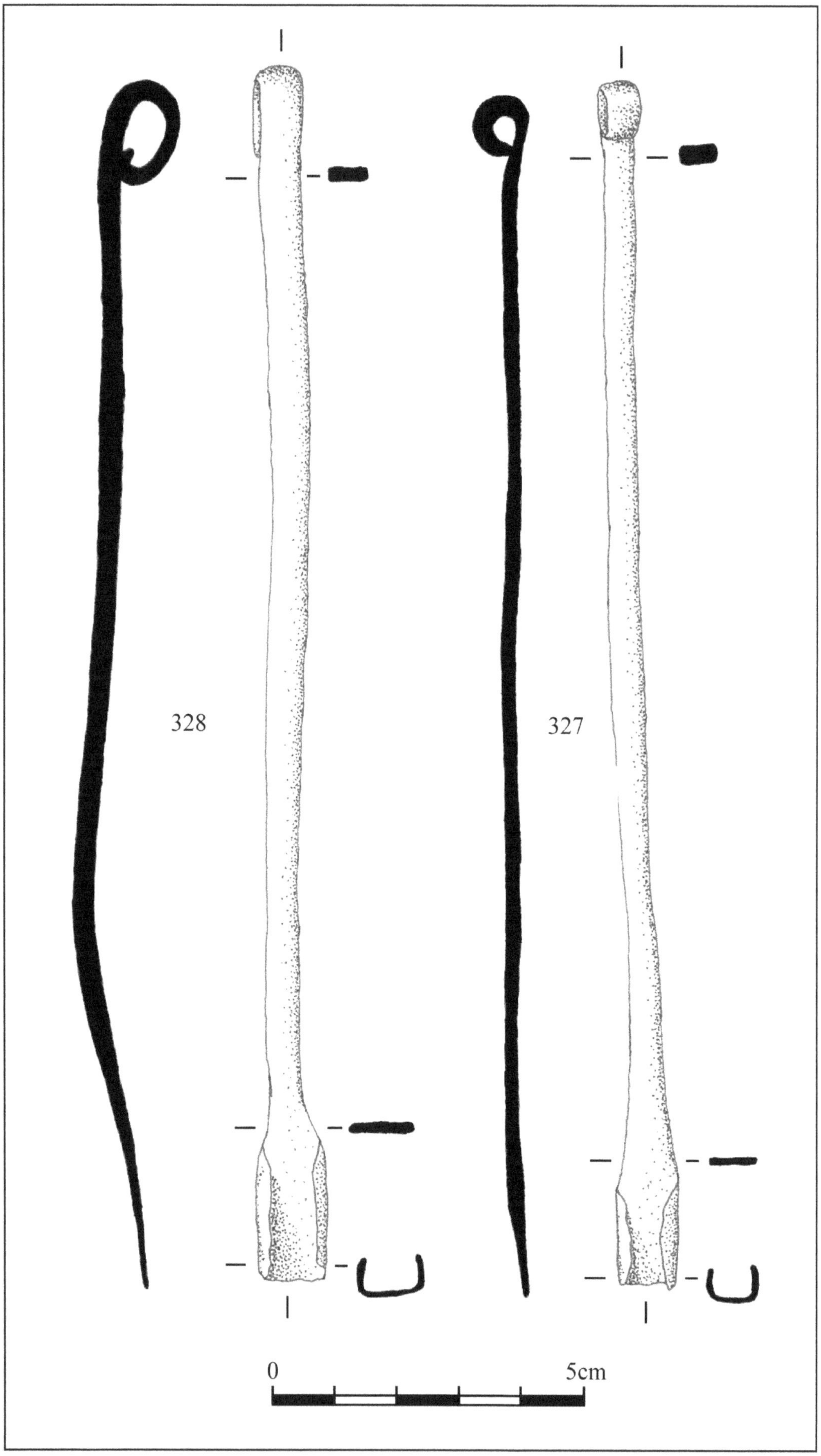

Fig. 181

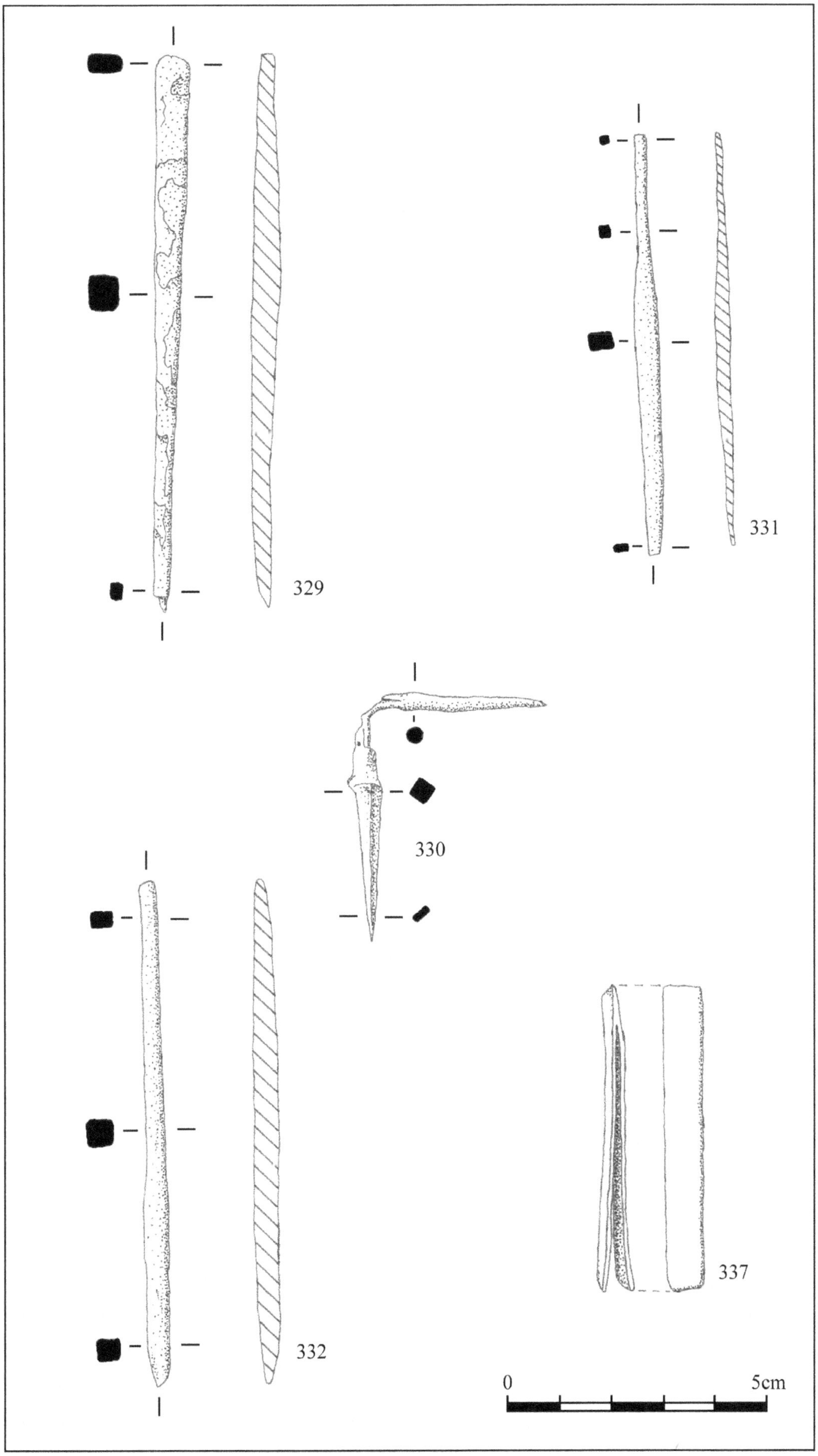

Fig. 182

330. Iron Tool (E14:1111 / 1166)
Fig. 182
Iron awl bent in two, with one half now at a right angle to the other. The surface so badly corroded that it is impossible to determine whether or not the piece is complete. One half of the tool is circular in section, the other rectangular. E area, outside palisade.
Total L 79mm.
Circular section; Diam. 3mm.
Rectangular section; L 3–5mm. B 1–4mm. C17:15.

331. Iron Tool (E14:1112 / 1207)
Fig. 182
A simple iron awl, rectangular in section. The tool is in relatively good condition. E area, south of river. Found outside palisade.
L 77mm. B 2–5mm. T 1–3mm.

332. Iron Tool (E14:1113 / 1578)
Fig. 182
Fragment of an awl with sub-rectangular section. W area, surface find.
L 93mm. B 4–5mm. T 3–5mm. C17:15.

333. Iron Tool (Wk.8, D.238) (Old Find)
Fig. 183
Made from a single piece of iron these shears are in good condition. The spring is of a simple open-ringed type, the exterior surface of which is decorated with incised lines in a criss-cross pattern. The stems or arms are circular in cross-section. The blades are roughly rectangular in shape with stepped bases joining them to the stems. The blades overlap near the base.
L 220mm. Stems; Diam. 5mm.
Blades; L 70mm. B 15mm. T 1mm. B24:4.

334. Iron Tool (Wk.10, D239) (Old Find)
Fig. 184
A well preserved iron shears made from one piece of metal. The spring used is of the open-ring type. The stems are circular in cross- section expanding into roughly triangular shaped blades, the bases of which take the form of a curved notch. There is no decoration present nor do the two blades overlap.
L 135mm. Stems; Diam. 2mm.
Blades; L 63mm. B 10mm. T 1mm. B24:4.

335. Iron Tool (Wk.36, P.912) (Old Find)
Fig. 184
Iron shears in good condition made from a single piece of iron. A large open-ringed spring joins the stems which are of circular cross-section. The blades are roughly triangular in shape and overlap at the base. There is no decoration present.
L 170mm. Stems; Diam. 4mm.
Blades; L 60mm. B 20mm. T 2mm. B17.

336. Iron Tool (Wk.37) (Old Find)
Fig. 184
A well preserved shears made from a single piece of metal. The spring is of a double-looped type. The stems are circular in cross-section while the blades are roughly triangular in shape and slightly overlap at the base. There is no decoration present on the tool.
L 170mm. Stems; Diam. 4mm.
Blades; L 47mm. B 15mm. T 1mm.

337. Bronze Tool (E14:124 / 246)
Fig. 182
Bronze tweezers made of a single strip of very thin bronze folded over. There are no visible markings or decoration. The two arms of the tweezers are blunt-ended. NE area. Found between palisades in brushwood.
L 50mm. B 7mm. T 1mm. C17:3.

338. Bronze Waste (E14:116 / 1380)
Fig. 185
A drop of solidified bronze, spilled while molten. E area, north of river. Found inside inner palisade in brown peat layer 3.
Diam. 2mm. C17:3.

339. Bronze Waste (E14:127 / 266)
Fig. 185
A small, irregularly shaped lump of bronze, spilled while molten. NE area, between palisades in brushwood.
L 50mm. B 17mm. T 15mm. C17:3.

340. Bronze Waste (E14:190–E14:196)
Fig. 185
Seven very small, corroded fragments of sheet bronze. 190–193 found in SE area, 194–196 found in W area. All approx. L 8mm. B 8mm. T 1mm. C17:3.

341. Bronze Waste (E14:204–E14:208 / 1262)
Fig. 185
Five triangular fragments of sheet bronze, in good condition. E area, north of river in black peaty material.
L 5mm–18mm. B 5mm–7mm. T 1mm. C17:3.

342. Bronze Wire (E14:177 / 814)
Fig. 186
Piece of round-sectioned bronze wire, bent into a 'C' shape. SE area.
L 150mm. Diam. 2mm. C17:3.

343. Bronze Wire (E14:178 / 41)
Fig. 186
Bronze wire, oval in section, bent into a 'C' shape. SW area.
L 77mm. Diam. 2mm. C17:3.

344. Bronze Wire (E14:179 / 1654)
Fig. 186
Straight piece of bronze wire, rectangular in section. W area.
L 87mm. B 2mm. T 1mm. C17:3.

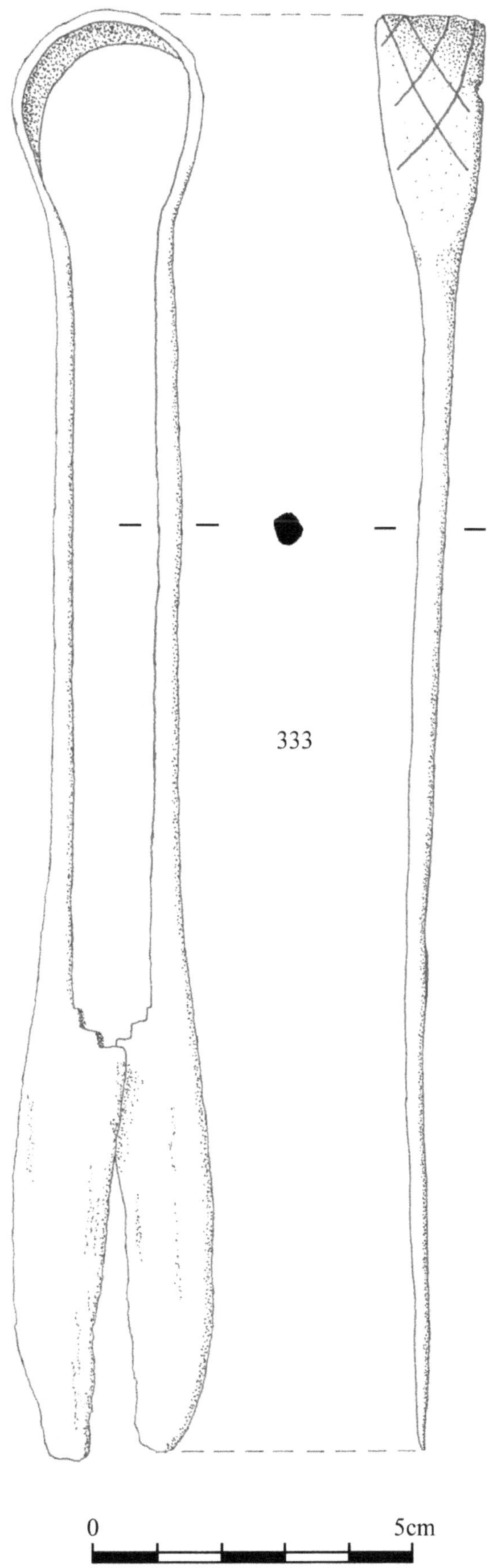

Fig. 183

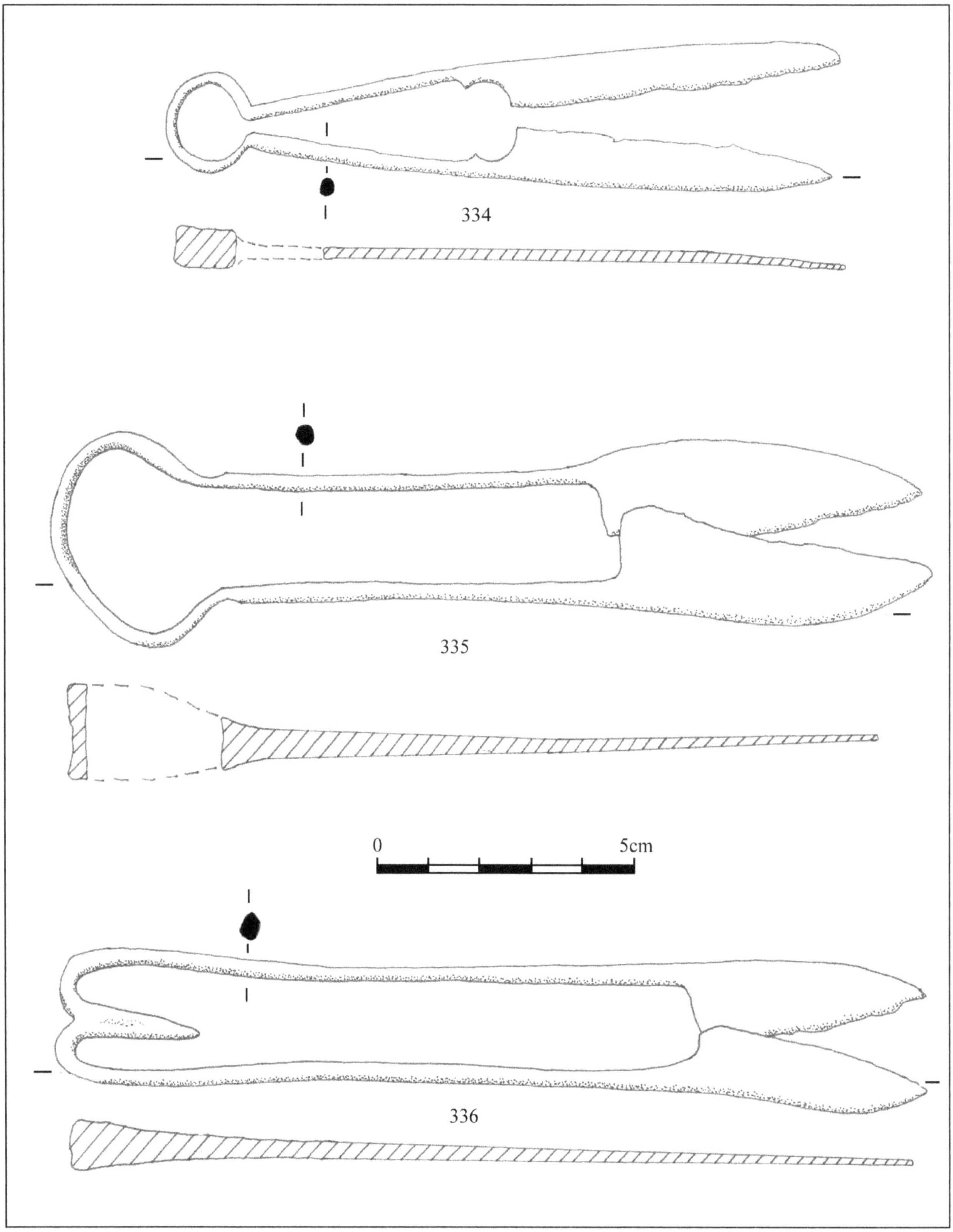

Fig. 184

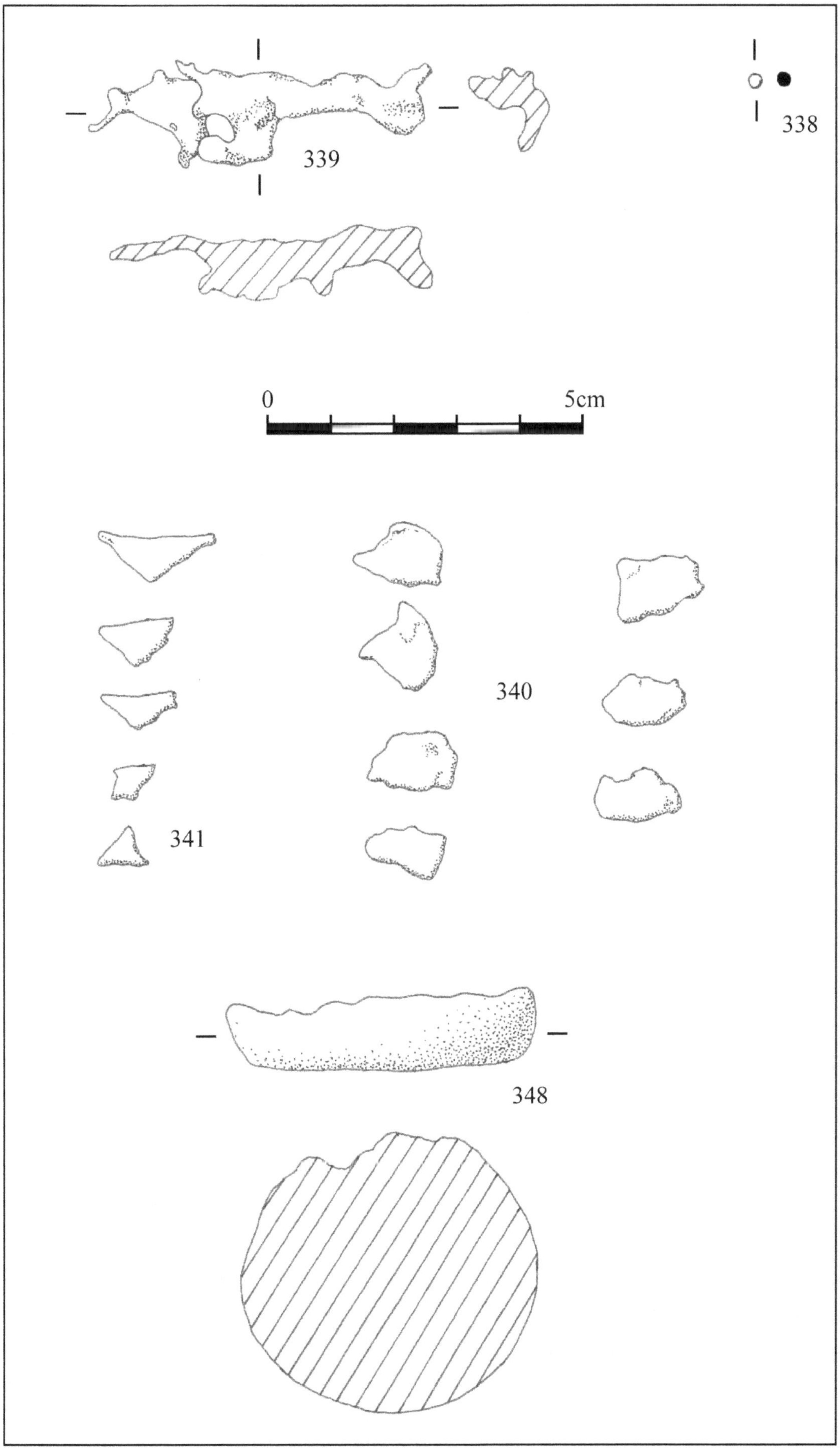

Fig. 185

345. Bronze Wire (E14:180 / 1628)
Fig. 186
Bronze wire, circular in section, looped twice. W area.
L 414mm. Diam. 1mm. C17:3.

346. Bronze Wire (E14:181 / 1628)
Fig. 186
Bronze Wire, circular in section, looped once. W area.
L 244mm. Diam. 1mm. C17:3.

347. Bronze Wire (E14:182 / 1628)
Fig. 186
Bronze wire, circular in section, bent into a 'C' shape. W area. L 95mm. Diam. 1mm. C17:3.

348. Bronze (Wk.110) (Old Find)
Fig. 185
A piece of bronze, squatly cylindrical in shape. The piece is now blue in colour and could possibly be a casting from a circular matrix like that visible on the stone mould Wk.22, W97.
Diam. 42mm x 48mm. T 11mm.

349. Bronze Rod (E14:158 / 945)
Fig. 187
Plain bronze rod, rectangular in section. NW area, in peat under brushwood layer 1.
L 59mm. B 3mm. T 1mm. C17:3.

350. Bronze Rod (E14:159 / 1472)
Fig. 187
Bronze rod, rectangular in section. The sides are slightly hammered up to form minute flanges. E area, south of river.
L 50mm. B 3mm. T 1mm. C17:3.

351. Bronze Rod (E14:160 / 744)
Fig. 187
Plain bronze rod, square in section. Disturbed SE area.
L 33mm. B 3mm. T 3mm. C17:3.

352. Bronze Rod (E14:161 / 1073)
Fig. 187
Plain bronze rod, rectangular in section. WSW area.
L 71mm. B 6mm. T 4mm. C17:3.

353. Bronze Rod (E14:162 / 367)
Fig. 186
Corroded bronze rod, oval in section. SW area.
L 41mm. Diam. 5mm x 1mm. C17:3.

354. Silver Fragment (E14:214 / 481)
Fig. 187
Small fragment cut from a silver Viking bracelet. Flat in section, the piece is decorated with a pattern consisting of a repeating incised x. Disturbed SE area.
L 21mm. B 16mm. T 2mm. Safe 3:12.

355. Lead Model (unrecorded)
Fig. 187
A small lead ring for a ringed pin bearing two head-like bosses, one on either side of the projection. In perfect condition.
External Diam. of ring 20mm. Internal Diam. of ring 12mm. T of metal 4mm. Viking Exhibit.

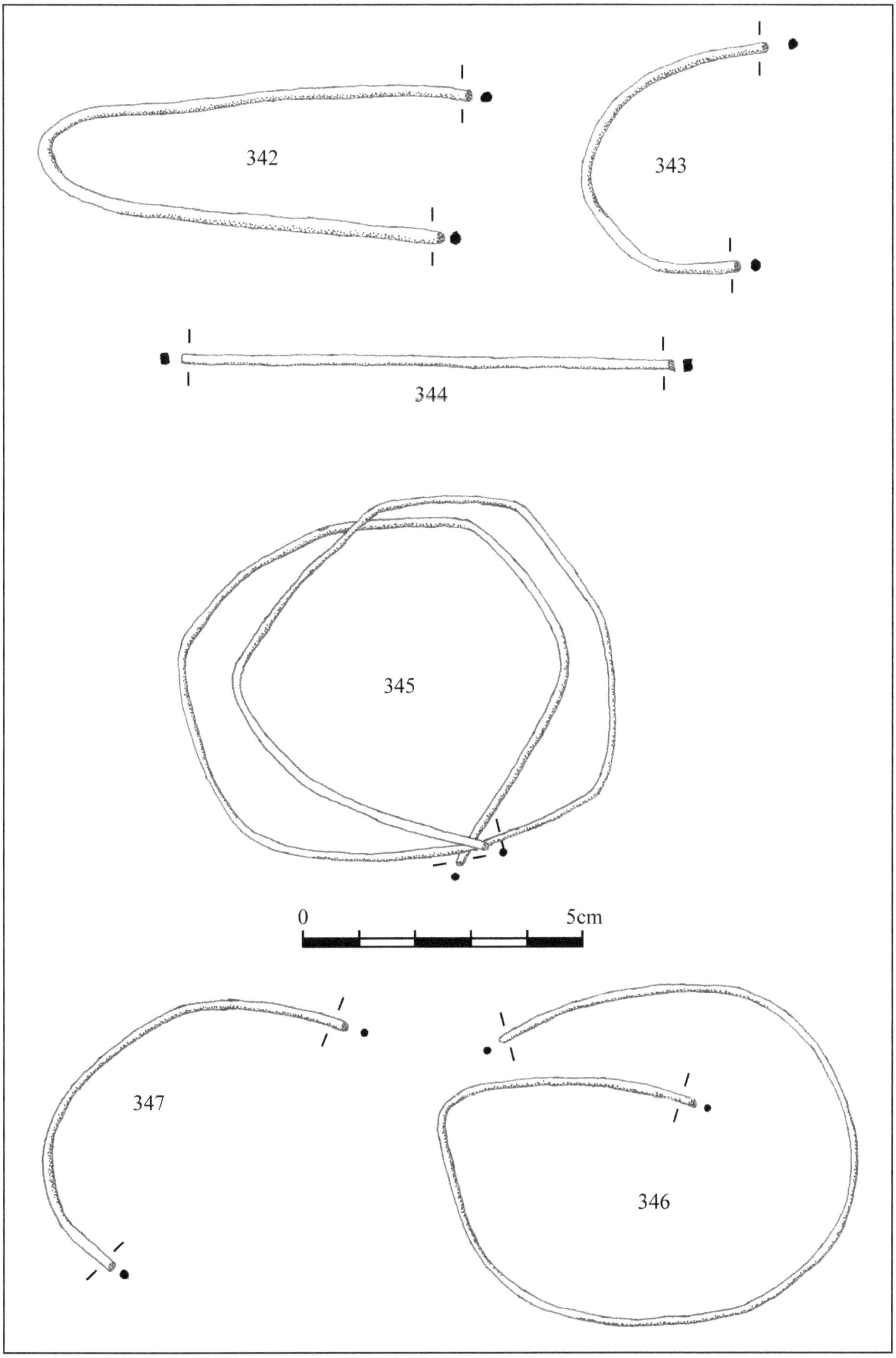

Fig. 186

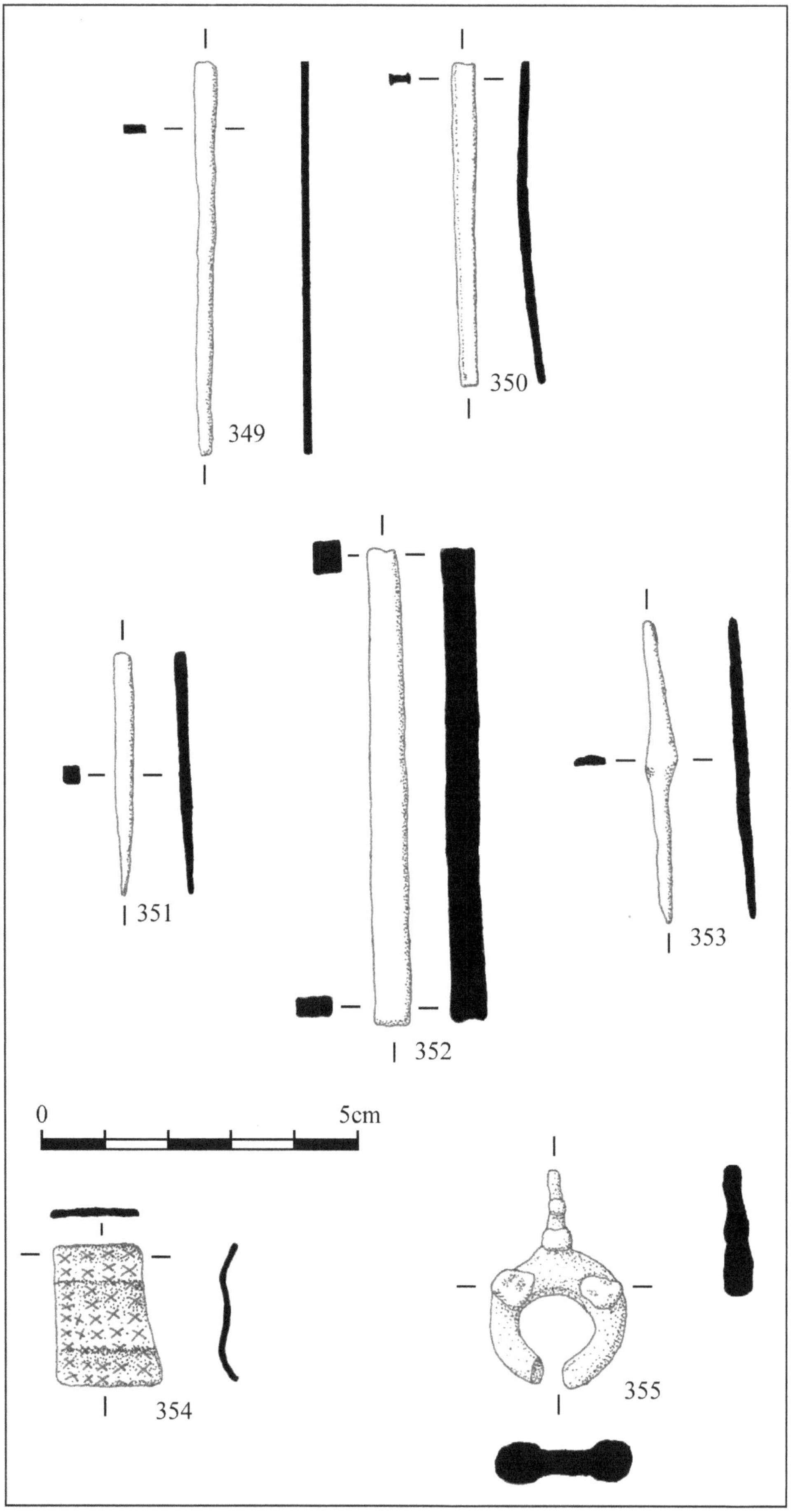

Fig. 187

GLOSSARY OF RELEVANT TERMS

AMBER
A fossilised tree resin, translucent, variable colour.

ALLOYING
The mixing together of two, or more, metals to produce an alloy with certain desired properties.

BIVALVE MOULD
A mould consisting of two halves or 'valves'.

BLOCK-TWISTING
The twisting of a metal rod and its subsequent rolling between two plane surfaces, in the manufacture of wire.

CASTING
The manufacture of a metal object through the shaping of molten metal in a mould.

CASTING-ON
The manufacture and joining of a new piece to a metal object in one operation, by constructing the mould around the existing item.

CHAPLETS
Small cylindrical pieces of metal reaching from the surface of the core to the inner surface of the mould, securing the core in place.

CHASING
The modelling of metal from the front using a hammer and punch to form oblique indentations, though removing no metal from the object being worked. Often used in conjunction with *repoussé*.

CIRE PERDUE
Literally meaning 'lost wax', the term describes the use of a wax model in the casting process. The desired item is modelled in wax, and this then coated in clay and baked. The wax melts and is run off, leaving behind its impression in the clay.

CONCENTRATION of ore
The crushing and sorting of ore to obtain the finest quality mineral for smelting. Also termed 'dressing' or 'beneficiation'.

CORE
A clay piece, inserted into a mould to enable the hollow casting of such items as socketed tools and weapons.

CORE-PRINTS
Small holes made in the walls of a mould to receive clay projections from a core, securing the core in place.

ENAMEL
A vitreous material, the colour of which is defined by various metallic oxides.

CHAMPLEVÉ The use of enamel to fill large background areas, cut or cast into an object.

CLOISONNÉ The use of enamel in gem-like fashion by clasping fragments or studs in a prepared setting.

ENGRAVING
The incising of lines of decoration by removing metal using a hammer and chisel.

FILIGREE
A type of decoration formed by soldering metal wires onto a metal background.

FIRE-GILDING or 'Mercury-gilding'
The application of a mixture of gold and mercury to an object followed by the heating of that object to evaporate off the mercury, leaving a gilded surface.

FLUXING
The introduction of an additional substance into a process to promote the separation of impurities through slag formation.

GATE
An opening or hole left in a mould, through which the molten metal is poured.

GRANULATION
The use of small individual beads of metal to form lines or areas of decoration.

INGOT
A cake of smelted metal, of varying shapes including ox-hide, bar-shaped or circular.

JET
Part of a casting consisting of the metal which solidified in the gate of the mould.

MILLEFIORI
Literally meaning 'thousand flowers' this type of decoration originally consisted of flower motifs. The ornament is formed by fusing and drawing out coloured glass rods and then using a cross-section as decoration, usually floated in a bed of red enamel.

MINING
The deliberate extraction of a mineral from the ground

OPEN-WORK
A form of decoration applied to sheet metal, entailing the cutting away of the background of a design.

PRESSBLECH
A type of ornament involving the use of a carved die or pattern upon which sheet metal was pressed or beaten.

PRESSURE-WELDING
The hammering together of two pieces of heated, though solid, metal.

PROSPECTING
The deliberate and organised search for a specific mineral.

PUNCHING
A form of decoration created using a hammer and punch, removing no metal from an object.

REFINING
The removal of unwanted or impure elements from a substance.

REPOUSSÉ
The production of relief motifs in sheet metal by hammering from the back. Sometimes finished from the front by chasing.

RISER
Excess part of a casting which formed in an air vent of a mould.

ROASTING
The open-air heating of sulphide ore to oxidise it prior to smelting.

SLAG
Waste material produced during smelting, consisting of the impurities from the ore being smelted.

SMELTING
The heating of ore under reducing conditions in a furnace, producing molten slag and metal.

SOLDERING
The joining of two pieces of metal using an alloy of lower melting point.

STITCHING
A method of securing decorative panels in place by cutting tabs in the walls of the setting, tabs which automatically curl over to rest on top of the panel.

TINNING
The application of tin-rich layers to the surface of an object to alter its appearance.

TRICHINOPOLY
The weaving or knitting of metal wire to produce a seamless, chain-like, tube. Cut and laid flat to be used as a decorative panel or employed in its original form.

TRIPLE BAND
The use of three parallel filigree wires, the centre one usually different to those on either side.

BIBLIOGRAPHY

Anderson, A.O. and Anderson, M.O. (eds.) 1961 *Adomnán's Life of Columba.* Edinburgh.

Alcock, L. 1963 *Dinas Powys. An Iron Age, Dark Age and Early Medieval Settlement in Glamorgan.* Cardiff, University of Wales Press.

Alcock, L. 1969 'Excavations at South Cadbury Castle 1968.' *Antiquaries Journal* 49, 30–40.

Alcock, L. and Alcock, E.A. 1987 'Reconnaissance excavations on Early Historic fortifications and other royal sites in Scotland, 1974-84: 2, Excavations at Dunollie Castle, Oban, Argyll, 1978.' *Proceedings of the Society of Antiquaries of Scotland* 117, 119–148.

Alcock, L., Alcock, E.A. and Driscoll, S.T. 1989 'Reconnaissance excavations on Early Historic fortifications and other royal sites in Scotland, 1974-84: 3, Excavations at Dundurn, Strathearn, Perthshire, 1976-77.' *Proceedings of the Society of Antiquaries of Scotland* 119, 189–226.

Almqvist, B. and Greene, D. (eds.) 1976 *Proceedings of the 7th Viking Congress.* London, Viking Society for Northern Research.

Anon. 1952 'Crannóg finds at Lough Gara.' *Journal of the Royal Society of Antiquaries of Ireland* 82, 182–3.

Ball, V. 1892 'On a block of red glass said to have been found at Tara Hill.' *Transactions of the Royal Irish Academy* 30, 277–293.

Bamberger, M. 1992 'The working conditions of the ancient copper smelting process.' In Craddock and Hughes, 151–159.

Bateson, J.D. 1973 'Roman material from Ireland: a re-consideration.' *Proceedings of the Royal Irish Academy* 73C, 21–97.

Bayley, J. 1988 'Non-ferrous metalworking, continuity and change.' In Slater and Tate (eds.), 193–208.

Bayley, J. 1995 'Crucibles and Moulds.' Historical Metallurgy Society, Datasheet 1.

Bayley, J. 1995 'Precious Metal Refining.' Historical Metallurgy Society, Datasheet 2

Bennett, I. (ed.) 1987-2002 *Excavations 1986–2000.* Wicklow, Wordwell.

Bersu, G. 1947 'The rath in townland Lissue, Co. Antrim.' *Ulster Journal of Archaeology* 10, 30–58.

Best, R.I. and O'Brien, M. 1967 *The Book of Leinster.* Dublin. 5 vols.

Bimson, M. and Oddy, W.A. 1983 'Coloured glass and millefiori in the Sutton Hoo grave deposit.' In R. Bruce-Mitford (ed.).

Binchy, D.A. 1941 *Críth Gablach.* Dublin, Medieval and Modern Irish series 11.

Binchy, D.A. 1955 'Bretha Nemed.' *Ériu* 17, 4–6.

Binchy, D.A. 1958 'The Fair of Tailtiu and the Feast of Tara.' *Ériu* 17, 113–138.

Binchy, D.A. 1966a 'Ancient Irish Law.' *Irish Jurist* 1, 84–92.

Binchy, D.A. 1966b 'Bretha Déin Checht.' *Ériu* 20, 1–66.

Binchy, D.A. 1973a 'Distraint in Irish Law.' *Celtica* 10, 22–71.

Binchy, D.A. 1973b 'A text on the forms of distraint.' *Celtica* 10, 72–86.

Binchy, D.A. 1978 *Corpus Iuris Hibernici.* Dublin. 5 vols.

Bradley, J. 1983 'Excavations at Moynagh Lough, Co. Meath 1980-81.' *Ríocht na Midhe* 7, 2, 12–32.

Bradley, J. 1984 'Excavations at Moynagh Lough, 1982 - 83.' *Ríocht na Midhe* 7, 3, 86–93.

Bradley, J. 1986 'Excavations at Moynagh Lough, 1984.' *Ríocht na Midhe* 7, 4, 79–92.

Bradley, J. 1991 'Excavations at Moynagh Lough, Co. Meath.' *Journal of the Royal Society of Antiquaries of Ireland* 121, 5–26.

Bradley, J. 1993 'Moynagh Lough: an insular workshop of the second quarter of the eighth century.' in R.M. Spearman and J. Higgitt (eds.), 74–81.

Bradley, J. 1995 'Excavations at Moynagh Lough, Co. Meath.' *Ríocht na Midhe* 9, 1, 158–169.

Bradley, J. 1997 'Archaeological excavations at Moynagh Lough, Co. Meath 1995-6.' *Ríocht na Midhe* 9, 3, 50–61.

Brannon, N.F. 1982 'A rescue excavation at Lisdoo Fort, Co. Fermanagh.' *Ulster Journal of Archaeology* 44/45, 53–59.

Brash, R.R. 1870 'The precious metals and ancient mining in Ireland.' *Royal Historical and Archaeological Association of Ireland* 1, 509–534.

Bruce-Mitford, R. 1983 *The Sutton Hoo Ship Burial.* Vol. III London, British Museum.

Buick, R. 1893 'The crannóg of Moylarg.' *Journal of the Royal Society of Antiquaries of Ireland* 23, 27–43.

Buick, R. 1894 'The crannóg of Moylarg. 2nd Paper.' *Journal of the Royal Society of Antiquaries of Ireland* 24, 315–331.

Burgess, C. 1979 'The background of early metalworking in Ireland and Britain.' In M. Ryan (ed.), 207–214.

Byrne, F.J. 1987 *Irish Kings and High-Kings.* London, B.T.Batsford.

Campbell, E. and Lane, A. 1989 'Llangorse: a tenth century royal crannóg in Wales.' *Antiquity* 63, 675–681.

Campbell, E. and Lane, A. 1993a 'Celtic and Germanic interaction in Dalriada: the seventh-century metalworking site at Dunadd.' In R.M. Spearman and J. Higgitt (eds.), 52–63.

Campbell, E. and Lane, A. 1993b 'Excavations at Longbury Bank, Dyfed, and Early Medieval settlement in south Wales.' *Medieval Archaeology* 37, 15-77.

Charles, J.A. 1979 'From copper to iron - the origin of metallic materials.' *Journal of Metals* 30, 7, 8-13.

Charles, J.A. 1992 'Determinative mineralogy and the origins of metallurgy.' In Craddock and Hughes (eds.), 21–28.

Close-Brooks, J. 1986 'Excavations at Clatchard Craig, Fife.' *Proceedings of the Society of Antiquaries of Scotland* 116, 117–184.

Coatsworth, E. and Pinder, M. 2002 *The Art of the Anglo-Saxon Goldsmith.* Woodbridge, Boydell Press.

Coghlan, H.H. 1951 *Notes on the Prehistoric Metallurgy of Copper and Bronze in the Old World.* Oxford, Pitt-Rivers Museum Occasional Paper 4.

Coghlan, H.H. and Case, H.J. 1957 'Early metallurgy of copper in Britain and Ireland.' *Proceedings of the Prehistoric Society* 23, 91–123.

Coghlan, H.H. and Raftery, J. 1961 'Irish prehistoric casting moulds.' *Sibrium* 6, 223–244.

Collins, A.E.P. 1955 'Excavations in Lough Faughan crannóg, Co. Down.' *Ulster Journal of Archaeology* 18, 45–80.

Collins, A.E.P. and Proudfoot, B. 1959 'A trial excavation in Clea Lakes crannóg, Co. Down.' *Ulster Journal of Archaeology* 22, 92–102.

Comber, M. 1996 *Native Evidence of Non-ferrous Metalworking in Early Historic Ireland.* NUI, Galway, unpublished MA thesis.

Comber, M. 1997 'Lagore crannóg and non-ferrous metalworking in Early Historic Ireland.' *Journal of Irish Archaeology* 8, 101–114.

Comber, M. 1998 'The position of the smith in Early Historic Ireland and the implications for Garranes ringfort, Co. Cork.' In M. Comber (ed.), 5–16.

Comber, M. (ed.) 1998 *Proceedings of the A.Y.I.A. Annual Conference 1998.* Galway, Association of Young Irish Archaeologists.

Comber, M. 2002 'M.V. Duignan's excavations at the ringfort of Rathgurreen, Co. Galway, 1948–9.' *Proceedings of the Royal Irish Academy* 102C, 137–197.

Cotter, C. (ed.) 1986–1987 *Excavations 1985–1986.* Dublin, Irish Academic Publications.

Craddock, P.T. 1989 'Metalworking techniques.' in S. Youngs (ed.), 170–213.

Craddock, P.T. 1995 *Early Metal Mining and Production.* Edinburgh, Edinburgh University Press.

Craddock, P.T. and Hughes, M.J. (eds.) 1992 *Furnaces and Smelting Technology in Antiquity.* British Museum Occasional Paper 48.

Craw, J.H. 1929/30 'Excavations at Dunadd and at other sites on the Poltalloch Estates, Argyll.' *Proceedings of the Society of Antiquaries of Scotland* 64, 111–146.

Crew, P. and Crew, S. 1990 *Early Mining in the British Isles.* Porthmadog, Snowdonia Press.

Crew, P. and Rehren, T. 2003 'High temperature workshop residues from Tara: iron, bronze and glass.' In Roche, 83–102.

Crone, A. 1991 'Buiston Crannóg.' *Current Archaeology* 127, 295–297.

Curle, A.O. 1913/14 'Report on the excavation of ... the Mote of Mark.' *Proceedings of the Society of Antiquaries of Scotland* 48, 125–168.

D'Arcy, S.A. 1900 'An account of the excavations of two lake-dwellings in the neighbourhood of Clones.' *Journal of the Royal Society of Antiquaries of Ireland* 30, 204–236.

Davies, O. 1942 'Contributions to the study of crannógs in south Ulster.' *Ulster Journal of Archaeology* 5, 14–30.

Davies, O. 1950 'Excavations at Island MacHugh.' *Supplement to the Proceedings and Reports of the Belfast Natural History and Philosophical Society.*

Delaney, T. (ed.) 1971–77 *Excavations 1970–1976.* Association of Young Irish Archaeologists.

De Paor, M. and De Paor, L. 1967 *Early Christian Ireland.* London, Thames and Hudson.

Dickinson, C.W. and Waterman, D.M. 1959 'Excavation of a rath with motte at Castleskreen, Co. Down.' *Ulster Journal of Archaeology* 22, 67–82.

Dickinson, C.W. and Waterman, D.M. 1960 'Excavations at Castleskreen, Co. Down.' *Ulster Journal of Archaeology* 23, 63–77.

Dillon, M. 1954 *Early Irish Society.* Dublin, Cultural Relations Committee of Ireland.

Dillon, M. 1962 *Lebor na Cert.* Dublin, Irish Texts Society 46.

Doherty, C. 1980 'Exchange and trade in early Medieval Ireland.' *Journal of the Royal Society of Antiquaries of Ireland* 110, 67–89.

Doyle, I. 1998 'The Early Medieval activity at Dalkey Island, Co. Dublin: a reassessment.' *Journal of Irish Archaeology* 9, 89–104.

Doyle, I. *et al.* 2000 *Unpublished Excavations Survey 1930–1997.* 1st Draft Report. Heritage Council, Oxford Archaeological Unit.

Dutton, L.A. 1990 'Surface remains of early mining on the Great Orme.' In Crew and Crew (eds.), 11–14.

Edwards, N. 1990 *The Archaeology of Early Medieval Ireland.* London, B.T. Batsford.

Eogan, G. 1974 'Report on the excavations of some passage graves, unprotected inhumation burials and a settlement at Knowth, Co. Meath.' *Proceedings of the Royal Irish Academy* 74C, 11–112.

Eogan, G. 1977 'The Iron Age-Early Christian settlement at Knowth, Co. Meath.' In V. Markotic, (ed.), 69–76.

Eogan, G. 1994 *The Accomplished Art.* Oxford, Oxbow Monographs 42.

Evans, E. 1966 *Prehistoric and Early Christian Ireland.* London, B.T. Batsford.

Fairhurst, H. 1939 'The galleried dun at Kildonan Bay, Kintyre.' *Proceedings of the Society of Antiquaries of Scotland* 73, 185–228.

Fanning, T. 1981 'Excavation of an Early Christian cemetery and settlement at Reask, Co. Kerry.' *Proceedings of the Royal Irish Academy* 81C, 67–172.

Foster, J. 1980 'Metalworking at Gussage All Saints, Dorset: A review of recent work.' In Oddy (ed.).

Frank, S. 1982 *Glass and Archaeology.* London, Academic Press.

Fredengren, C. 1998 'Lough Gara through time.' *Archaeology Ireland* 12, 1, 31–33.

Gaskell Brown, C. and Harper, A.E.T. 1984 'Excavations on Cathedral Hill, Armagh 1968.' *Ulster Journal of Archaeology* 47, 109–161.

Gelling, P.S. 1969 'A metalworking site at Kiondroghad, Kirk Andreas, Isle of Man.' *Medieval Archaeology* 13, 67–83.

Gibbons, E. 1995 'Cathair Fionnúrach, Ballnavenooragh.' In *Excavations* 1994, 42–3.

Gibson, B. 1982 *Craft Production and Consumption in Ireland, 200–1200 A.D.* University College Dublin, M.A. thesis.

Gilles, W. 1981 'The craftsman in early Celtic literature.' *Scottish Archaeological Forum* 11, 70–85.

Gray, E.M. 1982 *Cath Maige Tuired.* Dublin, Irish Texts Society 52.

Harper, A.E.T. 1973/74 'The excavation of a rath in Crossnacreevy Townland, Co. Down.' *Ulster Journal of Archaeology* 36/37, 32–41.

Hartnett, P.J. and Eogan, G. 1964 'Feltrim Hill, Co. Dublin: a Neolithic and Early Christian site.' *Journal of the Royal Society of Antiquaries of Ireland* 94, 1–38.

Hawthorne, J. and Stanley-Smith, C. 1963 *On Divers Arts - the treatise of Theophilus.* New York, Dover.

Hencken, H. O'N. 1942 'Ballinderry Crannóg no.2.' *Proceedings of the Royal Irish Academy* 47C, 1–76.

Hencken, H. O'N. 1950 'Lagore Crannóg: an Irish royal residence of the 7th to 10th centuries AD.' *Proceedings of the Royal Irish Academy* 53C, 1–248.

Hennessy, W.M. and McCarthy 1887–1901 *Annála Uladh, Annals of Ulster.* Dublin. 4 vols.

Hodges, H. 1964 *Artifacts. An Introduction to Early Materials and Technology.* London, John Baker.

Hoover, H.C. and Hoover, L.H. 1950 *De Re Metallica – Georgius Agricola (1556).* New York, Dover.

Howard, H. *The Bronze Casting Industry in Later Prehistoric Southern Britain.* Unpublished Doctor of Philosophy Thesis, University of Southampton.

Hughes, K. 1972 *Early Christian Ireland: Introduction to the Sources.* London, Hodder and Stoughton.

Hughes, K. 1980 'Introduction to A History of Medieval Ireland.' In A. Otway-Ruthven (ed.).

Hughes, K. 1987 *Church and Society in Ireland AD 400–1200.* London, Variorum.

Hull, V. 1964 'Early Irish find(b)ruine.' *Zeitschrift fur Celtische Philologie 29, 101–103.*

Hurley, V. 1982 'The early church in the southwest of Ireland: settlement and organisation.' In Pearce (ed.), 297–339.

Ivens, R.J. 1984a 'Killyliss Rath, Co. Tyrone.' *Ulster Journal of Archaeology* 47, 9–35.

Ivens, R.J. 1984b 'Movilla Abbey, Newtownards, Co. Down: excavations 1981.' *Ulster Journal of Archaeology* 47, 71–108.

Ivens, R.J. 1987 'The Early Christian monastic enclosure at Tullylish, Co. Down.' *Ulster Journal of Archaeology* 50, 55–121.

Ivens, R.J. 1989 'Dunmisk Fort, Carrickmore, Co. Tyrone: excavations 1984-86.' *Ulster Journal of Archaeology* 52, 17–110.

Ivens, R.J., Simpson, D.D.A., Brown, D. 1986 'Excavations at Island MacHugh 1985 – interim report.' *Ulster Journal of Archaeology* 49, 99–102.

Jessup, R. 1950 *Anglo-Saxon Jewellery.* London, Faber and Faber.

Joyce, P.W. 1968 *A Social History of Ancient Ireland.* New York, B. Blom.

Keeley, V. 1990 'Marlinstown, Co. Westmeath.' In I. Bennett (ed.), 55.

Keeley, V. 1999 'Iron Age discoveries at Ballydavis.' In Lane and Nolan (eds.), 25–34.

Kelly, F. 1988 *A Guide to Early Irish Law.* Vol. 3. Dublin, Institute for Advanced Studies.

Kenney, J.F. 1929 *Sources for the Early History of Ireland.* Vol. 1: Ecclesiastical. New York, Columbia University Press.

Kilbride-Jones, H.E. 1980 *Zoomorphic Penannular Brooches.* London, Thames and Hudson.

Kilbride-Jones, H.E. 1994 'Metalworking practices in early Ireland.' *Archaeology Ireland* 8, 3, 19–22.

King, H. 1992 'Excavations at Clonmacnoise.' *Archaeology Ireland* 6, 3, 12–14.

King, H. (ed.) 1998 *Clonmacnoise Studies – vol. 1 Seminar Papers 1994.* Dublin, Dúchas.

King, H. (ed.) 2003 *Clonmacnoise Studies – vol. 2 Seminar Papers 1998.* Dublin, Stationery Office.

Knowles, W.J. and Coffey, G. 1907 'Craigywarren Crannóg.' *Proceedings of the Royal Irish Academy* 26C, 109–118.

Laing, L. 1973 'The Mote of Mark.' *Current Archaeology* 39, 121–125.

Laing, L. 1975 *Late Celtic Britain and Ireland c.400–1200 AD.* London, Methuen.

Laing, L. and J. 1995 *Celtic Britain and Ireland: Art and Society.* Dublin, Gill and Macmillan.

Lane, A. and Campbell, E. 2000 *Dunadd. An Early Dalriadic Capital.* Oxford, Oxbow Books.

Lane, P.G. and Nolan, W. (eds.) 1999 *Laois: History and Society.* Dublin.

Lang, J. and Hughes, M.J. 1980 'Joining techniques.' In Oddy, W.A. (ed.), 169–178.

Lawlor, H.C. 1925 *The Monastery of Saint Mochaoí of Nendrum.* Belfast, Belfast Natural History and Philosphical Society.

Liversage, G.D. 1968 'Excavations at Dalkey Island, Co. Dublin 1956–59.' *Proceedings of the Royal Irish Academy* 66C, 53–233.

Lowery, P.R., Savage, R.D. and Wilkins, R.L. 1971 'Scriber, graver, scorper, tracer: notes on experiments in bronze working technique.' *Proceedings of the Prehistoric Society* 37, 167–182.

Lynn, C.J. 1978 'A rath in Seacash townland, Co. Antrim.' *Ulster Journal of Archaeology* 41, 55–74.

Lynn, C.J. 1982 'The excavation of Rathmullan, a raised rath and motte in Co. Down.' *Ulster Journal of Archaeology* 44/45, 65–171.

Lynn, C.J. 1983 'Some 'early' ring-forts and crannógs.' *Journal of Irish Archaeology* 1, 59–62.

Lynn, C.J. 1985a 'Excavations on a mound at Gransha, Co. Down, 1972 and 1982 – an interim report.' *Ulster Journal of Archaeology* 48, 81–90.

Lynn, C.J. 1985b 'The excavation of Rathmullan, Co. Down: Addenda.' *Ulster Journal of Archaeology* 48, 130–131.

Lynn, C.J. 1986 'Lagore, Co. Meath and Ballinderry no.1, Co. Westmeath crannógs: some possible structural reinterpretations.' *Journal of Irish Archaeology* 3, 69–74.

Lynn, C.J. 1988 'Excavations at 46–48 Scotch Street, Armagh, 1979–80.' *Ulster Journal of Archaeology* 51, 69–84.

Manning, C. and Hurl, D. (eds.) 1990 'Summary account of archaeological excavations in Ireland 1980–1984.' *Journal of Irish Archaeology* 5, 65–80.

Markotic, V. (ed.) 1977 *Ancient Europe and Mediterranean Studies*. Warminster, Aris and Phillips.

Melia, D.F. 1982 'The Irish church in the Irish laws.' In Pearce (ed.), 363–378.

Meyer, K. (ed.) 1906 *The Triads of Ireland*. Royal Irish Academy Todd Lecture Series 13.

Meyerdirks, V., Rehren, Th. and Harvey, A. 2004 'Reconstructing early Medieval copper smelting at Ross Island.' In W. O'Brien, 651–664.

Mitchell, G.F. 1940 'Studies in Irish Quaternary deposits: some lacustrine deposits near Dunshaughlin, Co. Meath.' *Proceedings of the Royal Irish Academy* 46B, 2, 13–37.

Mitchell, G.F. 1945 'The relative ages of archaeological objects recently found in bogs in Ireland.' *Proceedings of the Royal Irish Academy* 50C, 1, 1–19.

Mortimer, C. 1995 'Chemical analysis of metalwork and metalworking debris.' *Historical Metallurgy Society: Archaeology Datasheet* no.12.

Moss, R.J. 1927 'A chemical examination of the crucibles in the collection of Irish antiquities of the Royal Irish Academy.' *Proceedings of the Royal Irish Academy* 37C, 175–193.

Mytum, H. 1992 *The Origins of Early Christian Ireland*. London, Routledge.

MacNeill, E. 1923 'Ancient Irish law: law of status or franchise.' *Proceedings of the Royal Irish Academy* 36C, 265–316.

MacNiocaill, G. 1971 'Tír Cumaile.' *Ériu* 22, 81–86.

MacNiocaill, G. 1972 *Ireland before the Vikings*. Dublin, Gill History of Ireland 1.

Northover, J.P. 1985 'The complete examination of archaeological metalwork.' *The Archaeologist and the Laboratory*. Council for British Archaeology Report 58, 56–59.

O'Brien, W. 1994 *Mount Gabriel: Bronze Age Mining in Ireland*. Galway, Galway University Press.

O'Brien, W. 1995a 'Ross Island - the beginning.' *Archaeology Ireland* 9, 1, 24–27.

O'Brien, W. 1995b 'Ross Island and the origins of Irish-British metallurgy.' In J. Waddell and E. Shee Twohig (eds.), 38–48.

O'Brien, W. 1996 *Bronze Age Copper Mining in Britain and Ireland*. Shire Archaeology 71. Bucks. Shire Publications.

O'Brien, W. 2203a 'Early metalworking and myth on the lakes of Killarney.' In J. Fenwick (ed.), 185–196.

O'Brien, W. 2003b 'The Bronze Age copper mines of the Goleen area, Co. Cork.' *Proceedings of the Royal Irish Academy* 103C, 13–59.

O'Brien, W. (with contributors) 2004 *Ross Island: Mining, Metal and Society in Early Ireland*. Bronze Age Studies 6, National University of Ireland, Galway.

Ó Córráin, D. 1974 'Aspects of early Irish history.' in Scott (ed.), 64–75.

Ó Cróinín, D. 1995 *Early Medieval Ireland, 400–1200*. London, Longman History of Ireland.

Oddy, W.A. 1980 'Gilding and tinning in Anglo-Saxon England.' in Oddy (ed.), 129–134.

Oddy, W.A. (ed.) 1980 *Aspects of Early Metallurgy*. London, British Museum Occasional Paper 17.

O'Donnell, M. 1998 'Scrahane cropmark.' In I. Bennett (ed.), 87.

O'Donnell, M. 2000 'Scrahane 1: enclosure and smelting site.' In I. Bennett (ed.), 95–96.

O'Donovan, J. and O'Curry, E. 1901 *The Ancient Laws of Ireland.* London. 5 vols.

Ó Faoláin, S. and Northover, J. 1998 'The technology of Late Bronze Age sword production in Ireland.' *Journal of Irish Archaeology* 9, 69–88.

Ó Floinn, R. and King, H. 1998 'Archaeological investigations at St. Ciarán's National School.' In H. King (ed.), 119–132.

O'Kelly, M.J. 1952 'St. Gobnet's house, Ballyvourney, Co. Cork.' *Journal of the Cork Archaeological and Historical Society* 57, 18–40.

O'Kelly, M.J. 1961 'The ancient Irish method of smelting iron.' *Bericht űber den V Internationalen Kongress fűr vor-und Frűhgeschichte*, 459–462. Hamburg.

O'Kelly, M.J. 1963 'The excavations of two earthen ringforts at Garryduff, Co. Cork.' *Proceedings of the Royal Irish Academy* 63C, 17–150.

O'Kelly, M.J. 1965 'The belt-shrine from Moylough, Sligo.' *Journal of the Royal Society of Antiquaries of Ireland* 95, 149–188.

O' Kelly, M.J. 1967 'Knockea, Co. Limerick.' In Rynne (ed.), 72–101.

O'Kelly, M.J. 1989 *Early Ireland - An Introduction to Irish Prehistory.* Cambridge, Cambridge University Press.

O'Mahony and Richey 1873 *Ancient Laws of Ireland.* Dublin, vol. 3.

Ó Meadhra, U. 1979 *Early Christian, Viking and Romanesque Art: Motif Pieces from Ireland – An Illustrated and Descriptive Catalogue.* Stockholm, International North European Archaeology 7, Almqvist and Wiksell.

Ó Meadhra, U. 1986 'Preliminary observations on the manufacture of relief backplates in Hiberno-Saxon filigree work.' *Laborativ Arkeologi* 1, 81–94.

Ó Meadhra, U. 1987a 'Irish, Insular, Saxon and Scandinavian elements in the motif-pieces from Ireland.' In M. Ryan (ed.), 159–165.

Ó Meadhra, U. 1987b *Motif Pieces from Ireland – A Discussion.* Stockholm, International North European Archaeology 17, Almqvist and Wiksell.

O'Rahilly, C. 1970 *Táin Bó Cualnge from the Book of Leinster.* Dublin, Institute for Advanced Studies.

O'Rahilly, T.F. 1984 *Early Irish History and Mythology.* Dublin, Institute for Advanced Studies.

Organ, R.M. 1973 'Examination of the Ardagh Chalice – a case history.' In Young (ed.), 238–271.

Ó Ríordáin, A.B. and Rynne, E. 1961 'A settlement in the sandhills at Dooey, Co. Donegal.' *Journal of the Royal Society of Antiquaries of Ireland* 91, 58–64.

Ó Ríordáin, A.B. 1971 'Excavations at High Street and Winetavern Street, Dublin.' *Medieval Archaeology* 15, 73–86.

Ó Ríordáin, A.B. 1973 'The High Street Excavations.' In B. Almqvist and D. Greene (eds.), 135–140.

Ó Ríordáin, S.P. 1942 'The excavation of a large earthen ringfort at Garranes, Co. Cork.' *Proceedings of the Royal Irish Academy* 47C, 77–150.

Ó Ríordáin, S.P. 1949 'Lough Gur Excavations, Carraig Aille and the 'Spectacles'.' *Proceedings of the Royal Irish Academy* 52C, 6–111.

Ó Ríordáin, S.P. and Hartnett, P.J. 1943 'The excavation of Ballycatteen fort, Co. Cork.' *Proceedings of the Royal Irish Academy* 49C, 1–44.

O'Sullivan, A. 1998 *The Archaeology of Lake Settlement in Ireland.* Dublin, Discovery Programme Monographs 4.

O'Sullivan, J., Hannon, M. and Tierney, J. 1998 'Excavation of Lisnagun ringfort, Darrara, Co. Cork (1987-89).' *Journal of the Cork Historical and Archaeological Society* 103, 31–66.

Otway-Ruthven, A.J. 1980 *A History of Medieval Ireland.* London, Ernest Benn.

Pearce, S.M. (ed.) 1982 *The Early Church in Western Britain and Ireland.* Oxford. British Archaeological Report, British Series 102.

Penhallurick, R.D. 1986 *Tin in Antiquity.* London, Institute of Metals.

Phillips, C.W. 1934 'The excavation of a hut group at Pant-y-Saer in the parish of Llanfair-Mathafarn-Eithaf, Anglesey.' *Archaeologia Cambrensis* 89, 1–36.

Pittioni, R. 1951 'Prehistoric copper mining in Austria, problems and facts.' London, Institute of Archaeology 7th Annual Report, 16–43.

Plummer, C. (ed.) 1910 *Vitae Sanctorum Hiberniae.* Oxford, 2 vols.

Proudfoot, V.B. 1961 'The economy of the Irish rath.' *Medieval Archaeology* 5, 94–122.

Proudfoot, V.B. 1970 'Irish raths and cashels, some notes on origin, chronology, and survival.' *Ulster Journal of Archaeology* 33, 37–48.

Raftery, B. 1994 *Pagan Celtic Ireland.* London, Thames and Hudson.

Raftery, B. 1995 'The excavations of a crannóg at Rathtinaun, Lough Gara.' Lecture given to N.U.I., Galway Archaeological Society 10/11/95.

Raftery, J. 1952 'Crannóg finds at Lough Gara.' *Journal of the Royal Society of Antiquaries of Ireland* 82, 182–183.

Rahtz, P.A. 1962/63 'The Saxon and medieval palaces at Cheddar, Somerset - an interim report of excavations in 1960–62.' *Medieval Archaeology* 6/7, 53–66.

Roche, H. 2003 'Excavations at Ráith na Ríg, Tara, Co. Meath, 1997.' *Discovery Programme Reports 6*, 19–165. Dublin, Royal Irish Academy.

Rolleston, T.W. 1994 *Celtic Myths and Legend.* Guernsey, Guernsey Press.

Rothenberg, B. and Blanco-Freijeiro 1981 *Studies in Ancient Mining and Metallurgy in South-West Spain.* London, Institute of Archaeo-Metallurgical Studies.

Ryan, M. (ed.) 1979 *The Origins of Metallurgy in Atlantic Europe.* Dublin, Proceedings of the Fifth Atlantic Colloquium.

Ryan, M. 1980 'An Early Christian Hoard from Derrynaflan, Co. Tipperary.' *North Munster Antiquarian Journal* 22, 9–25.

Ryan, M. 1981 'Some archaeological comments on the occurrence and use of silver in pre-Viking Ireland.' In B. Scott (ed.), 45–50.

Ryan, M. 1983a *The Derrynaflan Hoard I. A Preliminary Account.* Dublin, National Museum of Ireland

Ryan, M. (ed.) 1983b *Treasures of Ireland, Irish Art 3000BC-1500 AD.* Dublin, Royal Irish Academy.

Ryan, M. (ed.) 1987 *Ireland and Insular Art AD 500-1200.* Dublin, Royal Irish Academy.

Ryan, M. (ed.) 1991 *The Illustrated Archaeology of Ireland.* Dublin, Country House.

Ryan, M. 1993 *Metal Craftsmanship in Early Ireland.* Dublin, Criterion Press.

Rynne, E. 1956 'Excavation of a ring-fort at Ardcloon, Co. Mayo.' *Journal of the Royal Society of Antiquaries of Ireland* 86, 203–214.

Rynne, E. (ed.) 1967 *North Munster Studies.* Limerick, Thomond Archaeological Society.

Rynne, E. 1987 'The date of the Ardagh chalice.' In M. Ryan (ed.), 85–89.

Savory, H.N. 1960 'Excavations at Dinas Emrys, Beddgelert, Caernarvonshire 1954–56.' *Archaeologia Cambrensis* 109, 13–77.

Scott, B.G. (ed.) 1974 *Perspectives in Irish Archaeology.* Belfast, Association of Young Irish Archaeologists.

Scott, B.G. 1981a 'Early Irish Caer, Iarn Aithlegtha, Credumae.' *Éiru* 32, 153–157.

Scott, B.G. 1981b 'Goldworking terms in early Irish writings.' *Zeitschrift fur Celtische Philologie* 38, 242–254.

Scott, B.G. 1981c 'Some conflicts and correspondences of evidence in the study of Irish archaeology and language.' In B. Scott (ed.), 115–119.

Scott, B.G. (ed.) 1981d *Studies on Early Ireland: Essays in Honour of M.V. Duignan.* Belfast, Association of Young Irish Archaeologists.

Scott, B.G. 1983 'An early Irish law tract on the blacksmith's forge.' *Journal of Irish Archaeology* 1, 59–62.

Scott, B.G. 1990 *Early Irish Ironworking.* Belfast, Ulster Museum.

Shee, E. 1974-1975 'Sluggary, Co. Limerick.' In T. Delaney (ed.), 22–23.

Sheehan, J. 1984 *Viking Age Silver Arm-rings from Ireland.* University College Galway, Unpublished M.A. Thesis.

Shennan, S. and Beck, C.W. 1991 *Amber in Prehistoric Britain.* Oxford, Oxbow Books.

Sheridan, A. 1983 'A reconsideration of the origins of Irish metallurgy.' *Journal of Irish Archaeology* 1, 11–19.

Slater, E. and Tate, J.O. (eds.) 1988 *Science and Archaeology.* Oxford, British Archaeological Report, British Series 196.

Spearman, R.M. and Higgitt, J. (eds.) *The Age of Migrating Ideas.* Edinburgh, National Museum of Scotland.

Stenberger, M. 1967 'A ring-fort at Raheennamadra, Knocklong, Co. Limerick.' *Proceedings of the Royal Irish Academy* 65C, 37–54.

Stevenson, J. 1838 *Nennii Historia Britonum*. London.

Stevenson, J. 1983 'Further notes on the Hunterston and Tara brooches, Monymusk Reliquary and Blackness bracelet.' *Proceedings of the Society of Antiquaries of Scotland* 113, 469–477.

Stokes, W. 1887 *Tripartite Life of St. Patrick*. London. 2 vols.

Stokes, W. 1890 *Lives of the Saints from the Book of Lismore*. Oxford.

Stokes, W. 1891 'The Second Battle of Moytura.' *Revue Celtique* 12, 52–130, 306–308.

Stokes, W. 1909 *In Cath Catharda*. Leipzig, Irische Texte IV.

Swan, L. 1973 'Kilpatrick, Co. Westmeath.' In T. Delaney (ed.), 26–27.

Swan, L. 1994/95 'Excavations at Kilpatrick, Killucan, Co. Westmeath.' *Ríocht na Midhe* 9, 1, 1–21.

Swindells, N. and Laing, L. 1980 'Metalworking at the Mote of Mark, Kirkcudbright, in the 6th–7th Centuries AD.' in W. Oddy (ed.), 121–128.

Talbot, J. 1849 'Arms and Implements...found at Lagore, near Dunshaughlin, Co. Meath.' *Archaeological Journal* 6, 101–109.

Twohig, D.C. 1990 'Excavation of three ringforts at Lisduggan North, Co. Cork.' *Proceedings of the Royal Irish Academy* 90C, 1–32.

Twohig, D. and O'Kelly, M. 1973 'Lisduggan North, Co. Cork, ringforts 1 and 2.' In T. Delaney (ed.), 7–8.

Twohig, D. and O'Kelly, M. 1974 'Lisduggan North, Co. Cork, ringforts 1 and 2.' In T. Delaney (ed.), 7–8.

Tylecote, R.F. 1962 *Metallurgy in Archaeology*. London, Edward Arnold.

Tylecote, R.F. 1986 *The Prehistory of Metallurgy in the British Isles*. London, Institute of Metals.

Tylecote, R.F. 1987 *The Early History of Metallurgy in Europe*. London, Longman.

Tylecote, R.F. 1992 *A History of Metallurgy*. London, Institute of Materials (2nd ed.).

Waddell, J. and Clyne, M. 1995 'M.V. Duignan's excavations at Kiltiernan, Co. Galway, 1950–1953.' *Journal of the Galway Archaeological and Historical Society* 47, 149–204.

Waddell, J. and Shee-Twohig, E. (eds.) 1995 *Ireland in the Bronze Age*. Proceedings of the Dublin Conference, April 1995. Dublin, Stationery Office.

Wakeman, W.F. 1879/82 'On certain discoveries of ancient crannóg structures.' *Journal of the Royal Historical and Archaeological Association of Ireland* 5, 324.

Wallace, P. 1992 *The Viking Age Buildings of Dublin*. Series A, Vol. 1. Part 1, Text. Part 2, Illustrations. Dublin, National Museum of Ireland and Royal Irish Academy.

Warhurst, C. 1969 'Excavations at Rathbeg, Co.Antrim.' *Ulster Journal of Archaeology* 32, 3–100.

Warner, R.B. 1973 'The excavations at Clogher and their context.' *Clogher Record* 8, 1, 5–12.

Warner, R.B. 1985/85 'The date of the start of Lagore.' *Journal of Irish Archaeology* 3, 75–77.

Waterman, D.M. 1954 'Excavations at the Cathedral Hill, Downpatrick, Co. Down, preliminary report on excavations in 1953.' *Ulster Journal of Archaeology* 17, 97–102.

Waterman, D.M. 1956 'Excavations at the Cathedral Hill, Downpatrick, Co. Down, preliminary report on excavations in 1954.' *Ulster Journal of Archaeology* 19, 57–72.

Waterman, D.M. and Collins, A.E.P. 1952 'The excavation of two raths at Ballywillwill, Co. Down.' *Ulster Journal of Archaeology* 15, 71–83.

Whitfield, N. 1987 'Motifs and techniques of Celtic filigree: are they original?' In M. Ryan (ed.), 75–84.

Whitfield, N. 1993 'The sources of gold in Early Christian Ireland.' *Archaeology Ireland* 7, 4, 21–23.

Wilde, W. 1836-1840 'The animal remains and antiquities recently found at Dunshaughlin in the county of Meath.' *Proceedings of the Royal Irish Academy* 1, 420–426.

Williams, B.B. 1985 'Excavation of a rath at Coolcran, Co. Fermanagh.' *Ulster Journal of Archaeology* 48, 69–80.

Wilson, D.M. (ed.) 1981 *The Archaeology of Anglo-Saxon England*. London, Cambridge University Press.

Wilson, D.M. and Hurst, D.G. 1966 'Medieval Britain in 1965.' *Medieval Archaeology* 10, 168–219.

Wood-Martin, W.G. 1886 *The Lake-Dwellings of Ireland*. Dublin.

Yates, M.J. 1983 'Preliminary excavations at Movilla Abbey, Co. Down, 1980.' *Ulster Journal of Archaeology* 46, 53–66.

Young, J. (ed.) 1973 *Application of Science in Examination of Works of Art.* New York/Museum of Fine Arts Boston.

Youngs, S. (ed.) 1989 *The Work of Angels, Masterpieces of Celtic Metalwork, 6th–9th Centuries AD*. London, British Museum.

www.ingramcontent.com/pod-product-compliance
Lightning Source LLC
LaVergne TN
LVHW071205210926
848482LV00015B/731